Manufacturing in the New Urban Economy

I0891186

In large cities in developed countries, the share of manufacturing has declined drastically in recent decades, whilst the share of services has grown. Many manufacturing plants have closed or moved to lower-cost locations. Manufacturing is pushed out of cities because of space constraints, rising costs of land and real estate, and increasing environmental awareness; at the same time it is attracted to other regions because of cost advantages, availability of space and better transportation possibilities. The ongoing process of globalisation speeds up the shift of manufacturing employment to lower-cost locations. 'Routinised' activities tend to disappear more rapidly than complex manufacturing.

This book discusses the new role of manufacturing in the emerging knowledge-based economies of cities. A central issue addressed in the book is how manufacturing activity relates to typically urban 'knowledge-based' activities, such as design and R&D. Is a manufacturing base a necessary condition to develop and expand R&D and other high-level services? How important is the link between manufacturing on the one hand, and R&D, design and command/control functions on the other, and to what extent do these functions benefit from co-location? What strategic options are open to urban and regional policymakers?

Taking a comparative approach to answering these questions, the book unravels the complex interaction between manufacturing and knowledge-based activity in cities and identifies some typical patterns. It shows how manufacturing industries have undergone a process of fundamental changes, with far-reaching consequences for regions. Combining insights from economic geography, industrial organisation and urban studies, this book contains extensive practical examples, cases and illustrations. It will prove a valuable source of information and analysis for both researchers and policymakers alike.

Willem van Winden works as a professor at Amsterdam University of Applied Sciences and is affiliated with the European Institute for Comparative Urban Research (Euricur) and the Institute for Housing and Urban Development Studies (IHS).

Leo van den Berg is a professor in regional and urban economics at Erasmus University Rotterdam, scientific director of the Euricur and member of the academic committee of the IHS.

Luis Carvalho is a doctoral candidate at Erasmus University Rotterdam and is affiliated with the Euricur and the IHS. He is an expert in urban management, regional development and economic geography.

Erwin van Tuijl is a researcher specialising in regional development and economic geography at Erasmus University Rotterdam. He works for the Euricur and the IHS.

Regions and Cities

Series editors: Ron Martin, *University of Cambridge, UK*; Gernot Grabher, *University of Bonn, Germany*; Maryann Feldman, *University of Georgia, USA*; Gillian Bristow, *University of Cardiff, UK.*

Regions and Cities is an international, interdisciplinary series that provides authoritative analyses of the new significance of regions and cities for economic, social and cultural development, and public policy experimentation. The series seeks to combine theoretical and empirical insights with constructive policy debate, and critically engages with formative processes and policies in regional and urban studies.

1. Beyond Green Belts
Managing urban growth in the 21st century
Edited by John Herrington

2. Retreat from the Regions
Corporate change and the closure of factories
Stephen Fothergill and Nigel Guy

3. Regional Development in the 1990s
The British Isles in transition
Edited by Ron Martin and Peter Townroe

4. Spatial Policy in a Divided Nation (April 1993)
Edited by Richard T. Harrison and Mark Hart

5. An Enlarged Europe
Regions in competition?
Edited by Louis Albrechts, Sally Hardy, Mark Hart and Anastasios Katos

6. The Regional Imperative
Regional planning and governance in Britain, Europe and the United States
Urlan A. Wannop

7. The Determinants of Small Firm Growth
An inter-regional study in the United Kingdom 1986–90
Richard Barkham, Graham Gudgin, Mark Hart and Eric Hanvey

8. The Regional Dimension of Transformation in Central Europe
Gorzelak Grzegorz

9. Union Retreat and the Regions
The shrinking landscape of organised labour
Ron Martin, Peter Sunley and Jane Wills

10. Regional Development Strategies
A European perspective
Edited by Jeremy Alden and Philip Boland

11. British Regionalism and Devolution
The challenges of state reform and European integration
Edited by Jonathan Bradbury and John Mawson

12. Innovation Networks and Learning Regions?
James Simmie

13. Regional Policy in Europe
S.S. Artobolevskiy

14. New Institutional Spaces
TECs and the remaking of economic governance
Edited by Martin Jones and Jamie Peck

15. The Coherence of EU Regional Policy
Contrasting perspectives on the structural funds
Edited by John Bachtler and Ivan Turok

16. Multinationals and European Integration
Trade, investment and regional development
Edited by Nicholas A. Phelps

17. Unemployment and Social Exclusion
Landscapes of labour inequality and social exclusion
Edited by Sally Hardy, Paul Lawless and Ron Martin

18. Metropolitan Planning in Britain
A comparative study
Edited by Peter Roberts, Kevin Thomas and Gwyndaf Williams

19. Social Exclusion in European Cities
Processes, experiences and responses
Edited by Judith Allen, Goran Cars and Ali Madanipour

20. Regional Development Agencies in Europe
Edited by Charlotte Damborg, Mike Danson and Henrik Halkier

21. Community Economic Development
Edited by Graham Haughton

22. Foreign Direct Investment and the Global Economy
Corporate and institutional dynamics of global-localisation
Edited by Jeremy Alden and Nicholas F. Phelps

23. Restructuring Industry and Territory
The experience of Europe's regions
Edited by Anna Giunta, Arnoud Lagendijk and Andy Pike

24. Out of the Ashes?
The social impact of industrial contraction and regeneration on Britain's mining communities
Chas Critcher, Bella Dicks, David Parry and David Waddington

25. Regional Innovation Strategies
The challenge for less-favoured regions
Edited by Kevin Morgan and Claire Nauwelaers

26. Geographies of Labour Market Inequality
Edited by Ron Martin and Philip S. Morrison

27. Sustainable Cities
Graham Haughton and Colin Hunter

28. Regions, Spatial Strategies and Sustainable Development
David Counsell and Graham Haughton

29. Clusters and Regional Development
Critical reflections and explorations
Edited by Asheim Bjorn, Philip Cooke and Ron Martin

30. Regional Competitiveness
Edited by Ron Martin, Michael Kitson and Peter Tyler

31. Regional Development in the Knowledge Economy
Edited by Philip Cooke and Andrea Piccaluga

32. The Rise of the English Regions?
Edited by Irene Hardill, Paul Benneworth, Mark Baker and Leslie Budd

33. Geographies of the New Economy
Critical reflections
Edited by Peter W. Daniels, Andrew Leyshon, Michael J. Bradshaw and Jonathan Beaverstock

34. European Cohesion Policy
Willem Molle

35. Creative Regions
Technology, culture and knowledge entrepreneurship
Edited by Philip Cooke and Dafna Schwartz

36. Devolution, Regionalism and Regional Development
The UK experience
Edited by Jonathan Bradbury

37. Intelligent Cities and Globalisation of Innovation Networks
Nicos Komninos

38. Whither Regional Studies?
Edited by Andy Pike

39. Business Networks in Clusters and Industrial Districts
The governance of the global value chain
Edited by Fiorenza Belussi and Alessia Sammarra

40. China and Europe
The implications of the rise of China as a global economic power for Europe
Edited by Klaus Kunzmann, Willy A. Schmid and Martina Koll-Schretzenmayr

41. Globalizing Regional Development in East Asia
Production networks, clusters and entrepreneurship
Edited by Henry Wai-chung Yeung

42. Manufacturing in the New Urban Economy
Willem van Winden, Leo van den Berg, Luis Carvalho and Erwin van Tuiji

43. The Impacts of Automotive Plant Closures
A tale of two cities
Edited by Andrew Beer and Holli Evans

44. The Futures of the City Region
Edited by Angela Hull and Michael Neuman

45. Migration in the 21st Century
Rights, outcomes, and policy
Kim Korinek and Thomas Maloney

Manufacturing in the New Urban Economy

Willem van Winden, Leo van den Berg, Luis Carvalho and Erwin van Tuijl

Routledge
Taylor & Francis Group
LONDON AND NEW YORK

First published 2011
by Routledge
2 Park Square, Milton Park, Abingdon, Oxon, OX14 4RN

Simultaneously published in the USA and Canada
by Routledge
711 Third Avenue, New York, NY 10017

Routledge is an imprint of the Taylor & Francis Group, an informa business

Typeset in Times by
HWA Text and Data Management, London

First issued in paperback 2013

British Library Cataloguing in Publication Data
A catalogue record for this book is available from the British Library

Library of Congress Cataloging-in-Publication Data
Manufacturing in the new urban economy / by Willem van Winden ... [et al.].
p. cm.
1. Manufacturing industries. 2. Urban economics. 3. Urban policy.
I. Winden, Willem van.
HD9720.5.M362 2010
338–dc22 2010022513

ISBN 978-0-415-71009-1 (pbk)
ISBN 978-0-415-58607-8 (hbk)
ISBN 978-0-203-84773-2 (ebk)

Contents

Illustrations

Figures

Tables

Boxes

Acknowledgements

The idea to start an inquiry into the urban future of manufacturing was born at an OECD conference on urban competitiveness and globalisation, held in Santa Cruz, Spain, in early 2005. Keynote speakers from several cities independently expressed their concern about the eroding manufacturing base and the rapid structural change affecting their urban economies and societies. Euricur (European Institute for Comparative Urban Research) picked up the signal, and launched a proposal to start an international comparative study, to analyse the changing face of manufacturing in different types of cities.

Four years, thousands of air miles, and countless interviews with CEOs, policymakers, and other experts later, we are proud to present an analysis of the dynamics of manufacturing industries in ten cities: Dortmund, Eindhoven, Munich, Ostrava, Paris, Porto, Rotterdam, São Paulo, Shanghai and Turku.

Many people have contributed to this study, and we are deeply grateful for their help.

First, we are highly indebted to our contact persons in the case cities, who helped us to arrange the interviews. Without them, it would not have been possible to have in-depth discussions with policymakers and key informants from the local industry and knowledge institutes. We thank Rob Bagchus, Rasmus Beck, Frank Grützenbach, Pavel Bednar, Petr Wilam, Sirpa Simola, Seppo Torikka, Thomas Fischer, Thierry Petit, Hilda Pereira, Isabel Santos, Emídio Gomes, António Lacerda, Jan Smeekens and Erik van Merrienboer for their support during the study. In particular, we want to thank Anass Selmani and Rachel Feng: not only did they arrange a number of interviews, they are also co-authors of the case studies on Rotterdam and Shanghai respectively. Furthermore, this study would have been impossible without the support of two organizations: the Academic Centre TransPort (ACTP) in Rotterdam, and the Institute for Housing and Development Studies (IHS) in Rotterdam. The latter institute made it possible to include three 'emerging economy' case studies in our inquiry: Shanghai, Ostrava and São Paulo. This greatly contributed to our understanding of the global dimension of developments in manufacturing industries and the impact on cities.

Second, we are grateful for the administrative support of Ankimon Vernède, Marlies Vossen and Lenneke Wester. Without them, we would not have been able to organise the meetings and finish this book.

Finally, we want to thank the Economic Development Agency ('Wirtschaftsförderung') of the City of Dortmund, for their trust and willingness to co-organise an international conference on this theme, together with Euricur. Dortmund, with its rich industrial past and diversified present, was the perfect place to host a conference on this topic!

1 Cities and manufacturing

Setting the scene

1.1 Introduction

In large cities in developed countries, the share of manufacturing has declined drastically in the last decades, and the share of services has grown. Many manufacturing plants have closed or moved to lower-cost locations. Manufacturing is pushed out of cities because of space constraints, rising costs of land and real estate, and an increasing environmental awareness; at the same time it is pulled to other regions because of cost advantages, availability of space, and/or better transportation possibilities. The ongoing process of globalisation seems to speed up the shift of manufacturing employment to lower-cost locations. 'Routinised' activities tend to disappear more rapidly than complex manufacturing.

Adam Smith and Karl Marx believed that wealth is created by producing real things, using capital and labour as inputs. They considered services as non-productive. This idea corresponds to the popular belief that service providers such as traders, bureaucrats and financial experts are parasites (Illeris, 1996), and that the manufacturing of tangible products is the source of 'real' wealth. In this view, the service sector is considered as a laggard that follows the growth of manufacturing. Indeed, productivity increases in manufacturing are high while in service industries they are hardly observable.

There are strong arguments against the idea that manufacturing is the only ultimate source of growth, however. The main one is that services help to create productivity improvements and thus make the economic system more productive. Thus, the border between manufacturing and services is blurred. Illeris (1996) argues that some business service activities (for instance R&D activity and marketing) may show slow productivity growth in itself, but the effect of it may translate into high increases in productivity, for instance when a new, more efficient production process is invented. This explains why countries in which the service sector is best developed and takes the highest share are relatively rich (Noyelle, 1994). A second point is that not only manufacturing but also services may be independent motors of economic growth. If, through rising incomes, new services are demanded that satisfy a need (for instance gardeners, butlers or kindergartens), this creates new economic sectors and circuits (Bell, 1973). Knowledge-intensive business services (KIBS) have been growth engines in many large metropolitan areas (Wood, 2009).

Many commentators see the process of deindustrialisation as part of the inevitable shift towards a 'knowledge-based economy'. In this new economy, not the production of tangible, physical goods but rather the creation of abstractions such as knowledge and information would form the economic 'raison d'être' of modern metropolitan areas. Urban economies come to rely on research and development, headquarters functions, financial and other business services, and tourism. Recently, the 'creative industries' have been added to the list as sources of growth.

Is the decline of manufacturing a natural process and should the market be allowed to do its job, or is it a potentially dangerous or threatening tendency that should be addressed by specific urban policy? The latter point seems to be gaining some weight among policymakers in recent years. The argument is that manufacturing cannot and should not be de-linked from typically urban 'knowledge-based' activities such as design and R&D. Or, to put it more strongly, a manufacturing base is a necessary condition to develop and expand R&D and other high-level services. Production facilities are needed to produce small batches of innovations and new products, and test whether the concepts work in practice, and researchers need to stay in touch with the production process. Unfortunately, there has been hardly any research into the relation between manufacturing and other urban activities. This leads to another key question: how important is the link between manufacturing on the one hand, and R&D, design and command/control functions on the other, and to what extent do these functions benefit from co-location?

In some cities, policy makers are worried about the decline of the urban manufacturing base and develop explicit strategies to keep these activities in the city. In some thriving and successful cities (Munich and New York are prime examples), interventions are designed to promote manufacturing in the city and counter the strong market pressure to reconvert urban sites into commercially viable offices, housing estates or retail. For other cities, the employment aspect is more important. This is especially the case in cities that have tried in recent years to create new jobs in R&D and knowledge-based industries to compensate for the heavy loss in traditional manufacturing and/or mining. Many of these cities now realise that although new jobs have been created, it is not enough to replace the lost jobs, and there is a big skills mismatch. Another issue is how policy could contribute to a better link between manufacturing on the one hand, and R&D, design and services on the other. To which extent can urban managers facilitate regional clusters? What is the place of local manufacturing plants in (inter)national, regional and local value chains? How to engage in strategic partnerships with other regions in order to exploit possible complementarities in the knowledge-based economy?

This book contains the results of an international comparative study that focuses on the issues raised above. To get a deeper understanding of the new role of manufacturing in the 'urban knowledge economy', we have conducted case studies in the following cities: Dortmund, Eindhoven, Munich, Paris, Porto, Rotterdam, São Paulo, Shanghai and Turku.

This book is organised as follows. Chapter 2 presents an overview of existing views and visions on the changing role of manufacturing in urban economies. It is based on a review of existing academic literature from the fields of organisational studies (focusing on new organisational forms, notably inter-firm networking) and from the regional economic literature. By the end of the chapter, we unfold our own approach, and present a framework of analysis that is applied in the case-study work. Chapters 3–12 contain the results of our case studies. In Chapter 13, the case studies are put into perspective: here, we compare and 'synthesise' the case material and draw conclusions.

Bibliography

Bell, D. (1973) *The coming of the post-industrial society*, New York: Basic Books.

Illeris, S. (1996) *The service economy: a geographical approach*, Chichester: John Wiley & Sons.

Noyelle, T. (1994) 'Services et mutations urbaines aux Etats-Unis', in J. Bonamy and N. May (eds.) *Services et mutations urbaines: Questionnements et perspectives*, Paris: Anthropos.

Wood, P. (2009) 'Service competitiveness and urban innovation policies in the UK: the implications of the 'London Paradox'', *Regional Studies*, 43:8, 1047–60.

2 Manufacturing and urban development

A frame of analysis

2.1 Introduction

This chapter develops a 'framework of analysis' to analyse the role of manufacturing in cities. First, we will discuss the difference between manufacturing and services, and whether it makes sense to distinguish the two. We stress that the line is not always that clear (section 2.2). Next, we present basic data on the development of manufacturing activity in the world economy (section 2.3). In the sections that follow, we describe some recent trends in global production systems (section 2.4) and the changing geography of production networks (section 2.5). In section 2.6, we unfold our 'frame of analysis' that is used in the remainder of the book to analyse the changing face of manufacturing in ten urban regions. Finally, in section 2.7, we describe the case study methodology we adopted in this study.

2.2 Defining 'manufacturing'

It is important to define what manufacturing is and what it is not. In most official statistics, manufacturing sectors appear as industries that produce tangible goods. A number of sub-categories are discerned, defined on the basis of the type of product (like food products, textiles and chemical products). Distinctions can also be based on the complexity of production process, or the size of the series/batches that are produced (for example mass manufacturing versus the production of small series). The Organisation for Economic Co-operation and Development (OECD, 2005) has defined manufacturing according to the technological sophistication. Manufacturing industries are grouped into four categories according to their R&D intensity: high, medium-high, medium-low and low technology. In our study, we define manufacturing as an economic activity in which the output is a tangible, physical product that needs not to be consumed immediately. Unlike services, users do not participate in the production (Illeris, 1996).

However, it is hard to draw a sharp line between manufacturing and services. In the first place, all productive activities consist of services applied to pre-existing physical materials (Greenfield, 1966). Increasingly, the production of commodities involves combinations of manufacturing and service functions (Daniels and Bryson, 2002). The profitability of companies depends not just on the manufacturing part of the production process, but on a total package including design, branding, advertising, financing, additional services, etc. By now, many such 'packages' are sold in the market, for instance cars that are sold with maintenance contracts, or computers with service agreements.

Hansen (1994) estimates that 75–85% of the costs of producing a TV are service costs, and those costs are even higher to produce a personal computer. Giarini (1997) finds that services now account for 70–80% of the production costs in most manufacturing companies. He distinguished the following services:

- Before manufacturing: research, finance;
- During manufacturing: quality control, finance, safety;
- Selling: logistics, distribution, marketing;
- During product and system utilisation: maintenance, insurance, after-sales services, repairs;
- After product and system utilisation: recycling, waste management.

So, a manufacturing plant may be classified statistically as 'manufacturing', but in fact, many service activities happen there, too. Another indicator for the 'service intensity' of manufacturing is the percentage of service workers within the manufacturing industry. These workers are legislators, senior officials and managers, professionals and associate professionals, clerks, service workers, shop and market sales workers, as well as drivers, sales and services primary occupations and transport workers. These occupations account for between 30% (Portugal) and 58% (The Netherlands) of total employment in manufacturing (see Figure 2.1).

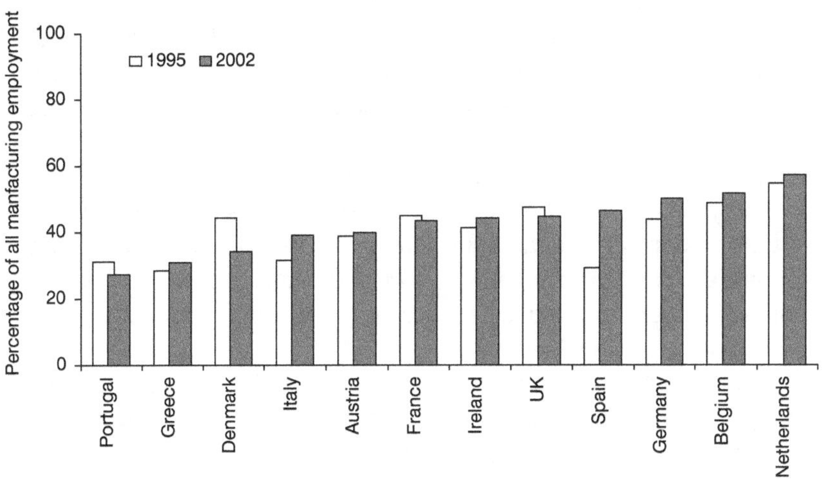

Figure 2.1 Share of employment in services-related occupations in the manufacturing sector, as a percentage of total manufacturing employment, 1995 and 2002 (source: OECD, 2005)

It is often argued that the growth of the service sector is a statistical illusion: it would reflect the trend of outsourcing producer services (like logistics, catering and legal/financial advice) to external service providers. However, there is little empirical evidence to support this claim overall, although for some types of services there is a clear tendency. One reason why the service sector has grown strongly is that services have high income elasticity: the richer households are, the bigger the share of service products in their total consumption. This is known as Engel's Law (a Saxon statistician of 1857). Very poor societies can only afford basic primary products (food and drink); when they get richer, they will consume more manufactured products (furniture, cars, etc.); in the third stage, there is an increasing consumption of services (such as culture, travel, education, entertainment and health). Furthermore, the productivity of manufacturing increases (with new technologies and machines, more goods can be produced by fewer people), whereas the service sector is characterised by low productivity growth. This explains the declining share of manufacturing in richer societies. It is not only consumers that spend more on services: the same holds for the business sector. The increased complexity of business operations coupled with a more dynamic and unpredictable environment leads to a higher service component in the production of goods. Illeris (1996) gives a perfect illustration of how this works: see Box 2.1.

Box 2.1 Why are business services related to manufacturing growth?

Think of a manufacturing firm some decades ago and now.

Previously, the production itself absorbed most of the man-hours of the firm. The planning and marketing of the products were relatively simple affairs, since there were a few standardised products and methods of production. The inputs of raw materials, capital and labour force were fixed, too. The markets, mainly domestic, demanded well-known goods and services for mass-consumption, and the distribution system was simple. Legal regulations and collective wage bargaining formed a standardised framework. Management largely consisted in combining these factors in such a way as to minimize costs and sending the orders down the hierarchy.

Today, the direct production of goods has largely been taken over by machines, though other activities still require many man hours. But all the other functions of the firm have become much more complex, unstable and risky. This is true of relationships both between departments and actors within the firms, and between the firm and the surrounding world.

Many types of mediation are required:

1 Markets are more segmented and turbulent, in many cases international. The marketing, advertising and distribution functions have become more demanding, and it has become essential to find out what these diversified markets want, to gather and analyse relevant information, and to communicate conclusions back to management.
2 Product innovation (R&D and design) is necessary in order to meet these diversified and changing demands. Innovation of production processes and internal organisation with the required flexibility subsequently becomes an essential function, too. The inputs of producer services increase productivity.
3 Inputs such as finance and raw materials can be obtained in an increasing number of forms, and specialists are required to find the forms that are best adapted to the different product lines.
4 The regulatory framework becomes more complex – for instance individual contract bargaining, environmental requirements, physical planning and public subsidies – in spite of government efforts to deregulate some areas. Legal, accounting and environmental services expand.
5 Firms increasingly need to communicate to the outside world, not only to buy inputs, sell outputs and adapt to regulations, but also to keep themselves informed about new developments in technology, markets and other factors, and to optimise public relations in general. In a world with so much information, middlemen are needed to follow, select and interpret these flows of information.
6 In order to perform those operations that are not bought externally, it becomes absolutely essential to employ people with high, specialised and changing qualifications, including motivation and social skills. Recruitment, personal management and up-skilling services become crucial.
7 The coordination of all these factors has become immensely more complex and requires the use of new information technology. Most of the personnel are preparing production, rather than producing themselves. Top management must concentrate on strategic planning while many integrating decisions must be made on decentralised levels, thus obliging all units to devote resources to overhead in the form of administrative functions.

Source: citation, slightly adapted, from Illeris (1996), pp. 69–70

2.3 Manufacturing in the world economy: some statistics

The role of manufacturing in Western economies has changed in a number of respects. Most strikingly, the share of manufacturing employment in total employment has declined. Figure 2.2 shows data for the G7 countries, but a similar pattern also holds for other OECD members. In most countries, the absolute number of workers (FTEs) in manufacturing has fallen. Figure 2.3 clearly shows this pattern. In the UK, manufacturing employment dropped by more than 20%. However, not all industrial sectors have faced a decline. Figure 2.4 shows sector data for the period 1990–2003. The reduction of employment is strongest in textiles, followed by computers, aircraft, petroleum refining, shipbuilding and chemicals. Some industries are stable or even growing slightly (food, plastics, motor vehicles, pharmaceuticals). Despite the negative employment trend, there is a continuing growth of output and value added in manufacturing. Again, Figure 2.5 shows the development for the G7 countries, but a similar pattern can be observed for other developed countries.

Whereas employment in manufacturing is declining in Western economies, the opposite is true in countries in central and Eastern Europe and East Asia, where manufacturing is a driver of economic growth. Figure 2.6 shows the rising share of Asia (and China in particular) in value added manufacturing. Despite fast growth in these regions, the developed countries remain dominant in absolute terms. Figure 2.7 shows that 14 of the 20 largest manufacturing countries (in terms of value added) still belong to this group.

Labour costs are assumed to be a key factor in the (re)location of manufacturing production. Figure 2.8 shows the large differences in hourly labour costs between countries, ranging from a little over $0.60 in China to more than $30 in Germany, Norway and Denmark. But evidently, when making location decisions, firms will not only consider labour costs but also the balance between costs and productivity (see Box 2.2). From this perspective, OECD (2006) suggests that China has

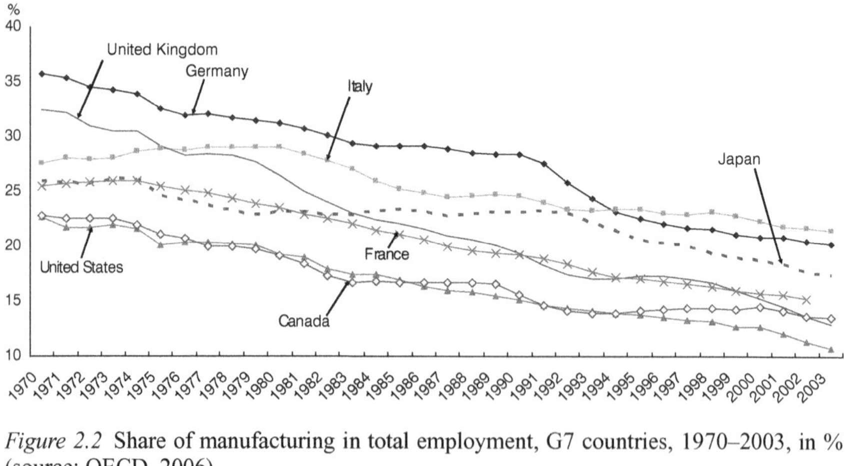

Figure 2.2 Share of manufacturing in total employment, G7 countries, 1970–2003, in % (source: OECD, 2006)

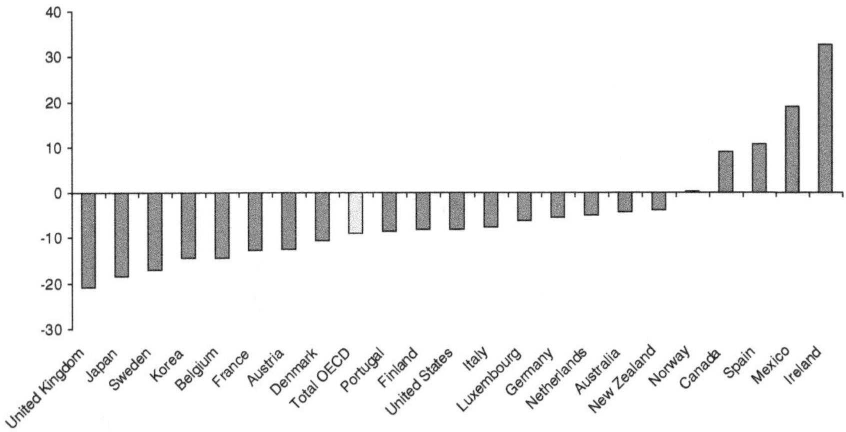

Figure 2.3 Percentage change in manufacturing employment, 1990–2003 (source: OECD, 2006)

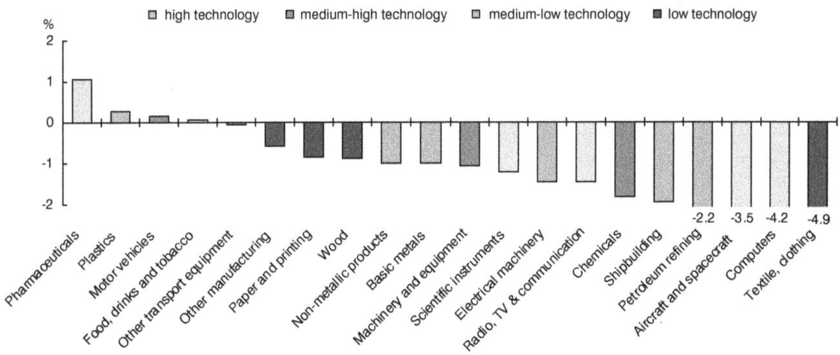

Figure 2.4 Manufacturing employment by key activity, average annual growth rates, in % (source: OECD, 2006)

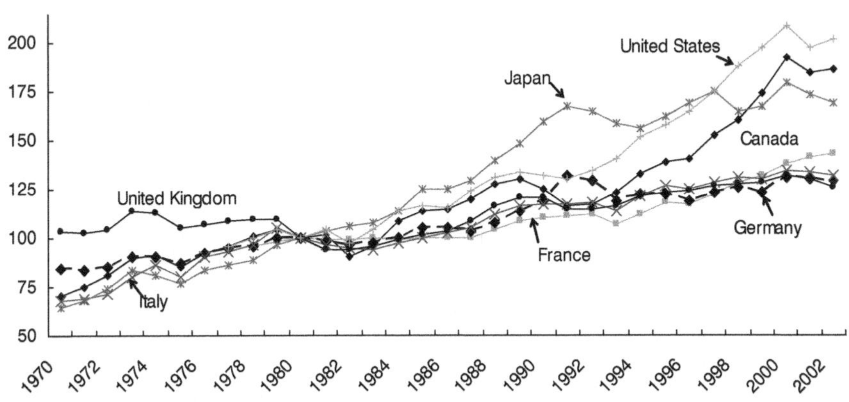

Figure 2.5 Index of manufacturing value added, G7 countries, 1970–2003 (source: OECD, 2006)

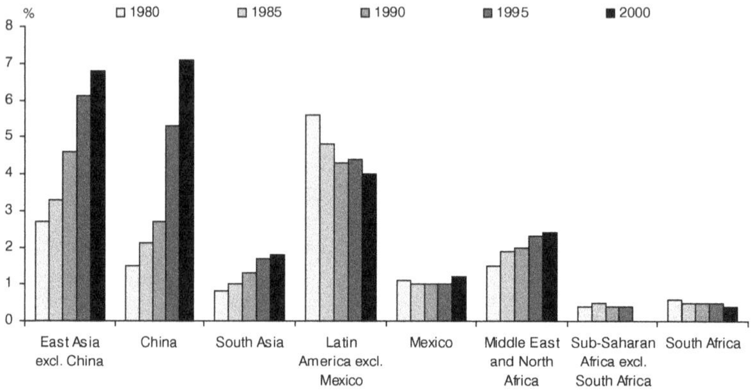

Figure 2.6 Share of major developing regions in global manufacturing value added (source: Unido, 2004)

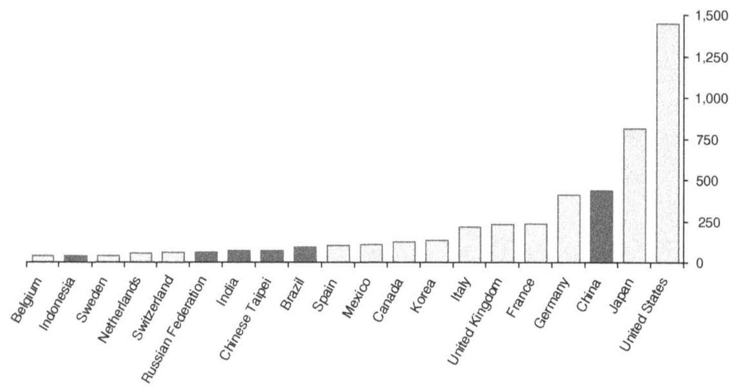

Figure 2.7 Top 20 manufacturing countries, 2002, in million dollars (source: OECD, 2006)

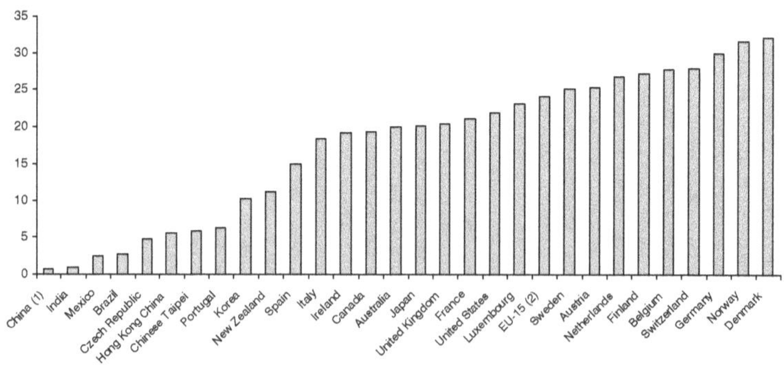

Figure 2.8 Hourly labour costs in manufacturing, 2003, in USD (source: OECD, 2006)

Box 2.2 Manufacturing in 'expensive' countries

Manufacturing activity can be located in developed countries with high wage levels only if productivity advantages outweigh the costs disadvantage. An example is Continental, a large German tyre and car parts producer. The company has production sites in China, Japan and Brazil but also a large factory close to Frankfurt, where it produces electronic braking systems. Labour costs amount to only 4.5% of the factory's costs, as 97% of the production is automated (Financial Times, 30-3-2005). Another example is Takaneka, a small Japanese producer of screws. This firm dominates the market in anti-corrosive screws that cost up to 10 times the price of a conventional screw. Also, it has a global monopoly on screws used by the nuclear industry. The secret is its superior and unrivalled quality control system. The company pays its workers more than 20 times the wages in China and Vietnam but it is highly profitable.

Source: own elaboration

relatively low level of unit labour costs, which explains its recent development as the 'workshop of the world'.

2.4 Trends in the global production system

In manufacturing industries, global competition has intensified due to the further decrease of trade barriers across the world. Tense competition has increased the pressure on industrial firms to cut costs and introduce new products more rapidly. Also, the pace of technological change has increased, thanks to the improvement in global communication networks. An important trend is the integration of 'emerging' economies in the world economy, with the BRIC countries (Brazil, Russia, India and China) as the most prominent ones. Because of their rapid economic growth, these countries have become important as new growth markets for established industrial firms. Fast demand growth and relatively low wages make these countries attractive for foreign direct investment. It is not only manufacturing activity that is 'offshored' to these countries: increasingly, multinationals also try to 'tap' from the rapidly developing knowledge base in these countries. Investments in R&D facilities have surged. Meanwhile, these countries are the home base of a rising number of 'new' multinationals that join the battle to conquer global markets.

The knowledge intensity of production processes has increased dramatically (Quinn, 1992), and this has major consequences for the organisation of industrial firms. The integrated 'bureaucratic' organisation is giving way to a networked type of organisation with more flexibility and capacity for learning. Many companies have undergone business process re-engineering (BPR), in which hierarchical structures have been replaced by horizontal structures with small, semi-independent units that concentrate on core activities for which they have developed a distinct competence. They are believed to be more 'agile' and better able to deal with fast-changing markets and technologies. These new 'horizontal' units still need to achieve economies of scale and scope, and to achieve that, they operate in networks with partners inside or outside the firm. The borders of the organisation become blurred, with more interactions with networks of other firms.

Networks make for flexibility: to benefit from chances, a firm has to be able to react fast, and to engage in partnerships with complementary strengths and capabilities. Networks are particularly important for innovation. Strong international competition and rapid technological development urge firms to produce new products or services, develop new processes and access new markets. Participation in a network enables a firm to concentrate on core capabilities and provides access to resources (such as specific know-how, technology, financial means, products, assets and markets) in other firms and organisations. This helps them to improve their competitive position.

The surge of intra- and inter-firm networks has provoked a debate about the boundaries of the firm: in the network economy, where does the firm end and the network begin? From this perspective, some scholars argue that the network rather than the firm is the appropriate unit of analysis to study value creation. Bahrami and Evans (1995) point at Silicon Valley to illustrate this point. The regional ecosystem of innovative firms operates in a constantly changing environment of incessant novelty and innovation. Dynamic networks are constantly 'recycling' knowledge and human resources to create new products that meet changing market demands.

Modularisation of industrial production

An important tendency in manufacturing industries is the modularisation of production. In this process, a final product is built up from modules, like building blocks that are produced by different networks of suppliers and sub-suppliers. Berger (2005) has dubbed this as the 'Lego model' of industrial organisation. In this organisational model, the original equipment manufacturers (OEMs) are at the top of a networked hierarchy. They specialise in the development, marketing and assembly of the final product (like cars, ships and machines). The production of core building blocks is outsourced to main suppliers (for a car, these are firms that fabricate the engine, the interiors, panels, etc.). These building blocks, in turn, are built up from smaller and less complex parts or components, created by lower-tier suppliers. In this system, a variety of companies and other actors are linked up, each of them playing a specific role in the network. The OEM takes a pivotal role: it assembles the final product from its parts, and typically is involved in the design of the product, as well as sales and marketing activity. In complex system development, OEMs cooperate intensively with other OEMs, and with a small number of 'system suppliers' that deliver the main building blocks.

Suppliers produce the largest part of the value added of an industrial product. In the relation between OEMs and suppliers, three main trends can be observed: 1) an ongoing pressure on suppliers to reduce costs and improve quality; 2) an increasing involvement of suppliers in product development; 3) a reduction in the number of suppliers. The second trend implies that main suppliers become increasingly an integrated part of the global knowledge networks of OEMs. Main suppliers, in turn, heavily rely on a network of lower tier suppliers. In these relationships, we see similar trends: a reduction in the number of suppliers, an increasing involvement of suppliers in development, shorter lead-times, and a larger share of purchasing from low cost countries.

2.5 The changing geography of production networks

Networks prevail in the knowledge-based economy. How do these networks develop in space? Several studies on the firm level shed light on this question. One reference study (Berger, 2005) reviews the strategies of more than 500 firms in Europe, Asia and America. It covers companies from diverse sectors like textiles, electronics, automotive and publishing. The author analyses the options for industrial organisation in a world of value chain fragmentation. The study focuses on two main questions: first, what components of their value chains do firms outsource versus 'keep in-house' (within the firm); and second, what components do they offshore versus 'keep at home'. Three findings of this study are particularly remarkable. First, there is 'resilience of variety': no single winning sourcing or location strategy could be identified. Across firms with strong results and performance, some have outsourced and offshored a large share of the production (like Gap), while others keep the largest share of the production within the home country (Zara). Second, there appear to be no winning or losing sectors. Despite the overall decline of slow technology (like textiles or metal), the pace of decline differs and many firms in those sectors thrive on the basis of their own advantages. Third, different firms construct and adapt their strategies on the basis of their own capabilities. In this context, firms in the same sector 'create their success in very different parts of the value chain', focusing in core capabilities and obtaining the other modules in different sources worldwide. A study of Doz *et al.* (2001) points in a similar direction. According to the authors, the successful firm in the knowledge economy is a 'meta-national firm', meaning that it is able to access and mobilise different sources of knowledge spread worldwide, and on the basis of this capability construct competitive advantage. The successful firm does not rely anymore only on its home country or region to construct competitive advantage (through local production factors, support industries, local demand and rivalry) and then grow internationally, but the process tends to be much more spatially interactive.

An interesting approach to inter-firm networks is provided by March (1991). He makes the distinction between two types of firm activity: exploration (the development of new capabilities) and exploitation (the efficient use of current assets and capabilities). The success of companies depends on their ability to strike the right balance between exploration and exploitation. However, both types of activities require different types of networks. Exploration requires innovative networks between different firms with substantial 'cognitive distance' (Nooteboom, 2004). This will generate new ideas and produce innovation. Exploitation, however, requires a more stable organisational structure, a narrow focus and clear standards. Unit production costs are to be minimised; quality control and logistics are important issues. Table 2.1 provides an overview of the difference between networks for exploration and network for exploitation.

Networks for exploration (aimed to discover new products or processes) operate in a context of uncertainty of outcomes. The type of knowledge that is exchanged is tacit, the process is creative. Firms prefer to have many (and frequently changing) knowledge partners: you don't know beforehand what the most useful contact will be. Levels of strategic interaction will be high. Levels of trust are also typically high. To limit the relational risk, firms will invest in mutual understanding.

In the exploitation stage, conditions are different. Uncertainty has been reduced (a 'dominant design' has emerged); the focus of activity shifts towards cost-efficient production and distribution. This requires the utilisation of economies of scale and the search for cheap supply sources. These changing conditions have implications for the network: the number and scope of 'ties' in the network can be reduced. Strategic interaction is needed less (specifications are clearly set), and relations shift from developmental to transactional. The increased division of labour leads to more specialisation in the networks, with each node of the network focusing on specific knowledge on a narrow scope of issues. Control becomes more formal (contracts, monitoring of compliance), and trust is less important.

In a recent paper, Nooteboom (2004) elaborates on the distinct geographical dimensions of these networks. In this analysis, he makes a rather strict separation between exploration and exploitation in both time and space. In time, first, there is localised exploration, and next, there is disembedded exploitation, and the associated networks consist of different actors. The proliferation of electronic networks has enabled the spatial separation of exploration and exploitation. Specifications of a new product can be sent through a wire to the other side of the world, where production can take place. The electronics industry is a typical example. R&D and design (exploration) take place in technologically advanced nations, while mass production is outsourced to low-wage countries and sales are global. In the pharmaceutical industry, there is some division of labour between exploration and exploitation: exploration is done by small biotechnology firms (in close collaboration with universities), whereas exploitation is performed by large internationally orientated pharmaceutical linkages that take care of the efficient production and marketing of drugs and medicines.

In several industries, however, it is difficult to separate exploitation and exploitation in time and space, notably when exploration and exploitation are functionally highly related. This is the case when produced quantities are very

Table 2.1 Networks for exploration and exploitation

General	*Exploration*	*Exploitation*
Overall characteristic	Volatility	Consolidation
Type of uncertainty	Radical technical and market uncertainty	Market risk
Focus of activity	R&D, design, prototyping	Production, distribution
Competition	Technical and market viability	Price
Type of knowledge	More tacit	More codified
Diffusion of knowledge	Limited	Wide
Network characteristics	*Exploration*	*Exploitation*
Network density	High	Low
Network stability	Low	High
Scope of ties	Wide	Narrow
Investment in mutual understanding	High	Low
Frequency of strategic interaction	High	Low
Trust/openness	High	Generally low

Source: adapted from Noteboom (2004) and Nooteboom and Gilsing (2004)

low (as in the case of cruise vessels), or when time to market is highly critical (as in the case of 'fast fashion'). Many activities are explorative to some extent, and rely heavily on exploitation functions. This is the case in incremental innovations, or innovations in production processes. Thus, the 'border' between exploration and exploitation is often far from sharp.

Nevertheless, thinking in terms of exploration and exploitation networks can be useful to study the dynamics of regional networks (or clusters). Nooteboom (2004) argues that proximity and local embedding are 'enablers' for exploration networks, but not for exploitation networks. Proximity enhances trust (through shared norms and values); it facilitates frequent strategic interaction, and facilitates reputation mechanisms. However, if relations become too durable, and there are very few 'outside' contacts, this will be good for trust but bad for learning and innovation. From this perspective, the view that outside contacts are important sources for local cluster dynamics is gaining ground (for example Asheim and Gertler, 2005; Bathelt *et al.*, 2004; Amin and Cohendet, 2004; Coe and Bunnell, 2003). While still stressing the relevance of specific localised and regionally 'sticky' features for a firm's learning and innovation, Malmberg and Maskell (2006) acknowledge that there are increasingly 'spatial extensions of the learning process'. New multi-scalar geographies of interaction between local assets and other localised sources of learning and innovation have been emerging. The conceptualisation of innovative regions as dynamic containers of 'buzz' and 'pipelines' has been particularly influential (Bathelt *et al.*, 2004). The authors distinguish between community-based learning processes attained by co-location and 'side effect' knowledge spillovers, called 'buzz' (Storper and Venables, 2004), and the knowledge obtained through the external contacts and relations of the local actors worldwide, called 'global pipelines'. The virtuous interaction between a vibrant buzz with the presence of global pipelines would provide regions and clusters with strong advantages, as they would be able to mix the best of two different knowledge and learning processes, and reinforce their competitive advantages over time (Giuliani, 2005; Bathelt, 2007).

2.6 Manufacturing industries and urban regions: a frame of analysis

For a long time, regional economists considered the manufacturing industry as the single driver for regional economic growth. According to the long-dominant 'economic base theory', the manufacturing sector is the only sector that is able to sell its products outside the region (it is assumed that services can hardly be sold over longer distances). The money that is earned by the manufacturing sectors flows to the employees (through wages and salaries). These people demand services (like retail, barbers, banking and insurance), so they support the local service sector. People who work in the service sector also demand services. So, 'basic' industries create a multiplier effect, and the size of the regional service sectors depends on the purchasing power that the basic activities (manufacturing) bring to it.

This theory has been criticised on several grounds, however. First, an increasing number of services can now also be sold outside the region and even globally, due to information and telecommunications technology: think about online services such as computer network maintenance, banking or routinised service activities such as billing (India in particular has built a large service industry with clients in the West). Second, as we have seen, many producer services (technological, marketing and management advice) are strategic for manufacturing. In fact, much manufacturing would not function without the input of such services. Third, there are a number of non-manufacturing activities that can be considered basic (or at least crucial for the functioning of basic industries), for instance telecommunication services and educational facilities. Fourth, many services and amenities (such as good restaurants, cultural facilities) are important aspects of quality of life, and determine the attractiveness of a place. This is becoming increasingly relevant in the race to attract talent (for example Florida, 2002).

There is a rich and growing literature that seeks to understand the sources of urban economic growth in the context of a knowledge-based economy. Several theoretical and empirical studies have pointed at the role of the local knowledge base as a determinant for urban economic development (like Lawton Smith, 2003; Goldstein and Renault, 2004). Several studies discuss the importance of local knowledge spillovers, and analyse the relation between local and non-local linkages for innovation processes (see, for instance, Simmie, 2002; Simmie *et al.*, 2002; Howells, 2002). Florida (2002) has catalysed discussions on the crucial role of human capital in the urban economy and the ability of urban regions to attract talent. There is a large literature on the development of knowledge-based clusters in cities (for an overview, see Baptista, 2000; or van den Berg *et al.*, 2001) in urban regions. Other recent contributions focus on governance issues related to the local knowledge economy (Knight, 1995; Kearns and Paddison, 2000; Docherty *et al.*, 2004). This list can easily be extended. Van Winden *et al.* (2007) have used many of the recent insights to construct a coherent framework that helps to assess, compare and analyse the development of the knowledge-based economy in urban regions. Following Polese (2005), they recognise that the relevant margins for urban growth are set on the national or even international level. Macro-economic conditions, structural economic change, legal frameworks, national policy frameworks and cultural factors play important roles.

As previously mentioned, the key question of our current study concerns the role of manufacturing in the knowledge-based urban economy. To address this issue, we combine insights from industrial dynamics and regional development, and consider two main levels of analysis: 1) The *regional spatial-economic context* in which the local manufacturing activities take place, to what extent are regional economic assets conducive to the development of manufacturing activity? and 2) the *configuration of regional networks*, how is manufacturing activity linked to knowledge-based activities, functionally and spatially, and what dynamics can we observe?

Figure 2.9 shows our framework of analysis. The manufacturing industry in a region is represented as a house that rests on a number of 'foundations' (regional characteristics) and produces a certain 'performance'. Furthermore, the industry is influenced by external developments such as (supra)national policies, technological trends and market developments. Below, the elements of the framework are elaborated.

Regional foundations: the spatial economic context

Manufacturing activity develops and evolves in a specific local economic and social context. We unravel this context into five main elements. The first is the wider economic base of a region (like sector mix, types of firms and size distribution of firms). Industries partly rely on local inputs and will develop better if the 'economic soil' in a city is in better shape. (Note that for different types of industries, different types of economic base may be favourable.)

The *knowledge base* of a region is a second highly significant foundation, especially relevant for the upgrading potential of local manufacturing industries. The regional knowledge base includes the presence and quality of specialised knowledge institutions in the region, such as universities, laboratories, etc., and also the qualification level of the workforce. The knowledge infrastructure can be a key asset for a manufacturing cluster. Universities 'produce' new staff in the form of graduates, and research institutes can be a valuable source of new knowledge. This is only the case, however, when the knowledge infrastructure is not self-contained but rather well aligned with companies in the cluster. The quality of the knowledge base can be measured with available data on educational levels of the working population, patents, R&D expenditures and the share of high-tech industries and knowledge-intensive services in the economy.

Quality of life is the third aspect of the spatial economic context. The relation between manufacturing and quality of life is an intricate one. Manufacturing industries (especially of the heavy kind) are notorious for having a negative impact on quality of the local environment. In public opinion, they are considered as sources of pollution of air, water, soil and landscape. However, in the last decade, the environmental damage caused by manufacturing has decreased drastically (at least in Europe), thanks to a combination of stricter environmental regulations, technological progress and huge investments by industrial companies. Furthermore, it may be increasingly in the self-interest of manufacturing industries to keep the environment clean: they become more knowledge-intensive, and thus increasingly thrive on highly-skilled people. These people prefer attractive and non-polluted places as locations to work and live. In a more general perspective, Florida (2002) finds that quality of life in a place is a key determinant to attracting and retaining these people. 'Talented people do not simply select a place to work based on the highest salary, they are typically concerned with a whole series of place-based characteristics' (Florida, 2002). Besides the environment, attributes of quality of life are an attractively built environment, high-quality houses, attractive city parks, attractive natural surroundings, a rich variety of cultural institutions, high-quality hospitals and (international) schools.

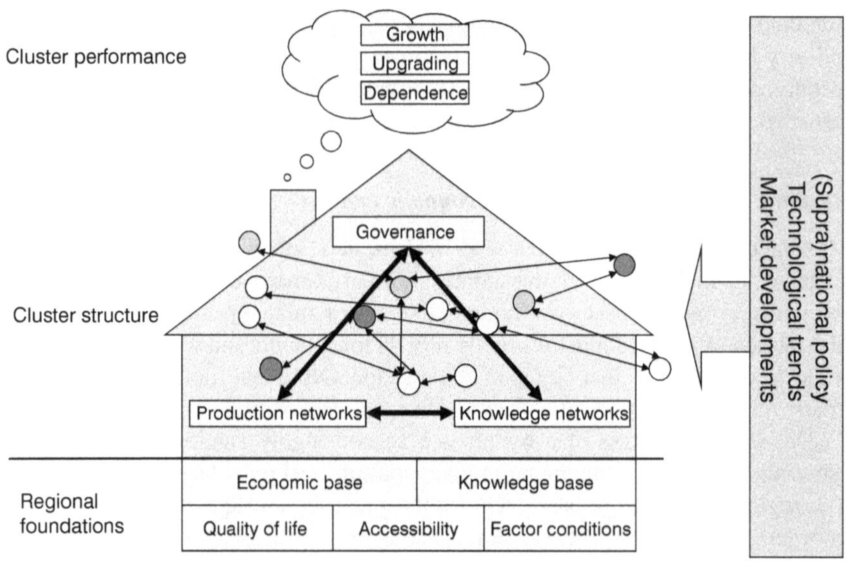

Figure 2.9 Framework of analysis (source: own elaboration)

Fourth, *accessibility* is a crucial regional context aspect for manufacturing industries. Manufacturing plants need good connections for goods transport (to tranship the manufactured goods to the customers, and for the timely delivery of inputs), and for passenger transport. The latter is becoming more significant, with the increased 'knowledge component' in many manufacturing industries and the increasing intensity and scope of networks in which they operate. For innovation and new knowledge development, manufacturing firms become more reliant on the inputs of foreign experts and knowledge centres abroad. Furthermore, the ongoing internationalisation of many (supplier) industries reinforces the importance of connections for passenger transport. Manufacturing companies increasingly become part of international conglomerates and need more frequent interaction with parent and sister companies.

Finally, the *factor conditions* matter for the dynamics of manufacturing. We discern two types of factor conditions: 'factors costs' in the region (prices of land, real estate, capital and labour), and 'soft factors' including social capital and cultural attitudes. Evidently, factor costs are highly significant for the development of manufacturing industries. Comparatively high labour costs will, ceteris paribus, reduce labour intensity of production, promote the substitution of labour for machines, and/or urge companies to compete on innovation and quality rather than cheap production costs. Likewise, high land or real estate costs will drive out space-intensive industries, or provoke innovations to make manufacturing less space intensive. Soft factors are important to explain cooperative behaviour within a region (or the absence of it!).

Cluster structure, networks and governance

The regional spatial economic context sets the margin for the development of manufacturing clusters. As a second step of our analysis, the industry itself (the middle part of the house) is described and analysed. In our different case studies, we select one specific manufacturing industry that we submit to an in-depth analysis. In particular, for a number of companies, we analyse how physical production ('manufacturing') is related to the more knowledge-intensive parts of the value chain (innovation, R&D), and how these relations are maintained in space.

Thus, in each case study, we will conduct a qualitative *analysis of networks*; this is the central part of each case study. We analyse the dynamics of inter-organisational networks, with a particular focus on the spatial dimension of these networks (regional, national, global). We put the manufacturing function (the production location) in the centre, and analyse how this manufacturing is linked up with other activities. In this respect, we make the distinction between two types of firm activity: exploration (the development of new capabilities) and exploitation (the efficient use of current assets and capabilities). Based on these activities we distinguish two types of networks: 'knowledge networks' which include exploration activities and 'production networks' that focus on exploitation activities.

Cluster performance

In this study, we assume that performance of the regional industry is dependent on the foundations described above, as well as industry-specific aspects (the size and composition of the cluster, the network structure and the governance of the cluster). On top of that, we assume that cluster performance is also affected by external developments such as market trends, technological developments and (supra)national government policies (the box at the right side of the house). It is not easy to measure the 'performance' of a manufacturing industry within a region. In this study, we identify three dimensions of performance. For each dimension, we will identify the direction of development of the cluster under consideration.

The first dimension is *growth* in quantitative terms. We will measure growth in terms of jobs, value added, turnover and number of units produced, as far as data are available, at least over a five-year period. The second performance dimension is *development*. This is a more qualitative measure, where we look at the upgrading of the cluster: does the cluster become more knowledge-intensive and technologically sophisticated, and does it move up in the value chain? For this aspect we will identify whether firms in manufacturing hire more highly skilled staff, and whether they increase their cooperation with knowledge partners. The third aspect of performance is *dependence/vulnerability*. Here, we will assess a) whether manufacturing clusters depend on one single firm, or just a few large companies; if so, they are highly vulnerable for an eventual downturn; b) whether manufacturing clusters depend very much on expertise and technology that comes from outside the region or abroad; if yes, their vulnerability is likely to be higher; and c) whether companies in the manufacturing cluster are steered from far away headquarters; if so, they are more vulnerable to decisions by these headquarters.

In many countries, politicians and citizens are deeply worried by the job losses in manufacturing industries. This is even more the case in countries that do not manage to compensate this loss with the creation of new jobs in other sectors, and subsequently experience a rise in unemployment. Policymakers and opinion leaders are sharply divided regarding the interpretation of de-industrialisation and propose a wide variety of policy answers, varying from protectionism to radical reforms to increase the flexibility in the labour market. For this study, however, the focus is not on national policy but on the scope for urban and regional policy.

In our cases, we address the role of urban management in the development of urban manufacturing industries. Based on an analysis of all the elements described above, we aim to identify options open to urban policymakers regarding the local manufacturing industry for the benefit of the entire local economy. From the outset, it is clear that policy interventions can be indirect (focusing on the foundations that underlie the development of the industry), or more direct (targeting the industry itself). In this study, we emphasise the last category without losing sight of the first one.

2.7 Research design and methodology

In this study, we apply the framework in a number of cities/urban regions by deploying a case study methodology (Yin, 2003; Dul and Hak, 2007). By doing so, we looked simultaneously at different levels of analysis, namely industry, networks and urban regions. Moreover, to unravel a large diversity of experiences, in each city we selected one specific manufacturing industry to analyse in-depth (see Table 2.2 and Figure 2.10).

The manufacturing industries were selected taking into consideration the history and structural relevance for each specific city/urban region. With the exception of Shanghai, all the cities/urban regions analysed have a long-standing history in the chosen sector. The automotive industry is slightly overrepresented in our cases, but there are noteworthy differences within. While in Paris and Ostrava the focus is on the mass-market segments, the Munich case focuses on the premium and luxury segment; in Eindhoven the focus is primarily on trucks and buses. Moreover, the case of São Paulo sheds particular light on the links between automotive and the bio-fuels industry.

On what concerns its 'manufacturing profile', our cities/urban regions can be divided into five groups:

1 'Traditional' industrial cities in transition (Dortmund, Porto, Rotterdam),
2 Post-communist industrial cities (Ostrava),
3 Advanced diversified cities (Munich, Paris),
4 Smaller industry towns (Eindhoven, Turku),
5 Emerging giants (São Paulo, Shanghai).

Each group is different in terms of 'regional foundations' and it is reasonable to assume that changes in global production systems have a different impact on

Table 2.2 Case studies: cities and industries

City/urban region	*Country*	*Manufacturing industry*
Dortmund	Germany	Metal–electro
Eindhoven	The Netherlands	Automotive
Munich	Germany	Automotive
Ostrava	France	Automotive
Paris	Portugal	Automotive
Porto	Czech Republic	Metal–electro
Rotterdam	The Netherlands	Food
São Paulo	Brazil	Automotive
Shanghai	China	Automotive
Turku	Finland	Shipbuilding

Source: author elaboration

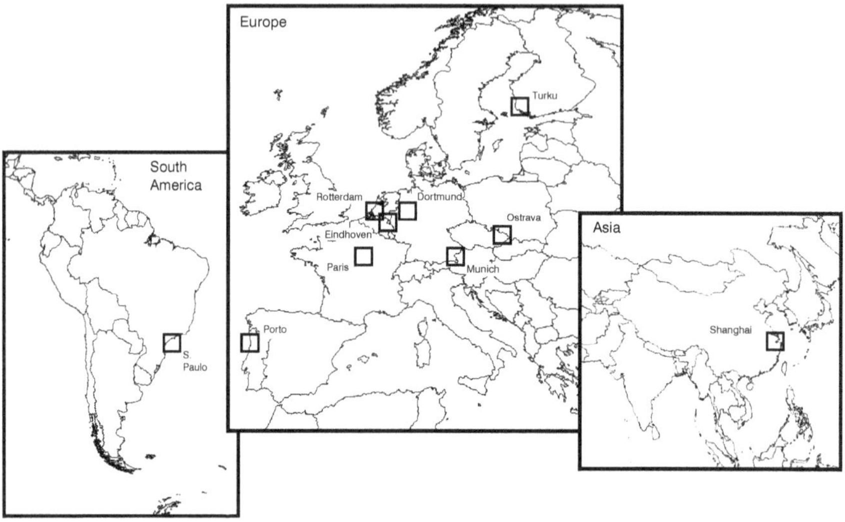

Figure 2.10 Geography of the case studies (source: own elaboration)

different types of cities. In our synthesis (Chapter 13), we systematically analyse whether this is the case. The cases also differ from each in terms of the 'weight' of industrial activities in their economy. Figure 2.11 maps our cases along two structurally distinctive factors: a) the 'industrial intensity' of the regional economy and b) the regional R&D intensity. All our cities and urban regions (except Paris) show a relatively high 'industrial intensity' in their economies, measured by the share of industrial gross added value generated in the total economy. Ostrava, Shanghai and Turku present the most 'industrial' economic bases, while Dortmund, Paris and Munich are the ones where the share of the service sector in the total economy is higher.

The cities differ substantially in terms of size, measured by number of inhabitants, and 'technological intensity', measured by the R&D spending as a share of regional GDP. Paris and Shanghai are 'world cities' with more than 10 million inhabitants, whereas the urban agglomerations of Turku and Eindhoven count less than 1 million inhabitants. In terms of R&D spending, Munich, Paris and Turku present values well ahead of the European average, while Porto, Rotterdam and Dortmund still lag in terms of 'technological' intensity. The present share of R&D investments in Shanghai is remarkable; it is already higher than the European average.

Methodological procedures

The field work and case studies were developed from February 2006 to December 2007, in the cities and urban regions under analysis.[1] For each case study we triangulated a diverse array of quantitative and qualitative secondary data (e.g. industrial reports, statistical sources, previous studies, press releases, etc.) with primary qualitative data collected through in-depth and semi-structured interviews with a dozen key players in each of the cities/urban regions under analysis. For each of the interviews, we used a semi-structured questionnaire, focusing on the different issues under analysis in our research framework, allowing for comparability and cross-case analysis. In total we held 123 interviews with high-level representatives of firms and corporations (CEOs, managing directors, and marketing, R&D, sales and procurement managers), universities and R&D institutes (university deans, directors of R&D centres, professors and engineers), external experts (in the region or in the specific industry), representatives of different government tiers (mainly local, supralocal and regional) and intermediate organisations (development agencies, sectors and cluster associations, etc.) – see Table 2.3.

Each interview took between 60–90 minutes and detailed interview reports were produced afterwards. Following the procedures of Merton and Kendall (1946) and Kincaid and Bright (1957), the interviews were conducted simultaneously by at least two of the authors, in order to reduce interpretation bias and systematically explore new emerging issues and relations. One preliminary case study report was sent to all the interviewees in each region for comments, to enhance the validity of the information and relations conveyed. During the field work process, the framework of analysis was adapted and fine tuned as new relations and important

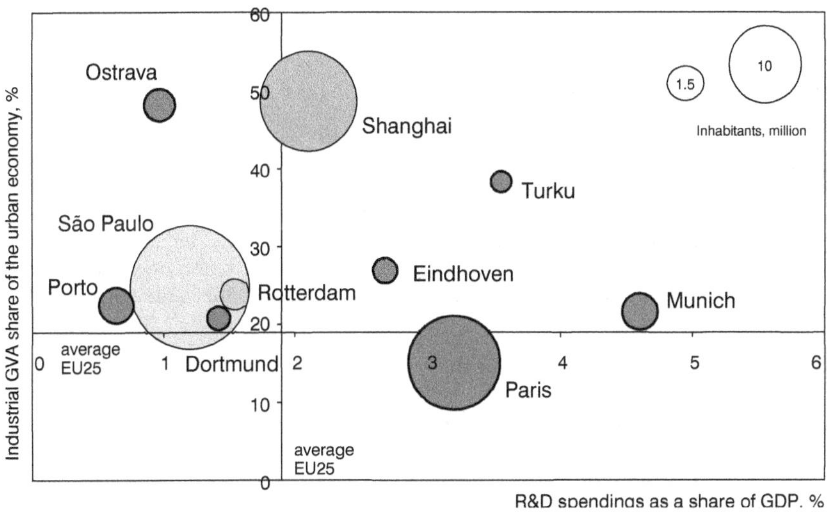

Figure 2.11 Mapping the urban regions (source: own elaboration; data from EUROSTAT and Shanghai Statistical Yearbook, 2006). Note: urban regions depicted correspond to the statistical NUTS3 level for Industrial Gross Added Value share and size, and NUTS2 level for R&D spending as a share of GDP. Exceptions are made for the largest and 'world cities' like Paris, São Paulo and Shanghai, which are based on national standards for delimitation of the urban agglomeration(s).

Table 2.3 Case studies: interview distribution

City/Urban Region	*Company*	*Expert*	*Govern-ment*	*HEI / R&D*	*Inter-mediaries*	*Total*
Dortmund	4	–	4	2	2	12
Eindhoven	6	–	–	2	–	8
Munich	4	1	5	2	–	12
Ostrava	5	–	2	2	5	14
Paris	4	1	–	–	2	7
Porto	4	–	2	2	2	10
Rotterdam	7	–	–	2	1	10
São Paulo	9	2	6	3	3	23
Shanghai	9	3	1	–	1	14
Turku	8	–	1	–	4	13
Total	60	7	21	15	20	123
%	49	6	16	12	17	100

Source: own elaboration
Note: HEI = higher education institutes

mechanisms were inductively revealed from the analysis (Eisenhardt, 1989). After the conclusion of the set of case studies, a cross-case analysis was deployed to find relevant similarities, differences and provide evidence of the mechanisms proposed in the framework. The final results of the study were presented in an international conference in the City of Dortmund on 17 January 2008.

Note

1 In our analysis we considered not a fixed administrative unit of analysis (a city or a region), but defined the unit of analysis (usually functional and larger urban regions) in a flexible fashion according to the territorial span of the sector concentration and of its main players, as well as their perception about what constituted the 'urban region'.

Bibliography

Amin, A. and P. Cohendet (2004) *Architectures of knowledge: firms, capabilities, and communities*, New York: Oxford University Press.

Asheim, B.T., and L. Coenen (2005) 'Knowledge bases and regional innovation systems: comparing Nordic clusters', *Research Policy*, 25:5, pp. 655–70.

Asheim B., and M. Gertler (2005) 'The geography of innovation: regional innovation systems', in Fagerberg, J., D. Mowery and R. Nelson (eds) *The Oxford Handbook of Innovation*, 291–317. Oxford: Oxford University Press.

Bahrami, H. and S. Evans (1995) 'Flexible recycling and high-technology entrepreneurship', *California Management Review*, 37:3, 62–89.

Baptista, R. (2000) 'Do innovations diffuse faster within geographical clusters?', *International Journal of Industrial Organization*, 18, 515–35.

Bathelt, H. (2007) 'Buzz-and-pipeline dynamics: towards a knowledge-based multiplier model of clusters', *Geography Compass*, 1, 1282–98.

Bathelt, H., A. Malmberg and P. Maskell (2004) 'Clusters and knowledge: local buzz, global pipelines and the process of knowledge creation', *Progress in Human Geography*, 28, 31–56.

Berg, L. van den, E. Braun and W. van Winden (2001) *Growth clusters in European metropolitan cities*, Aldershot: Ashgate.

Berger, S. (2005) *How we compete: what companies around the world are doing to make it in today's global economy*, New York: Doubleday.

Castells, M. (1996) *The rise of the network society*, Cambridge: Blackwell Publishers.

Coe, N. and T. Bunnell (2003) '"Spatializing" knowledge communities: towards a conceptualization of transnational innovation networks', *Global Networks*, 3, 437–56.

Daniels, P.W. and J.R. Bryson (2002) 'Manufacturing services and servicing manufacturing: knowledge-based cities and changing forms of production', *Urban Studies*, 39:5/6, 977–91.

Docherty, I., S. Gulliver and P. Drake (2004) 'Exploring the potential benefits of city collaboration', *Regional Studies*, 38, 445–56.

Doz, Y., J. Santos and P. Williamson (2001) *From global to metanational: how companies win in the knowledge economy*, Boston, MA: Harvard Business School Press.

Dul, J. and T. Hak (2007) *Case study methodology in business research, Rotterdam School of Management*, Rotterdam: Erasmus University Rotterdam.

Eisenhardt, K. (1989) 'Building theories from case study research', *The Academy of Management Review*, 4, 532–50.

Financial Times (2005a) 'Conti's formula for racing ahead', 30 March 2005.

—— (2005b) 'Made in Japan: Tide of history is halted in search for quality', 30 March 2005.

Florida, R. (2002) *The economic geography of talent*, Pittsburgh, PA: Carnegie Mellon University.

Giarini, O. (1997) 'Notes on economics, globalisation and insurance', *Information Letter 152*, Geneva Association.

Giuliani, E. (2005) 'Cluster absorptive capacity: why do some clusters forge ahead and others lag behind?', *European Urban and Regional Studies,* 12, 269–88.

Goldstein, H.A. and C.S. Renault (2004) 'Contributions of universities to regional economic development: A quasi-experimental approach', *Regional Studies*, 38:7, 733–46.

Greenfield, H.I. (1966) *Manpower and the growth of producer services*, New York: Columbia University Press.

Hansen, N. (1994) 'The strategic role of producer services in regional development', *International Regional Science Review*, 16:2, 187–95.

Howells, J.R.L. (2002) 'Tacit knowledge, innovation and economic geography', *Urban Studies*, 39, 871–84.

Illeris, S. (1996) *The service economy: a geographical approach*, Chichester: John Wiley & Sons.

Jarillo, J.C. (1993) *Strategic networks: creating the borderless organisation*, Oxford: Butterworth-Heinemann.

Kearns, A. and R. Paddison (2000) 'New challenges for urban governance', *Urban Studies*, 37, 845–50.

Kincaid, H. and M. Bright (1957) 'Interviewing the business elite', *American Journal of Sociology*, 63–3, 304–11.

Knight, R.V. (1995) 'Knowledge-based development: policy and planning implications for cities', *Urban Studies*, 32:2, 225–60.

Lawton Smith, H. (2003) 'Knowledge organisations and local economic development: The cases of Oxford and Grenoble', *Regional Studies*, 37, 899–909.

Malmberg, A. and P. Maskell (2006) 'Localized learning revisited', *Growth and Change*, 37, 1–18.

March, J. (1991) 'Exploration and exploitation in organisational learning', *Organization Science*, 2:1, 101–23.

Merton, R. and P. Kendall (1946) 'The focused interview', *American Journal of Sociology*, 51–56, 541–57.

Nooteboom, B. (2004) 'Innovation, learning and cluster dynamics', *Research Paper ERS ERS-2004-006-ORG*, Erasmus Research Institute of Management (ERIM), RSM Erasmus University.

—— (1999) *Inter-firm alliances: analysis and design*, London: Routledge.

—— (1994) 'Innovation and diffusion in small firms: theory and evidence', *Small Business Economics*, 6:5, 327–47.

—— (1992) 'Governance and competence: how can they be combined?', *Cambridge Journal of Economics*, 28, 505–25.

Nooteboom, B. and V.A. Gilsing (2004) 'Density and strength of ties: a competence and governance view', *ERIM Research Series ERS-2004-005-ORG*, Erasmus University Rotterdam, 29.

OECD (2005) *STI scoreboard 2005*, OECD Paris.

—— (2006) 'The changing nature of manufacturing in OECD countries', *OECD Science, Technology and Industry Working Papers*, 2006/9, OECD Publishing.

Polèse, M. (2005), Cities and national economic growth: a reappraisal, *Urban Studies*, vol. 42, pp. 1429–1451.

Porter, M.E. (2000) 'Location, competition, and economic development: local clusters in a global economy', *Economic Development Quarterly*, 14:1, 15–34.

Quinn, J.B. (1992) *Intelligent enterprise: a knowledge and service based paradigm for industry*, New York: Free Press.

Simmie, J. (2002) 'Knowledge spillovers and reasons for the concentration of innovative SMEs', *Urban Studies*, 39, 885–902.

Simmie, J., J. Sennett, P. Wood and D. Hart (2002) 'Innovation in Europe: a tale of networks, knowledge and trade in five cities', *Regional Studies*, 36, 47–64.

Storper, M. and A.J. Venables (2004) 'Buzz: face-to-face contact and the urban economy', *Journal of Economic Geography*, 4, 351–70.

Unido (2004) 'Inserting local industries into global value chains and global production networks: Opportunities and challenges for upgrading', Working Paper, Vienna: Unido

Winden, W. van, L. van den Berg and P.M.J. Pol (2007) 'European cities in the knowledge economy: towards a typology', *Urban Studies*, 44:3, 525–50.

Yin, R. (2003) *Case study research – design and methods*, Applied Social Research Methods Series. Thousand Oaks, CA: Sage.

3 Dortmund: metal-electro sector

3.1 Introduction

Dortmund is one of the largest cities in the polycentric industrial Ruhr area. The Ruhr area was an industrial core of Europe and one of the most prosperous regions. However, since the 1970s the dominant 'smokestack industries' (such as coal, iron and steel industries) have declined steeply, and the region has suffered from this de-industrialisation process resulting in high unemployment rates (16% in 2005), above the German average of 12.5% (RVR, 2006a). The de-industrialisation process and upgrading differ widely within the Ruhr area; some cities still have a strong industrial base and have successfully upgraded industrial sectors (the region has gained a competitive advantage in environmental research technology, energy supplies and waste disposal) (Hospers, 2004), while in other cities manufacturing nearly disappeared. Dortmund is an example of the latter, as it opted for a renewal strategy (the 'Dortmund project') focusing on three knowledge-intensive service sectors, information and communication technology (ICT), microsystems technology (MST) and logistics, while the pillar industries of the past (beer, steel and coal) nearly disappeared. Dortmund has the lowest share of workers in the manufacturing industries; only 13% of the workers are active in manufacturing, whereas in the Ruhr area on average this is 21%.[1]

Despite the low share of manufacturing in Dortmund's economy (in terms of employment), manufacturing is still important, especially in a regional perspective due to the still-large industrial base in other parts of the Ruhr area. Moreover, Dortmund's service industries and knowledge infrastructure might be linked with the manufacturing sector in order to upgrade old industries and to gain competitive advantage over other regions. In this study, we analyse the changing role of manufacturing in value chains, particularly the (geographical) link between manufacturing and other operational functions (such as logistics) on the one hand, and R&D, design and other knowledge-related activities on the other. In this chapter on Dortmund and the wider Ruhr area, we focus on the metal-electro sector, which accounts for nearly 70% of the employment in the manufacturing sector in Dortmund (Stadt Dortmund, 2007). The metal-electro sector, which produces various metal, electronic and plastic products, has strong linkages with classical manufacturing industries in the region (such as the steel industry), as well as with modern research fields, like new materials.

The rest of the chapter is structured as follows. In section 3.2, we put the metal-electro sector in the regional economic context of Dortmund and the Ruhr area. Next, in section 3.3 we analyse the metal-electro sector in Dortmund and the Ruhr region. In section 3.4 we discuss the sector networks in the region where we investigate the linkages of the cluster (within and outside the region), and in section 3.5 we discuss the role of the government and cluster organisations. In the last section (3.6) we draw conclusions.

3.2 Foundations of the Dortmund region[2]

Dortmund is a medium-sized city with nearly 600,000 inhabitants. It is the largest city in the Ruhr area, which, with 5.3 million inhabitants, is one of the most densely populated areas of Germany. One of the major sectors within manufacturing is the metal-electro sector, which has several linkages with other manufacturing and service industries. In order to understand the current dynamics of the metal-electro sector, this section puts the sector in the perspective of the regional economic context. This is done by describing the regional economic 'foundations' on which the metal-electro sector in the Dortmund region rests: economic base, knowledge base, accessibility, quality of life and factor conditions.

Economic base

Dortmund used to be known as the 'city of coal, steel and beer'. These industries formed the main economic activities of Dortmund until the 1980s. In 1951, the city had the largest steel output of Germany, and in 1955 it had a coal output of 13 million tons. In the 1960s Dortmund had eight breweries which produced 10% of the German beer consumption. However, since the 1970s, the pillar industries have been in crises that resulted in the closure of the last mine in 1987, the closure of the last blast furnace in 2001, and the last two independent breweries of the city were taken over by Oetker Group (Stadt Dortmund, 2005). The closure of the last blast furnace in 2001 concerns Thyssen Westfalen, which was dismantled in Dortmund and rebuilt in China, a literal example of the shift of manufacturing to low-cost countries (see Box 3.1).

The shift of the Westfalen plant is illustrative for the de-industrialisation process in the Ruhr area. Although the Ruhr area is still one of the largest industrial and mining areas of Europe, it is gradually transforming into a modern service economy. The decline of the eight largest manufacturing industries in the Ruhr area for the 1984–2000 period is depicted in Figure 3.1. The steel and the mining industries have been hit the most with a decline in employment of over 50%. Two industries of the metal-electro sector show different performances: the mechanical engineering industry has a position in the middle, with a decline of about 40%, while the electrical engineering industry performs the best of the eight largest manufacturing industries with a decline of less than 5%. While the manufacturing

Box 3.1 Movement of the Westfalen works

Westfalenhütte on its way to China.

The blast furnaces had burned at Thyssen's Westfalenhütte in Dortmund, went out last year, however, next year they will be built up again – and in China. For months, six to eight hundred workers from China have been busy dismantling the plant and soon they will establish the plant again on the west of Shanghai.

The Westfalenhütte belonged to the so-called 'dry smelting works'. It had no access to waterways, all raw material had to be delivered by rail and road. That was not only inconvenient, but also inefficient. This made the final product more expensive, and therefore it was not competitive anymore, particularly as there is enough cheap steel in Europe. Therefore, ThyssenKrupp was forced to close the plant at the beginning of 2001.

In October, the sales agreement was signed. And already in February 2001 the demolition work had started. Shagang has sourced out the deconstruction of the plant to three different companies. They dismantle the plant in Germany and build it up again in China. The works includes a sinter plant, which prepares the iron ore for the blast furnaces, two blast furnaces, one steel mill with two transit facilities and one hot rolling mill.

'One used to see workers all over the plant' according to Sigrid Kuhnke-Glöckner, employee of ThyssenKrupp Steel's sales division. 'Nowadays one sees only machines here. At most, there is somebody in the control office or in the control station. In the best times, we used to have 30,000 workers. Before the closure, we still had 2,400 and currently we have 1,700 workers. Nevertheless, this number will also be reduced. However, nobody would be fired; most are relocated to other parts of the company. Some workers used the social plan to retire and youths left voluntarily against compensation.' Regretfully, Miss Kuhnke-Glöckner adds: 'More generations of families worked at our company: grandfather, father and son. But this loyalty does not exist anymore. A pity!'

Source: cited from Die Tagepost (31-10-2002). Translated from German

industries have declined, most industries in the service sector have increased (Figure 3.2); business-orientated services especially show strong growth.

Within the Ruhr area, the de-industrialisation process differs between the cities and counties. The share of manufacturing employment in the total employment varies from 13% in Dortmund and Essen to 38% in Ennepe-Ruhr-Kreis, while

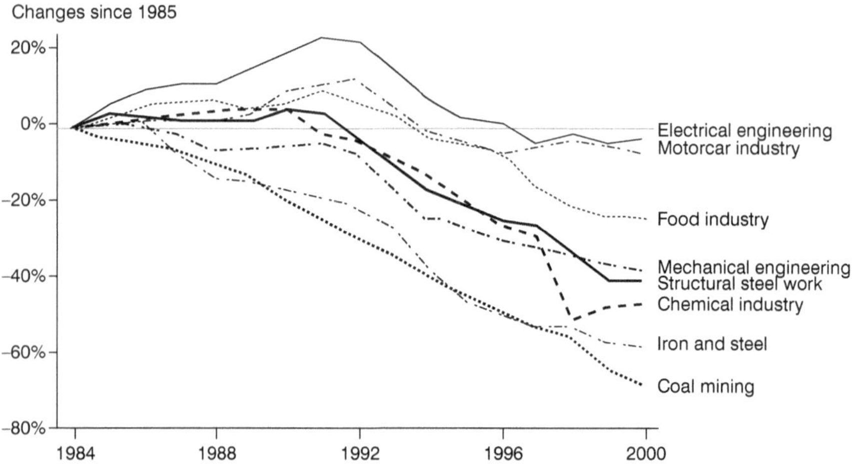

Figure 3.1 Development of employment in the eight largest manufacturing industries in the Ruhr area (%, 1984–2000) (source: Bömer, 2005)

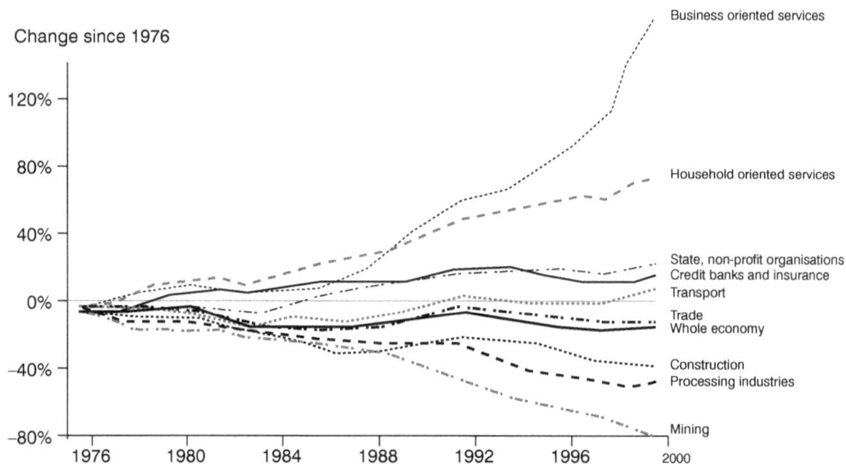

Figure 3.2 Development of employment in the ten largest economic industries in the Ruhr area (%, 1976–2000) (source: Bömer, 2005)

the average in the Ruhr area is 21% (see Figure 3.3). With a share of 86% of its employees in the service sector, Dortmund is one of the leading cities in the Ruhr area in the transformation process towards a modern service economy; by comparison, the average in the Ruhr area is 77%.

The Ruhr area and its cities suffer from the strong de-industrialisation, resulting in a relatively low GDP and high unemployment rates. Dortmund has a GDP per capita of €27,959 (in 2004), which is only slightly above the German average

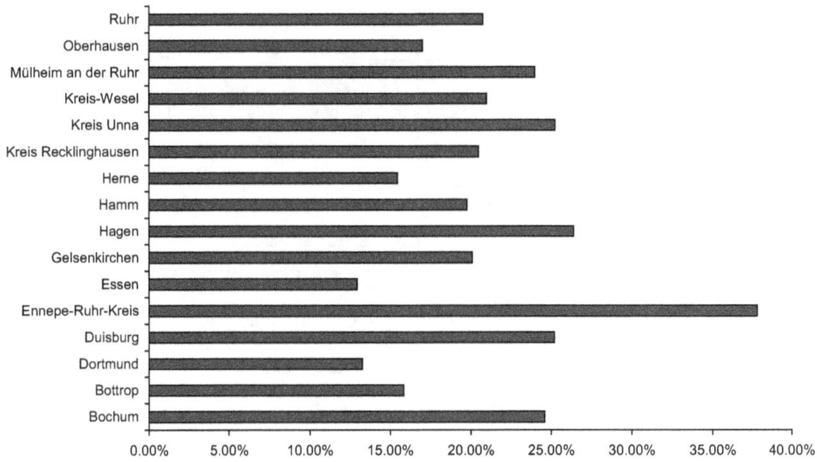

Figure 3.3 Share of manufacturing employment in the total employment in the Ruhr area, 2005 (source: www.Ruhrsite.de). Note that the supply of energy and water is excluded from the manufacturing sector.

(€26,754), which includes the values of eastern Germany. Of the ten largest cities (in number of inhabitants) in Germany, only Berlin has a lower GDP per capita (see Figure 3.4). Other indicators show that the performance of Dortmund is in line with the average of the Ruhr area; the purchasing power in Dortmund is a bit lower (€16,247, against €17,082 in 2004), while the GDP per employee is a bit higher (€57,247, against €55,855 in 2004). It should be noted that the GDP per capita in Dortmund increased from about €23,000 in 1995 to €27,959 in 2004 (Stadt Dortmund, 2006a). The unemployment rate increased from 15% in 1995 to 18.7% in 2006. Compared to other large German cities, the unemployment rate in Dortmund is high; only Berlin has a higher unemployment rate (see Figure 3.5). Also compared to the Ruhr area, which has an average unemployment rate of 15.1%, the unemployment rate in Dortmund is high. Only Gelsenkirchen has a higher unemployment rate of 20.6%.

Economic structure and manufacturing

From Figure 3.6 it becomes clear that Dortmund has a high share of workers in the modern service industries, such as banking and insurance, leasing and real estate. Within the Ruhr area, Dortmund is by far the most important location for the insurance industry; 70% of all employees in this industry work in Dortmund (RVR, 2006b).

Despite the relatively low share of manufacturing in the economy of Dortmund, manufacturing still has an important role in Dortmund's economy. A recent study by IW Consult (2007) on the future of manufacturing in Dortmund argues that the manufacturing sector functions as a central 'power wheel' in local economy. Many

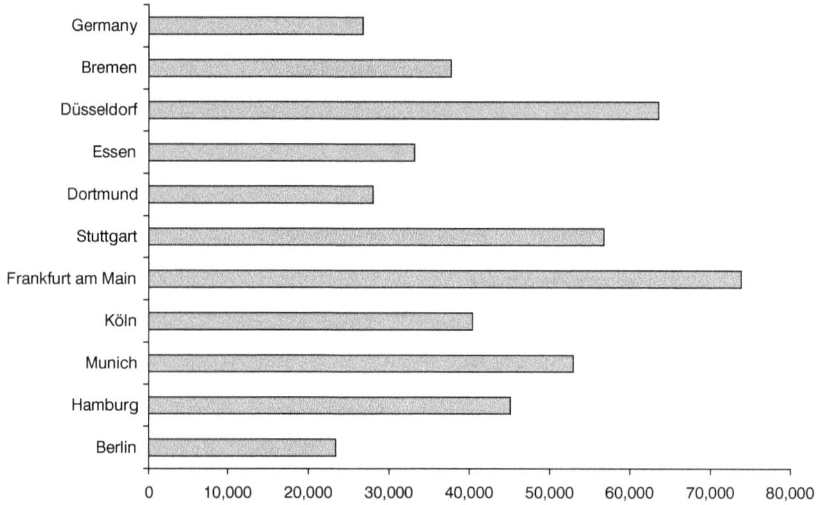

Figure 3.4 GDP per capita in Germany's ten largest cities in €, 2004 (source: Eurostat)

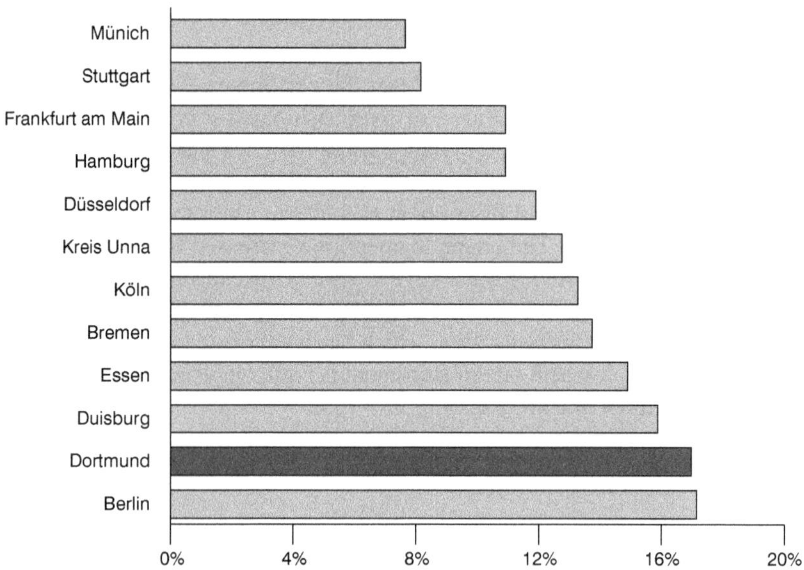

Figure 3.5 Unemployment rate in Dortmund compared to other German cities, 30 June 2006 (source: Stadt Dortmund, 2006a)

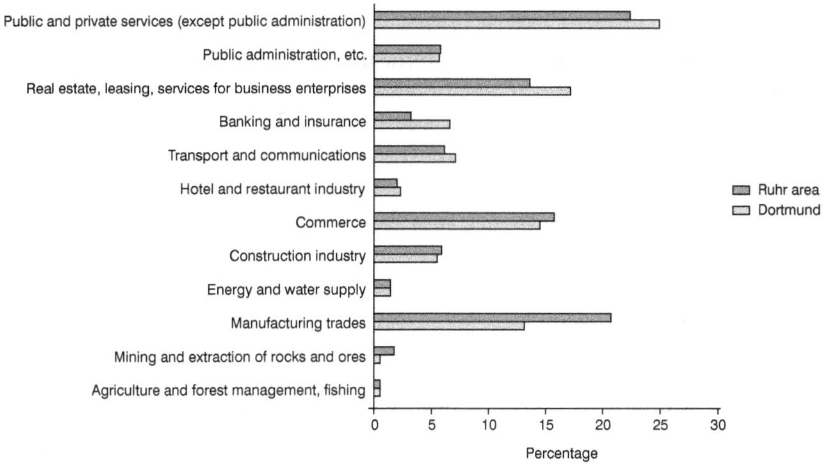

Figure 3.6 Economic structure in the Ruhr and Dortmund, percentage in total employment, 2005 (source: www.Ruhrsite.de)

services in Dortmund are related to manufacturing. For instance, ThyssenKrupp Schulte, a steel trading company, has narrow links with the steel industry in the Ruhr area. Furthermore, the services in Dortmund might serve the manufacturing industries in the surrounding areas of Dortmund. In the surrounding areas, the manufacturing sector still has a relatively high share; in Duisburg, Bochum, Hagen, Kreis Unna and Ennepe-Ruhr-Kreis the share of manufacturing is 25% or more (see Figure 3.3. Also, to the east of Dortmund, outside the Ruhr area, there are areas with a large share of manufacturing.

The study gives some additional reasons to show the importance of manufacturing in Dortmund: 1) it is a major contributor to exports, 2) it is a major source of innovation, and 3) it is the core of spatial clustering. The study also gives six branches that have the best growth chances in the future: chemical and pharmaceutical; rubber and plastic; automotive; machinery; radio, television and communication equipment; medical, precision and optical instruments. The last three industries are part of the metal-electro sector. Finally, the study makes clear that Dortmund has no large companies in the manufacturing sector. Apart from some establishments of ThyssenKrupp and Salzgitter AG, there are no companies with more than 500 employees (Stadt Dortmund, 2007). The structure of the manufacturing sector is dominated by many small and medium enterprises (SMEs), often owned via a family business structure. However, in the region (other parts of the Ruhr area) there are several large companies, such as ThyssenKrupp (headquartered in Duisburg/Essen), Opel and BP Germany in Bochum, and RWE AG and EON Ruhrgas AG in Essen (RVR, 2006a, b). This is in line with the metal-electro sector in general that is characterised by the presence of a few large companies and numerous SMEs.

Knowledge base

Dortmund has a good knowledge infrastructure with many research and educational facilities.

- *Technical University of Dortmund* (11,914 students; 2005) (Stadt Dortmund, 2006a);
- The *University of Applied Sciences Dortmund* (5526 students; 2004) and some smaller specialised universities of applied sciences (Stadt Dortmund, 2006a);
- *17 other scientific institutes*, including the Fraunhofer Institute for Software and Systems Engineering (ISST) (Stadt Dortmund, 2005);
- *Competence centres* offering several facilities and services (such as business support, qualification and 'soft infrastructure') for small companies and start-ups in specific industries. A competence centre for the machinery industry, one of the industries in the metal-electro sector, is the Centre for Robotics and Automation Dortmund ('Robotik- und AutomatisierungsCenter Dortmund', RACe);
- *Research departments of companies*. There are a number of manufacturing companies with specialised R&D departments, such as Murtfeldt Kunststoffe, which has a research department for the development of new sorts of plastics. There are no really large private research centres, which is related to the absence of large companies in Dortmund.

The city has some spatial concentrations of research facilities and companies in certain industries. One example is TechnologyParkDortmund, founded in 1985 together with TechnologieZentrumDortmund. The technology park and centre are located next to the Technical University of Dortmund and the Fraunhofer Institute for Material Flow and Logistics, aiming to increase knowledge transfer between research institutes and industry. Entrepreneurs in the technology centre can relocate to the technology park after a successful start-up. Altogether, more than 230 firms and about 8,500 employees are currently in the centre and the park, and these numbers are still increasing. The centre and park cover different technology fields (GfW, 2002), including 'electronics', which is relevant for the metal-electro sector.

The knowledge infrastructure is not limited to the local level. On the regional level, the Ruhr area has the largest spatial concentration of research institutes in Germany (Stadt Dortmund, 2007) and the total number of students in the universities and the universities of applied sciences is 148,127 (in 2005) (RVR, 2006a). In March 2007, the rectors of the three largest universities in the Ruhr area (the universities of Bochum, Dortmund and Duisburg-Essen) signed a cooperation agreement, the University Alliance Metropolis Ruhr, to combine various disciplines and to increase the critical mass (Ruhr University Bochum, 2007). Also, in other parts of the state of North-Rhine Westphalia (NRW), outside the Ruhr area, there are universities and research institutes, such as the University of Aachen. Several of these universities offer education for and are doing research in fields that are of import for the metal-electro sector, such as machinery/ mechanical engineering and ICT.

Spending on R&D and patent registrations are other relevant aspects for the knowledge base. If we compare the relative spending on R&D as share of the GDP, we see that the state of NRW has a relatively low share. NRW spent 1.78% of its GDP on R&D (in 2004), which is below the German average of 2.49% and below the Lisbon Treaty target of 3%. In addition, the R&D spending per employee (€113,810 in 2004) is below the German average (€116,880). The majority of the R&D spending (61% in 2003) is done by the private sector. After Bavaria and Baden-Württemberg, NRW has the most patent registrations with 16.9% of all German patents (in 2005). Within NRW, the majority of the patents (53%) are registered in Rhineland, followed by Westphalia (29%) and the Ruhr area (18%) (RWI, 2006).

Another part of the knowledge base is the ability of cities to attract highly skilled workers and to keep students. According to our interviewees, it is a positive factor for the labour pool that inhabitants are socially embedded to Dortmund. It is likely that students will stay after they have finished their study on the condition that they can find a job. Another positive development for the labour pool is that Dortmund has an increasing share of foreign students in the student population, from 4% in 1996 to 11% in 2005. Of other major cities in NRW, only Duisburg has a larger share of foreign students (about 16%) (Stadt Dortmund, 2006b). However, there are also some weaker points for Dortmund as a location for students and higher skilled workers. One is that Dortmund, just like the other cities in the Ruhr area, does not have a distinctive image and it is relatively unknown internationally. Another weak point is a lack of large companies that can offer (international) careers for students in Dortmund after they have finished their studies. Large, well-known companies from other regions, like Siemens and BMW in Munich and Daimler-Chrysler in Stuttgart, might attract students from Dortmund. However, as we have already noticed, there are several examples of well-known companies in the Ruhr area, such as ThyssenKrupp in Essen and Opel in Bochum. These companies may help to attract students to the region and keep students and workers in the region.

Just like many other European cities, Dortmund has some problems in the labour market. An important problem for the manufacturing sector is the negative image of manufacturing. Employees, especially young people and women, associate working in a plant with dirty and heavy work, and not as 'sexy'. This could lead to a labour shortage in the near future, especially if the ageing problem is taken into account. Currently, the dominant age group in the manufacturing sector is 42–45 years (interview). Youths are needed to follow this group. In order to address the problem, the government and other actors (such as Dortmund business associations) have taken several measures. For instance, the Chamber of Commerce in Dortmund organises meetings between older workers and youths in order to show the youths what happens in the manufacturing industry. Another problem that Dortmund and many other cities face is a shortage of trainee posts, despite that the gap between demand for and supply of trainees in Dortmund has decreased during the last years (Stadt Dortmund, 2006a). A last problem in the labour market is a qualitative mismatch: there is an oversupply of low or unskilled labour (mainly from the three old key industries) but there is a shortage of highly skilled knowledge workers in the three new pillar industries. But also there is a shortage of low-educated workers in the current manufacturing industries.

Accessibility[3]

The Ruhr area has always been a main corridor in Europe and has good connections with all major cities in Europe via all transport modes. It is located relatively close to some heavily populated areas in Europe, such as the Randstad area in the Netherlands and the greater Paris region in France. New markets and industrial areas in Central and Eastern Europe can also be reached quickly. In total, about 40% of the European population lives within a radius of 500km. The Ruhr area is directly connected to important highway, rail, air and inland waterway networks. Relevant for freight transport (for instance for the metal-electro sector) is a 272km long inland waterway network, with Duisburg as the largest inland port and Dortmund as the largest inland dock port of Europe. Both ports are multi-modal and are connected with important North Sea ports, such as Rotterdam and Antwerp. This is relevant for intercontinental transport, which gains importance due to the increasing globalisation and the rise of new growth markets outside Europe. Furthermore, relevant for freight transport is that logistics is a key industry in Dortmund (see economic base), which gives the city an additional asset for the input and output of goods for the metal-electro sector. Dortmund is also accessible by air and by high-speed train (HST) lines, which is particularly of interest for knowledge workers who increasingly work on international projects. The Ruhr area has 13 HST connections to the largest German and European cities, and via the intercontinental airport in Düsseldorf (16.6 million passengers; 2006) and the regional airport in Dortmund (2 million passengers) it has air connections to several European and intercontinental destinations.

Despite the good connections for both freight (of interest for the input and output of the metal-electro sector) and passenger transport, there are some problems with accessibility. As with most other polycentric regions, the Ruhr area has considerable bottlenecks in the road network leading to congestion. Especially on the B1, the main east–west axis in the region, and on the ring roads of cities, there are a lot of traffic jams during peak hours. Around Dortmund, however, the congestion is relatively low (Schönharting *et al.*, 2003).

Quality of life

The quality of life in Dortmund and the Ruhr area is generally improving. The manufacturing industries in the region have become cleaner through new production processes. Furthermore, heavy manufacturing industries are declining and a lot of former industrial sites are transformed for other functions or stay intact as industrial heritage. Despite the industrial image of the region with heavy industries and 'black coal mines', Dortmund has a lot of green spaces (19% of the land area in the city consists of forest land, green areas and public water areas). Outside the city, the Münsterland region and the Sauerland area offer possibilities for different recreational functions, such as cycling. Air pollution declined sharply in the 1994–2005 period and the crime rate declined in 2005 (compared to 2004) to a value of about 11.7 crimes per 100 inhabitants. This was lower than the neighbouring cities of Düsseldorf and Cologne, but higher than in Duisburg and Essen (Stadt Dortmund 2006a, b). In a city monitor of the 20 largest cities, Dortmund performed average (10th place) in terms of quality of life, according to a survey among 100 spokesmen or owners of companies in each city (Ernst & Young, 2006). Thus, Dortmund and the region have various possibilities for a good quality of life to attract highly skilled knowledge workers, who are also relevant for the metal-electro sector. A drawback, however, is that the city is relatively unknown internationally (it has no distinct image) and it faces competition from surrounding cities (such as Essen).

Factor conditions

In research on labour costs and labour productivity (2004, in Becker, 2006), it turns out that Germany is the most expensive location of 12 investigated countries in Western Europe, Japan and North America in terms of unit labour costs. Only Norway has comparably high wages. In terms of labour productivity, however, Germany ranks eighth out of the twelve investigated countries. Within Germany, NRW has gross annual wages in the manufacturing industry of €38,602 (in 2005), which is just below the Western German average of €39,474 (Bavarian Ministry of Economic Affairs, Infrastructure, Transport and Technology, 2007). In European perspective, a study in Manshanden *et al.* (2005) shows that the labour productivity in the Ruhr area is above the average of EU-15 countries (in 2004) and is comparable with the average of the 20 best performing European metropolitan regions. However, the growth of the labour productivity in the 1995–2004 period is 1%, just below the EU-15 average and the 'top 20' which both are about 1.5%.

An important factor that distinguishes Dortmund from many other cities in Germany and Western Europe is the fact that the city has many vacant spaces within the city borders and even within the central districts. Many of these 'brownfield' sites, former industrial areas for traditional industries, can be used for new industrial functions or can be transformed to new (urban) functions. For instance, the Westfalenhütte, with 460 hectares is the largest fallow land area in Europe,[4] and is planned to be redeveloped for mainly logistical functions, but there is also some space for industrial functions. The vacant spaces can also be an important asset to attract large manufacturing companies that are space intensive (such as certain companies in the metal-electro sector). A problem, however, is that many brownfields are (heavily) polluted and need to be cleaned and levelled before they can be transformed. This incurs high costs, which may lead to higher prices than competing sites in the Ruhr area.

3.3 The metal-electro sector in the Dortmund region

In this section, we describe the metal-electro industry in Dortmund and in the Ruhr area. First, we analyse some general data/statistics of the metal-electro industry. Second, we describe two kinds of actors: companies and educational and research institutes. Finally, we describe the geography of the metal-electro cluster in Dortmund.

Key features of the metal-electro sector in the Dortmund region

The metal-electro sector is by far the most important sector in the manufacturing sector in Dortmund; two-thirds of the employees in the manufacturing sector are working in the metal-electro sector (see Figure 3.7). In the 2000–2005 period, the total number of employees in the manufacturing sector declined by 24% to 25,035 workers.

The metal-electro sector showed a comparable decline of 26% to 16,801 workers. However, within the metal-electro industry there are large differences (see Figure 3.8). Outstanding is the halving of the basic metals industry in 2001–2002. This is largely caused by the closure of the Westfalen plant in 2001 (see Box 3.1). In total, the basic metals industry declined by nearly 60% to 3,262 workers, and its share in the total metal-electro sector dropped from 34% to 19%. Apart from the medical, precision and optical instruments (+6.2%) and office,

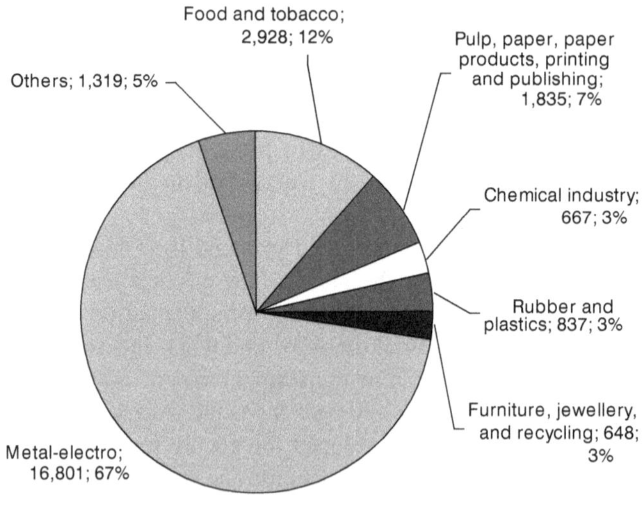

Figure 3.7 Employment structure in Dortmund's manufacturing sector, number of employees and relative share, 2005 (source: database city of Dortmund)

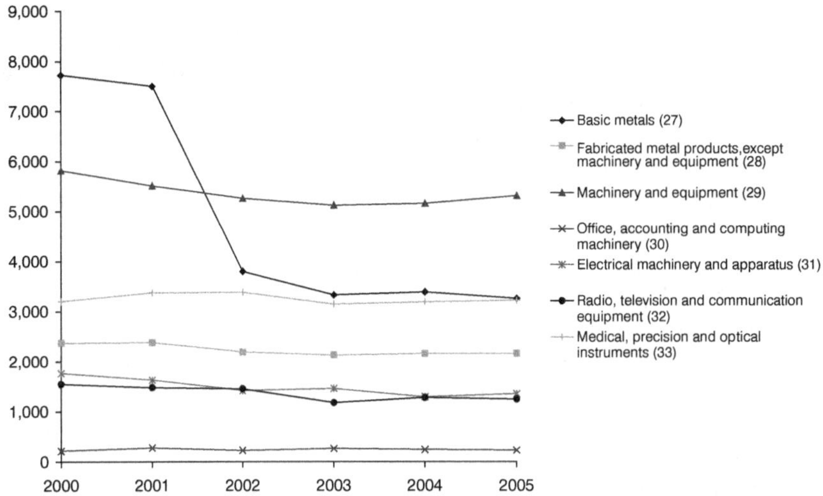

Figure 3.8 Employment development of industries in the metal-electro sector, number of employees (source: database city of Dortmund)

accounting and computing machinery industries (+6.9%), all other industries in the metal-electro sector had a declining employment. This can be explained by the fact that most of the industries are related to the basic metals industry, since basic metals forms an important input for many other industries.

Also on the Ruhr area level, the industries in the metal-electro sector have declined. However, there are large differences between the individual industries. The largest decline can be observed in the ‘manufacturing of machines for the creation and distribution of energy’ industry by 23.9% between 1999–2004. Also, in other industries of the metal-electro sector, a lot of jobs have disappeared; basic metals (–19.8%), radio, television and communication equipment (–17.8%), metal products (–16.6%) and machinery (–12.3%). Only the medical, precision and optical instruments industry has a relatively small decline in employment, of only 2.1%. A large part of the decline in employment in the industries in the metal-electro sector is probably related to the de-industrialisation in the Ruhr area. It should be noted that the job losses, in the absolute and relative sense, differ widely by location. In some locations, there are even increases in employment in certain industries of the metal-electro sector, like the machinery industry in Duisburg that has increased by 60.8% to 3,508 employees in 2004 (RVR, 2006a).

Companies in the metal-electro sector

Metal-electro is a very broad industry. Companies in the sector vary widely, from simple standardised metal-part manufacturers (with mainly lower educated workers) to highly innovative chipmakers (with mainly engineers). In total, about 600 companies in Dortmund are active in the metal-electro sector and related industries (Stadt Dortmund, 2007). Most of the enterprises are small and medium sized (family) enterprises, which are located in Dortmund for historical reasons. Typically, the founder lived in Dortmund, or the company used to be located close to the suppliers of steel, the major input in the area.

Some of the smaller companies in Dortmund and the region are specialised, niche players, and some are even among the market leaders in their market. These companies deliver highly innovative products, often customised for their customers. An example is Murtfeldt Kunststoffe, which produces, among others, highly innovative plastic chain guides for their customers in the machinery industry. These highly innovative manufacturing companies seem to have the best chances of surviving the de-industrialisation process, particularly because they also deliver services, such as R&D. It is notable that some of the companies have successfully changed their scope from 'classical' sectors (such as mining or steel) to more 'modern' ones (like ICT, bio-chemical and medical industries). An example of this is given in Box 3.2.

Although large multinational organisations (MNOs) are absent in Dortmund, there are exceptions, such as a subsidiary of Siemens VDO (which is a major customer of the metal-electro sector), Miebach (a large company in the machinery industry with over 1,000 employees, with its headquarters and some production

Box 3.2 GFG – an example of a company that successfully changed its scope

Gesellschaft für Gerätebau GmbH (GfG) was founded in 1961. The company developed and produced the first electric gas detector, which was used in coal mines. It was located in Dortmund, in the middle of the Ruhr area, one of the largest mining areas in Europe. When the mining industry in Europe went down, GfG developed many other products related to gas measuring and warning techniques and it found many other markets for its products, including communities, chemical industry, fire and civil protection, mining, power supply, food processing, shipping, steel industry, agriculture, and research institutes and universities.

Currently, the company belongs to the top 10 leading companies in the field of gas measuring and warning techniques and it has establishments in the USA, Singapore, Switzerland and South Africa. The company is one of the fastest growing companies in the branch and it expects its global sales to be doubled by 2009 and to be tripled by 2011, up to a value of about €90 million. To realise this growth, the company wants to expand its capacity in Dortmund.

Source: http://www.gasmessung.de/index.html (accessed 4 May 2007) and WAZ, 24-03-2007.

facilities in Dortmund), and some remaining plants of the large steel companies ThyssenKrupp and Salzgitter.

ThyssenKrupp still plays a key role in the Ruhr area, despite the closure or restructuring of some large manufacturing plants, such as the Westfalen plant in 2001. It still has its headquarters in the Ruhr area (in Duisburg and Essen), as do several highly innovative production plants and basic research centres. The main site of the Steel and Stainless divisions is located in Duisburg, which offers the company logistical advantages for the procurement of raw materials and short distances to major customers (ThyssenKrupp, 2006c). This is an indication that proximity matters for certain products in the steel industry.

ThyssenKrupp is a global company which is represented in more than 70 countries. Two-thirds of its sales come from abroad and more than half of its employees work in one of the 600 foreign-associated shareholdings. The company is aware of new growth markets in Southeast Asia (especially India and China), Latin America and Central and Eastern Europe, and therefore the company invests in these countries.[5] Apart from commercial reasons, ThyssenKrupp also invests in low-cost countries for financial reasons (low wages) and because of the presence of raw materials; an example is an investment of $2.4 billion for the construction of an iron and steel mill in Brazil. By benefiting from regional cost advantages (low labour costs) and the proximity of raw material sources, the company plans to supply low-cost slabs from the new Brazilian plant to its plants in Germany and in other parts of Europe and the USA.

Besides investments in low-cost countries, ThyssenKrupp also invests in production facilities in its home region, the Ruhr area. Two examples are:

- An investment of €33 million for the modernisation of a cold rolling mill in Dortmund in 2003. This made the Dortmund plant one of the most innovative plants for customised steel plates (mainly used in the automotive industry) in the world (ThyssenKrupp, 2003).
- An investment of €340 million for the construction of a new blast furnace and improving a current one on one of the Duisburg sites in 2008. Another €400 million is invested in Duisburg to remove bottlenecks and to optimize the production flow (ThyssenKrupp, 2006c,f).

The investments in Europe are mainly replacement investments, in contrast with the investments in low-cost countries which are extension investments. However, the investments in Europe are also important to make plants more innovative (for example for upgrading the production process). It seems to be the case that more high-tech production takes place in the Ruhr area, while more low-tech production takes place in low-cost countries. The investments in the Ruhr area are also important for job reasons. For instance, the investment in Duisburg will secure 1,200 jobs directly and 3,600 indirectly (ThyssenKrupp, 2006c).

R&D is a key function for ThyssenKrupp. In 2005–2006, the company spent €743 million on R&D, which is 1.5% of the sales. The share of the R&D spending on basic research stayed stable at about 30% during 2001–2006. About two-thirds of the spending on basic research was carried out in its home country of Germany. Nevertheless, the company has a global knowledge network with 85 international R&D centres, employing over 3,300 people. The majority of the scientists and engineers are specialised in (new) materials, production, process and electrical engineering, which are also important for the metal-electro sector. The company cooperates intensively with universities for joint research, knowledge transfer and education. The Technical University of Dortmund and Ruhr University Bochum are the partner universities in the Ruhr area. Other German partner universities are universities in Aachen, Berlin, Dresden, Hamburg-Harburg and Freiburg. Outside Germany, ThyssenKrupp cooperates with Tongji University in Shanghai (ThyssenKrupp, 2007).

Salzgitter is another important steel company that has several linkages with and activities in the metal-electro sector. For instance, its steel is used by companies in the metal-electro sector and the tube division is active in the field of the metal-electro sector. It has various companies in the Ruhr area, like Hoesch Spundwand und Profil (a company in Dortmund that delivers metal products for the construction industry), and the Dortmund-based KHS (a company in the machinery industry that produces machines for the packaging industry). Besides its headquarters and a plant in Dortmund, KHS has seven other plants in Germany and six abroad. KHS wants to locate close to its customers for short service and transport routes. The largest growth markets of the company are Asia (especially India and China), Latin America and Eastern Europe. Therefore, major investments are made in these countries. In 2006, the company acquired a majority stake in a Chinese firm, and in February 2007, it opened a new plant in India. The markets in Western Europe and the USA are saturated, but demand is generated in these markets through replacement investments, aseptic technology and new product development (KHS AG, 2007). There are no signs that production of KHS in Western markets will decline by the closure or restructuring of manufacturing plants in Europe.

Just like ThyssenKrupp, R&D is a key function for Salzgitter. In 2006, it spent €67 million on R&D, which is 0.8% of its sales. Moreover, the company is involved in cooperation projects with other market participants and research institutes with an overall budget of about €40 million. About 688 employees were engaged in R&D or related activities. Most of the research is done by Salzgitter Mannesmann Forschung (SZMF), the central R&D company of Salzgitter, and by Salzgitter Magnesium-Technology (SZMT). SZMF and SZMT are located in Salzgitter, in proximity to other departments (production plants and headquarters) of the company. However, depending on the function or the needs of the customer, there are also establishments in other locations. For instance, SZMF has also an establishment in Duisburg close to manufacturing plants (Salzgitter AG, 2007). This example suggests that R&D activities of Salzgitter require proximity of manufacturing.

Besides Salzgitter and ThyssenKrupp, there are more large companies and several SMEs relevant to the metal-electro sector in other parts of the Ruhr area and NRW. Some examples of the other large firms are the car manufacturers Opel in

Bochum and Ford in Cologne, which are important customers for companies in the metal-electro sector. The combination of many SMEs and a few large companies, not only in the Ruhr area, but in the rest of NRW as well, is characteristic for most of the industries (or 'sub sectors') in the metal-electro sector.

Education and R&D in the metal-electro sector

In section 3.2, we have seen that Dortmund and the region have a good knowledge base. In this section we investigate which institutes are of interest for education and R&D in the metal-electro sector in Dortmund and the region.

The Technical University of Dortmund (Technische Universität Dortmund, TUD) and the University of Applied Sciences (Fachhochschule Dortmund, FhD) are important actors in the training of highly skilled employees for the metal-electro industry. Both institutes offer several Bachelors and Masters courses in the field of metal-electro. Figure 3.9 shows that the number of students in machinery (TUD) and in electro-technical (TUD and FhD) increased during the 1999–2003 period.[6] Also other universities in the Ruhr area offer studies in the field of the metal-electro industry, like Ruhr University in Bochum and the University of Duisburg/Essen.

TUD is not only important because it trains (future) employees for the metal-electro sector, it also conducts research. It has various institutes that are engaged in the fields of the metal-electro sector. The major ones are:

- *Robotics Research Institute* (Institut für Roboterforschung, IRF); focusing on robotics, and its research projects cover various fields of the metal-electro sector.

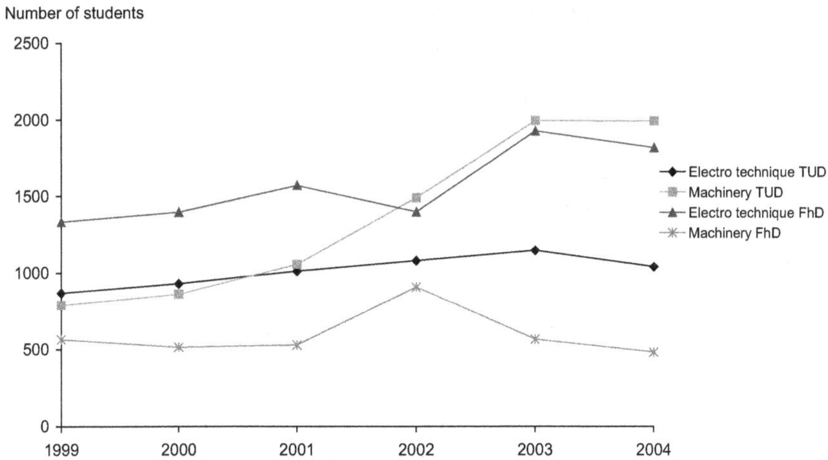

Figure 3.9 Students at TUD and FhD in metal-electro studies (source: Stadt Dortmund, 2006a)

- *Department of Machining Technology* (Institut für Spanende Fertigung, ISF), which emerged from the 'Chair of Machining Technology'. It offers various education and research programmes in different fields, which are all relevant for the metal-electro sector.
- *Institute of Forming Technology and Lightweight Construction* (Institut für Umformtechnik und Leichtbau, IUL) is organised into three departments (sheet metal forming, high-speed forming and profile forming) and it conducts research and offers courses for students.

The business association for the metal industry in Dortmund is an important actor for technical and vocational training. This association offers technical and vocational training for (potential) employees in the metal-electro sector, especially for youths.

Companies can also play an important role, since on-the-job training ('learning by doing') plays a large role for workers in the metal-electro sector. One example is ThyssenKrupp that, despite falling employee numbers, trains more youths than it can subsequently take on. The company does this to support youths in the difficult training market, and therefore it supports a training pact between the government and the industry. Furthermore, it increases the company's chances to recruit well-qualified personnel, before the expected shortage of workers in a few years as a consequence of the ageing problem. ThyssenKrupp employs about 1,130 apprentices (in 2006–2007) and it has training centres on its sites, including in Duisburg, Essen and Dortmund. It has various programmes to support youths. Often, ThyssenKrupp cooperates with its university partners to offer programmes for students.

ThyssenKrupp is also a major actor for R&D. It is involved in various R&D centres with several partners:

- *Dortmunder Oberflächen-Centrum (DOC),* a research institute that is set up as a joint venture between two companies in the metal-electro sector (ThyssenKrupp and SMS Debag) and the Fraunhofer Institute for Material and Beam Technology. The centre is active in the development of new corrosion systems and is located on the site of ThyssenKrupp in Dortmund. This location has been chosen, among other reasons, because ThyssenKrupp is one of the partners in the joint venture, and the Ruhr area offers many industries that can be customers of the centre, like automotive and steel;
- *Engineering Unit Ruhr (EUR)*, a partnership between the machinery faculties of TUD and Ruhr University in Bochum. The goal of the cooperation is to strengthen the machinery industry in the eastern part of the Ruhr area;
- *Interdisciplinary Centre for Advanced Materials Simulation (ICAMS)*, a research centre for material research and development based at Ruhr University Bochum. New material research and development is an important field for the metal-electro sector. Customers of the sector, such as shipbuilding and automotive industries, ask for increasingly stronger and lighter materials (plastics and steel). The centre is a public–private partnership between the

government, universities, other research institutes and industry. All actors are located in or near the Ruhr area, which makes quick information and knowledge transfer possible (ThyssenKrupp, 2006e).

Several other actors in Dortmund and other parts of the Ruhr area are engaged in R&D projects in fields of the metal-electro sector. One important facility is RACe, a competence centre for the machinery industry. It has a robotics and machine park that can be used by the companies in the centre. Another actor is the Fraunhofer Institute for Material Flow and Logistics (IML), which acts as a logistics consultant for companies from various industries, especially for companies from the metal-electro sector which have customers who are members of the institute. Finally, several companies in Dortmund have their own research departments. Especially for specialised niche players it is important to have their own research facilities because their specialised needs are not covered by local research institutes. Just like the companies in the metal-electro sector, R&D differs widely in the metal-electro sector in Dortmund. For instance, Murtfeldt Kunststoffe is interested in the development of new materials (plastics), while ETR is concerned with the development of highly innovative chips for the automotive industry.

To give an indication of the meaning of R&D in the metal-electro sector in the Dortmund region, we compared the spending on R&D in NRW with other states in Germany. NRW with €285 million (in 2003) is the largest spender on R&D in the basic metal industry, which means a share of 38% in total German spending. This top position can be explained by the presence of large metal companies, such as ThyssenKrupp. In the machinery industry and in 'electro-technical industries', the expenditures on R&D are €3,721 million and €7,590 million respectively, which means shares of 18% and 12% in German expenditures in these industries. This puts NRW in third place after the states of Bavaria and Baden-Württemberg, where large investments in R&D are made by large companies, such as Siemens and Daimler-Chrysler (RWI, 2006).

Geography of the metal-electro sector

The metal-electro sector is scattered across several districts and cities in the Ruhr area (see Figure 3.10). The major location for the basic metal industry is Duisburg, where 38% of the industry (based on employees in 2004) is concentrated. Other major locations for the basic metal industry are Kreis Unna and Bochum, each with a share of about 10%. For the metal products industry, Ennepe-Ruhr-Kreis is the major district, with a share of 29%, followed by Hagen (11%) and Essen (11%).The majority (18%) of the machinery industry is concentrated in Ennepe-Ruhr-Kreis. Dortmund and Mulheim have high shares of the machinery industry in the Ruhr area, with shares of 11% and 10% respectively. The high share in Dortmund can be explained by the presence of at least two major actors in this industry: Miebach and KHS.

Within Dortmund, most companies in the metal-electro sector are located in or around the central districts (see Figure 3.11). The majority of the companies in the

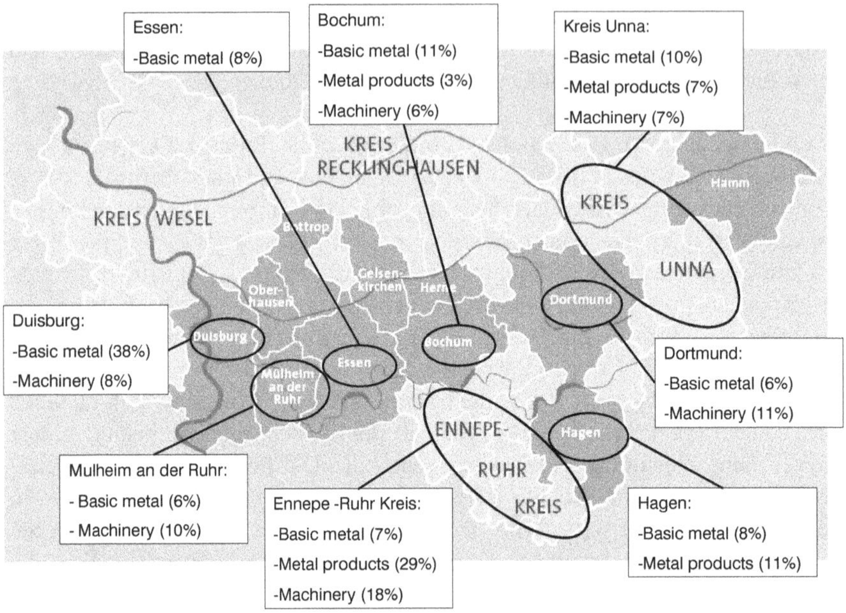

Figure 3.10 Important concentrations of the metal-electro industry in the Ruhr area with shares in employment (2004) in brackets (source: own elaboration, data from RVR, 2006b). Note that this figure is not comprehensive.

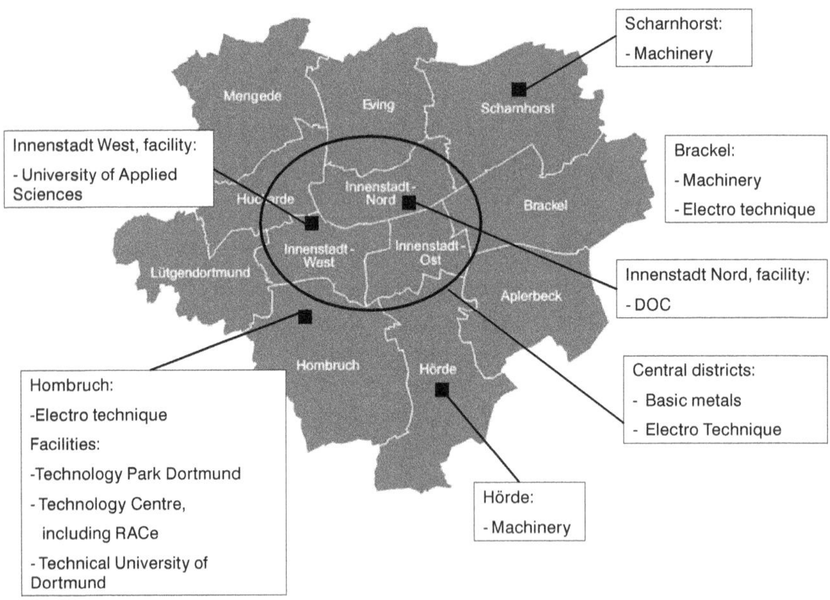

Figure 3.11 Important concentrations of the metal-electro sector in Dortmund and relevant R&D institutes (source: own elaboration). Note that this figure is not comprehensive.

basic metal industry are concentrated near old industrial sites. These old industrial sites used to be the location of the steel plants, which formed a major input for the basic metal industry. The companies in the electro-technical, fine mechanics and optics industry are, aside from being on old industrial sites, also located on (relatively) new sites, such as Technology Park Dortmund. The reason for this is that the electro-technical industry is more high-tech than the basic metal industry, and therefore, companies in this industry may need the physical proximity of research institutes on the technology park and of the Technical University Dortmund. Other parts of the metal-electro sector and the knowledge institutes are more spread across several locations in Dortmund. The knowledge institutes also are spread across various locations.

3.4 Local and global networks

A relevant question in this research concerns the extent to which proximity matters. What are the local relations, and conversely, what are the global ones? To answer these questions, we have set up a framework of analysis, distinguishing two types of networks: production networks (related to operational functions like manufacturing) and knowledge networks (related to R&D and design). This section discusses these networks in more detail.

Production networks

Steel is a major input in the metal-electro industry. The presence of steel companies in Dortmund used to be an important location factor for the metal-electro companies in Dortmund. Now that the steel industry has disappeared in Dortmund, steel and other inputs are sourced from the whole world. For instance, Anker Schroeder, an innovative metal parts producer, used to source steel locally from the steel industry in Dortmund, but now is importing steel from China. Furthermore, as we have already noted in section 3.4, ThyssenKrupp invested $2.4 billion to construct an iron and steel mill in Brazil, which will deliver cheap parts to production plants in Europe and in the USA.

Germany is an important market for the metal-electro sector in Dortmund; most of the companies deliver to clients all over Germany, since all destinations in Germany can be reached by truck in one day from Dortmund. This is especially of interest for spare parts (important products in the metal-electro industry), since spare parts require a quick delivery. There are also companies that deliver to clients in Europe or even other parts of the world. These are, for instance, the (highly innovative) companies that are market leaders in a certain niche. For example, ETR delivers highly innovative sensors to all major auto manufacturers in Europe, and Anker Schroeder delivers anchors for the building industry to construction sites across the whole world. Some companies also have establishments abroad. According to our interview partners, these establishments do not replace production in Dortmund, but mainly serve to produce for local markets. For instance, KHS invests in plants in Latin America and Asia to produce machines for the beverage industry for these new growth markets (see section 3.3). Another example is the automotive division of ThyssenKrupp, which expanded its activities in Brazil, China and India. The reason for this is the rising car production as a consequence of rising vehicle sales in these countries. Nevertheless, we have also seen that the production of certain labour-intensive products has been shifted to low-wage countries in Central and Eastern Europe. For instance, Murtfeldt Kunststoffe has a subsidiary in the Czech Republic for the production of labour-intensive standardised plastics. In Dortmund, in contrast, it produces customised plastics.

Knowledge networks

In the Ruhr area, the number of knowledge networks in the metal-electro industry with actors from industry, universities, other industries and various governments is increasing. ThyssenKrupp especially plays a major role in these networks; it has set up various partnerships with universities (like the Technical University of Dortmund and Ruhr University Bochum) and it is involved in other initiatives. New initiatives, like EUR and ICAMS, involve universities, other research institutes and partners from different industries. An older initiative is DOC, which is project based, with project teams involving different actors. Some projects of DOC illustrate this:

- Silver Ice® UV, a new coating which reduces steel's sensitivity to finger prints. This new coating is developed by DOC in cooperation with the Düsseldorf-based company Henkel. The product was brought to production readiness in conjunction with ThyssenKrupp Speciali in Terni, Italy. Currently, it is produced on the production line of this company (ThyssenKrupp, 2006a);
- European Solabs® project, a large, international project for the development of steel-based solar collectors with various international project members from the industry (for instance Interpane) and research institutes (like the University of Malaga, Spain, and a polytechnic from Lausanne, Switzerland). The project is financed by the EC, the consortium members and the Federal Office for Education and Science. ThyssenKrupp is involved in a part of the project by the development of steel for the 'solabs'. This is done by a project group with engineers from DOC and from ThyssenKrupp establishments in Dortmund and Kreuztal (ThyssenKrupp, 2006b);
- Research and development of various products for the automotive industry. DOC does this in cooperation with several establishments of ThyssenKrupp and its customers from the automotive industry. The car manufacturers give specifications and test the new products in the cars. The project teams for these new products consist of engineers from DOC, car manufacturers and establishments of ThyssenKrupp. Other actors also can be involved. For example, the development of a new coating for certain car parts is done by DOC, BMW and engineers from an establishment of the Fraunhofer Institute in Dresden (interview).

From these research projects of DOC, some conclusions can be drawn. First, R&D projects take place in cooperation with actors from various industries. This happens also in other research institutes, like ICAMS, which involves actors from the chemical industry, the automotive industry and the steel industry. The links with various industries is typical for the metal-electro sector, since the products from the metal-electro sector serve as input for various other industries (see also section 3.2). Second, the Silver Ice example, which is developed in Dortmund and produced in Italy, shows the possible separation between R&D and production. However, in the interviews we have also found indications that R&D and manufacturing cannot be separated. Therefore, some companies have production and R&D departments located on the same site. Murtfeldt Kunststoffe and ETR are two examples of this. These companies announced that the operational and production function can not be divided, since newly developed products have to be tested, which requires direct contact between R&D and production departments. Furthermore, the companies deliver R&D services to the clients as a 'physical product', in a sort of package. Often, the companies are developing new products in cooperation with their clients.

In contrast with the strong regional networks, we have found that there is limited cooperation between local companies and local research centres in Dortmund. This is caused by several factors. First, as we have noted, the industrial structure in Dortmund is dominated by SMEs. These enterprises often have no budget or capacity for (external) research. Note also that in other parts of the Ruhr area, there are SMEs which are not involved in R&D projects because they do not have resources for participation in research projects. Second, a part of the metal-electro sector is characterised by handwork, in which learning by doing is more important than R&D. Involvement of universities and other research companies is not necessary in these cases. Third, many companies specialise in a certain niche which is not covered by the local research institutes. These companies are conducting research in their own research centres or are trying to find research partners outside Dortmund that are active in the same niche. A fourth, and more general problem, is the 'language': companies ask for 'products' (more related to 'development'), while research institutes, especially universities, supply more 'basic research'.

An important aspect in our research is the role of (physical) proximity in knowledge networks. Do companies cooperate with local research partners or are they engaged in international research projects? DOC has research projects with partners from the Ruhr area, but also from other parts of Germany and even from abroad. This would suggest that proximity does not matter. However, the projects are often done in project teams in which engineers from other companies work temporarily in other locations. In this way, a sort of 'temporary proximity' is created, so that the actors can share information and transfer knowledge quickly. These projects can take place at the DOC site in Dortmund or on-site at a project partner elsewhere, like the Fraunhofer Institute in Dresden. IRF has research projects with various manufacturing companies inside and outside the Ruhr area, but from the interviews it becomes clear that cooperation with closer partners is better for social (informal) contacts, which improves the speed of knowledge transfer. Furthermore, ICAMS makes quick information transfer possible, since all actors in this research institute are located in the Ruhr area. All these examples show the importance of proximity for joint research. Nevertheless, companies have various partners outside the region. For instance, ETR cooperates with a technical university in Bavaria. This cooperation is based on personal contacts; a former manager of ETR works now as a professor at a technical university in Bavaria. Another reason for contacts with universities and other research institutes outside the Ruhr area is the research themes the institutes offer; just like companies, universities can also specialise in a certain niche.

3.5 The role of the government, cluster organisations and other networks

The manufacturing sector, including the metal-electro sector, is influenced by the actions of various governments and network organisations. In Dortmund, until now, the city government gave highest priority to the three key industries in the Dortmund project, which are actively supported. For these industries, specific locations have been selected as development zones, and specific centres have been set up as cluster and development organisations. However, currently, the city is giving increasing attention to other future industries with growth potential, mainly robotics (which is narrowly related to parts of the metal-electro sector) and the biomedical industry. Furthermore, as we have already noted in section 3.2, a recent study of IW Consult (2007) puts the role of the manufacturing sector forward as a central power wheel in the local economy. This report has advised Dortmund to focus in the future on three industries of the metal-electro sector (machinery; radio, television and communication equipment; medical, precision and optical instruments) and two other industries, which can act as an industrial axis with other industries and therefore have a high growth potential.

The Ruhr area, the regional level, is a typical example of a polycentric urban region (see for example Priemus and Zonneveld, 2004) which had a common economic specialisation (coal mining and steel), and as a consequence, a common image from the outside. However, from the inside, there are various individual cities with their own identity (Houtman and Lagendijk, 2001) and cooperation was low. To a certain extent, there was a lack of coordination and cities took similar competing initiatives (Van den Berg *et al.*, 2005). This has changed gradually; the cities organise joint events and under (financial) pressure of the state of NRW business agency Metropoleruhr was established in 2007, aiming to promote the Ruhr area as one region, support firms and to attract new business and inhabitants.[7] Moreover, various local bodies have joined forces at the regional level, such as the business associations in the Regional Business Association Ruhr and universities in the University Alliance Metropolis Ruhr. Intercity cooperation is important to realise the potential of various parts of the metal-electro industry that are scattered across several locations in the Ruhr area. Intercity cooperation may enlarge the critical mass and may link various individual sectors to 'new' combinations that can compete internationally.

On a larger spatial level, the state of NRW has selected 16 industries as future industries for a newly developed cluster policy (since March 2007). These industries play an important role for the innovation strategy, the location marketing and for the economic development of the state. The machinery industry, which is also a major industry in Dortmund, is one of these 16 industries.[8] An important programme on the national level is the excellence programme which aims to strengthen research and innovation in Germany by funding universities. The initiative encourages cooperation between research institutes and the industry by additional funding. Also, the city of Dortmund wants to strengthen the knowledge and information transfer between science and industry. Currently (2007), the city is developing a plan for doing that.

Cluster organisations and other networks

There is not a specific cluster organisation or platform for the metal-electro sector in Dortmund. This can be explained by the fact that the sector is too broad; the different industries within the metal-electro sector each have their own actors and characteristics. Furthermore, the actors have too few common interests and problems, which makes it hard to set up one single cluster organisation. Nevertheless, for individual industries in the sector, there are platforms, like 'Robotik- und AutomatisierungsCenter Dortmund' (RACe), a competence centre for the machinery industry. The goal of RACe is to support SMEs in machinery and related industries and to upgrade the cluster in this field. Another example is the Network of Industry RuhrOst (NIRO), which was set up by six companies (including KHS AG), the Technical University of Dortmund and the economic department of the Unna district in order to strengthen competitiveness of and innovation in the machinery and electro-technical industries in the eastern part of the Ruhr area. The network is funded by the EU, the state of NRW and the district of Unna. The scope of its activities includes: support SMEs, create a shared labour pool and support joint R&D projects.

Moreover, there are other actors in the Ruhr area that have networking functions and may be seen as cluster organisations. Chambers of commerce and business associations have traditionally played a strong role. One of these partnerships in Dortmund is the Dortmund Business Association, which plays an important role in the development of the labour pool. It provides training and education to managers and workers in the industry. Furthermore, the association represents its members in social and political issues and therefore it has frequent contacts with other actors, such as the city of Dortmund and IG Metall, the labour union.

3.6 Conclusion

Dortmund is one of the leading cities in Ruhr in the transformation process from a manufacturing dominated economy towards a modern services economy;

more than 80% of its workers are active in the service sector. Despite the low share of manufacturing in the economy of Dortmund, it is still relevant, for example due to its key role in innovation. One of the largest sectors in the manufacturing sector is the metal-electro sector, with a share of nearly 70% in the manufacturing employment. We have seen that the industries in the metal-electro sector in the Ruhr area and Dortmund have declined, which can be explained by the de-industrialisation process in the area, especially the closure of the Westfalen plant in 2001 that resulted in a strong decline in employment in the basic metal industry and related industries. Nevertheless, there are differences in individual industries and in each city or district. Some industries in certain cities have even grown.

In this chapter, we have analysed the development of the metal-electro sector in the Dortmund region. A specific aim of the study is to analyse the evolving relationship between physical production ('manufacturing'), and other activities in the value chain. In this section, we summarise our main findings and draw some conclusions.

Production and knowledge networks

Concerning production, we have seen that the companies in the metal-electro sector in the Ruhr area are increasingly investing in new facilities for standardised production in developing countries in Asia, Latin America and Central and Eastern Europe. For instance, Murtfeldt Kunststoffe has opened an establishment in the Czech Republic, ThyssenKrupp opened plants in Brazil and Asia, and KHS invests in Latin America and Asia. They do this to serve these growth markets or for cost reasons. We have indications that companies in the steel products industry (like Anker Schroeder and ThyssenKrupp) are increasingly sourcing from low-cost countries instead of the steel industry in the Ruhr area. In the Ruhr area, the remaining production is concerned with more customised manufacturing, higher up in the value chain. Companies are investing in these facilities to upgrade the production process. Another aspect relevant for upgrading of the cluster that we have found is an increasing number of knowledge networks. Major characteristics of these networks are:

- *Cooperation between industry and research institutes.* ThyssenKrupp especially plays a large role in this through university-based partnerships (in research and education) and involvement in other initiatives and research like DOC, EUR and ICAMS. Cooperation between industry and science is stimulated by the national government through the Excellence initiative, and the city of Dortmund is developing a plan to increase this cooperation. A drawback is that SMEs are often not involved in these networks;
- *Cooperation between industry and governments (public–private partnerships).* For example, ICAMS is funded by the state of NRW and by the involved companies;

- *Cooperation between actors from various industries*. For instance, DOC has R&D projects with ThyssenKrupp and with its customers, like companies from the automotive industry.

Moreover, relevant for upgrading the metal-electro sector in the region is an increasing number of (foreign) students in almost all studies related to the metal-electro sector at the Technical University of Dortmund and at the University of Applied Sciences. Increasing cooperation between universities in the Ruhr area, via the University Alliance Metropolis Ruhr, and specifically for the machinery industry via EUR, is also relevant for upgrading the cluster. This offers advantages for education (for instance, students have a broader choice) and for research (for example, there is more chance for additional funding via a third cash flow). Despite the increasing number of students, we have found two possible threats to attracting students to Dortmund or keeping them in the city after they have finished their studies. One is a lack of a distinct image on the local level, and the second is the absence of large companies in Dortmund that can offer (international) careers for students after their studies. Nevertheless, in other places in the Ruhr area, there are various large, well-known companies. ThyssenKrupp especially is an important employer for students.

Our case shows that proximity in knowledge networks seems to matter; companies cooperate with local research institutes (for instance, ThyssenKrupp and university partners), have joint research centres (like ICAMS) or create temporary proximity by working with project teams (DOC does this). Nevertheless, we have also seen examples of companies that cooperate with partners outside the region, based on personal contacts or on the specialisation of the research partner, since not all research themes are covered by partners in the region. In addition, the role of proximity between production and R&D departments differs. We have seen examples that production and R&D can not be separated geographically, for instance because the newly developed products have to be tested directly, which requires a direct contact between these departments. However, other examples show that R&D and production can be split geographically, like the Silver Ice project, where development takes place in the Ruhr area, while production has been done in Italy.

Different roles of regional foundations in the industrial economy and the knowledge economy

The case clearly shows the differences between the 'traditional' industrial economy, which was dominated by manufacturing and the 'modern' knowledge economy, which integrates services and manufacturing activities. The traditional industrial urban economy is characterised by a one-sided *economic base*, consisting of one or a few, often related, industries (coal, steel and beer in the case of Dortmund). The transition towards the knowledge economy requires a more diverse economic structure, with new (manufacturing) industries, linkages between several industries and possibilities for new combinations between industries. The metal-electro sector is a good example of a diverse sector; it includes various manufacturing industries, consists of some large MNOs and several local SMEs, and has both high-tech firms as well as simple parts makers. Also, in Dortmund and other parts of the Ruhr area, we have recognised this structure in the sector, with ThyssenKrupp and Salzgitter as major MNOs, and some well-performing niche players among the SMEs.

Furthermore, the *knowledge base* becomes important for the remaining manufacturing industries and new manufacturing industries in the knowledge economy. The R&D function is gaining importance in manufacturing plants, and thus plants need fewer 'blue collar workers' and more engineers. There are also more joint research projects between industry and partners from research institutes (for instance universities). The good knowledge infrastructure (TechnologyParkDortmund and Centre, Technical University of Dortmund and other institutes) in Dortmund is a key asset in the new knowledge-based economy.

A related difference regarding *factor conditions* is that manufacturing plants from new sectors are often located in the city, in contrast to plants in the traditional economy, which were pushed out of the cities because of the pollution and noise that the plants caused. Plants in the knowledge economy are smaller and cleaner, and are increasingly established in urban regions because of the proximity to research institutes and other urban facilities. In other words, there is a sort of 're-urbanisation of manufacturing'. Dortmund has a relatively unique asset for this: the vacant spaces in the city from former manufacturing plants, which can be used for new (manufacturing) industries.

In general, it can be said that the knowledge base and *quality of life* are gaining importance in the knowledge economy and the role of *accessibility* is changing. In the traditional economy, freight transport was the most important for manufacturing, and the truck, train and ship were the most important transport modes. Nowadays, passenger transport is gaining importance. Production and other functions are increasingly done in projects with actors on different locations. Despite new technologies (like the Internet), face-to-face contacts are important in these projects, and therefore there is an increase in passenger transport. Other modes, like airplanes and high-speed trains, are becoming more important. Dortmund has a good infrastructure for freight as an inheritance of its industrial past, which is still a key asset for certain manufacturing industries, like the metal parts industry. Dortmund also has a good infrastructure for the knowledge economy (high-speed train connections and two airports in the region), but the internal accessibility in Ruhr is a problem (congestion on the roads).

Another characteristic of the knowledge economy is an increasing competition between regions on an international level. This is also true for manufacturing. For instance, the metal-electro sector in the Ruhr area is competing with the metal-electro sector in the Ostrava region in the Czech Republic. Therefore, competition between the different cities in the Ruhr area should be prevented and cooperation should be stimulated. The cooperation between the universities in the region is a good step for this. In addition, events, like the Cultural Capital 2010, can act as catalysts to stimulate cooperation in the Ruhr area, which start with a 'Ruhr awareness'. Events are an important 'tool' to create this Ruhr awareness.

Outlook

From our study, two industries in the metal-electro sector in Dortmund seem to be promising: they combine the presence of a good knowledge base, key leading firms and an already existing strong cooperation between industry and knowledge institutes. The first is the machinery industry, because: 1) the decline of the machinery industry in Dortmund is relatively low, and in the last three years employment has even grown (see Figure 3.8); 2) Dortmund has IRF, ISF and IUL (all part of the Technical University of Dortmund) and IML some major research institutes and educational facilities (TUD and FhD) in this industry. It is also relevant that the machinery faculties of TUD and Ruhr University Bochum are cooperating in this industry via EUR; 3) Dortmund is home to two multinationals (KHS and Miebach) in this industry; 4) the machinery industry is selected as one of the 16 key industries in NRW's new cluster policy; 5) there are two cluster organisations (RACe and NIRO) for the machinery industry in the Dortmund region. The second promising industry, despite a strong decline in employment, is the basic metals industry. This industry is still important for Dortmund and the region since it serves as an important input for other manufacturing industries (like metal products). Moreover, the region has two important research institutes (DOC and ICAMS) for the basic metals industry, which are also conducting research in the field of new materials. New materials research and the basic metals industry is a promising 'new combination' for the region.

Notes

1 www.ruhrsite.de (accessed 25 May 2007).
2 Statistical data from www.ruhrsite.de, unless mentioned differently (accessed 25 May 2007).
3 Data from RVR (2006b), http://www.duesseldorf-international.de/e/index.php?type = index&path = 04_infos_service/lageplaene and http://www.flughafen-dortmund.de/flugplan.0.html?&L = 1 (accessed 3 May 2007).
4 It should be noted that in reality only 280 hectares of the Westfalenhütte will be converted.
5 http://www.krupp.com/en/konzern/zukunftsmaerkte.html (accessed 1 June 2007).
6 The year 2004 shows a decline that is caused by an adaptation in the education system. This has resulted in a decline in the total number of students at the university of about 15% and at the University of Applied Sciences of about 10%.
7 http://business.metropoleruhr.de/en/promoting_regional_business/responsibilities.html (accessed 1 June 2007).
8 http://www.wfgmetropoleruhr.de/templates/main.jsf?Select_id = c6a772ab-c4b1–4c30-b62b-8ffc922a619d&Message_id = a67024b8–8b7d-4ccc-a978-da99fdbb4cab&Filter_id = select (accessed 1 June 2007).

Bibliography

Bavarian Ministry of Economic Affairs, Infrastructure, Transport and Technology (2007) *Invest in Bavaria facts and figures*, Munich: Bayern International GmbH.

Becker, H. (2006) *High noon in the automotive industry*, Berlin: Springer-Verlag.

van den Berg, L., P.M.J. Pol, W. van Winden and P. Woets (2005) *European cities in the knowledge economy – the cases of Amsterdam, Dortmund, Eindhoven, Helsinki, Manchester, Munich, Münster, Rotterdam and Zaragoza*, Aldershot: Ashgate.

Bömer, H. (2005) *Modern local and regional development policies in times of mass-unemployment – the example of Dortmund*, 2nd edn, Dortmund: IRPUD.

City of Dortmund (n.d.) *Business location Dortmund*, available online: www.economicdevelopment-dortmund.de/upload/binarydata_wbfdod4cmsen/2689/Internet_engl.pdf.

Dortmund University/Ruhr University Bochum (2007) *Engineering unit Ruhr*, Essen: ThyssenKrupp.

Ernst & Young (2006) *Deutsche Großstädte: Zufriedenheit der Unthernehmen mit ihrem Standort 2006,* available online: www.ey.com/global/download.nsf/Germany/Studie_DeutscheGrossstaedte_2006/$file/DeutscheGrossstaedte_2006.pdf.

Etin (2005) *Quick scan industriële ketens*, available online: www.etin.nl/f_publ/ketensrotterdam.pdf (23 July 2007).

GfW (2002) *Welcome special – Dortmund IT centre of excellence in Nordrhein-Westfalen*, Düsseldorf: Gesellschaft für Wirtschaftsförderung Nordrhein-Westfalen GmbH (GfW).

Hospers, G.J. (2004) 'Restructuring Europe's rustbelt: the case of the German Ruhrgebiet', *Intereconomics*, 39:3, 147–56.

Houtman, H. and A. Lagendijk (2001) 'Contextualising regional identity and imagination in the constructuion of polycentric urban regions: the cases of the Ruhr area and the Basque country', *Urban Studies*, 38:4, 747–67.

IW Consult (2007) 'Wie sieht die Zukunft der Industrie in Dortmund aus?', Summary, Institut der deutschen Wirtschaft Köln Consult GmbH.

Iwd (2004) Schröder Ch., 'Productivität und Lohnstückkosten im internationalen Vergleich', *IW-Trends,* 3.

KHS AG (2007) *Annual report 2006.*

Manshanden, W.J.J., W. Jonkhoff, P.J.M. de Bruijn, S.P Newrly and R. en Koops (2005) *De top-20 van Europese stedelijke regio's 1995–2004; Randstad in internationaal perspectief*, Delft: TNO.

Priemus, H. and W. Zonneveld (2004) 'Regional and transnational planning: problems today, perspectives for the future', *European Planning Studies*, 12:3, 283–97.

Ruhr University Bochum (2007) 'Eine neue Instanz in der Wissenschaftslandschaft Universitäten Bochum, Dortmund und Duisburg-Essen gründen "Universitätsallianz Metropole Ruhr"', press release 12 March 2007.

RVR (2006a) *Kleiner Zahlenspiegel der metropoleruhr*, Essen: Regionalverband Ruhr.

—— (2006b) *Struktur und Entwicklung der Sozialversicherungspflichtig Beschäftigten im Ruhrgebiet 1980 bis 2004,* Essen: Regionalverband Ruhr.

RWI (2006) *Innovationsbericht 2006 – Zur Leistungsfähigkeit des Landes Nordrhein-Westfalen in Wissenschaft, Forschung und Technologie*, Essen: Rheinisch-Westfälisches Institut für Wirtschaftsforschung.

Salzgitter AG (2007) *Annual report 2006.*

Schönharting, J., A. Schmidt, A. Frank and S. Bremer (2003) 'Towards the multimodal transport of people and freight: interconnective networks in the RheinRuhr Metropolis', *Journal of Transport Geography*, 11, 193–203.

Stadt Dortmund (2005) *Dortmund 2005 Jahresmagazin*, Dortmund: Dortmund-Agentur.

—— (2006a) dortmunderstatistik, No. 180, Jahresbericht 2006, wirtschaft, Stadt Dortmund, Ambt für Statitik und Wahlen.

—— (2006b) dortmunderstatistik, No. 179, Jahresbericht 2006, lebensraum Dortmund, Stadt Dortmund, Ambt für Statitik und Wahlen.

—— (2007) *Branchenbericht 2007 – Trends und Analysen zum Wirtschaftsstandort Dortmund*, Dortmund: City of Dortmund, Department of Economic Promotion.

ThyssenKrupp (2003) 'ThyssenKrupp Stahl Dortmund: modernization of cold rolling mill completed', press release, 22 August 2003.

—— (2006a), *Technoforum*, No. 1, 2006.

—— (2006b), *Technoforum*, No. 2, 2006.

—— (2006c) 'German President Horst Köhler visits ThyssenKrupp Steel: dialogue with young people in education and initial training', press release, 21 June 2006.

—— (2006d) '24 neue Auszubildende bei ThyssenKrupp Steel in Dortmund', press release, 5 September 2006.

—— (2006e) 'Ruhr University Bochum and ThyssenKrupp establish research institute to develop new materials, State of NRW and industry to provide 24 million euros over five years', joint press release of the Ministry for Innovation, Science, Research and Technology of the State of North-Rhine Westphalia, Ruhr University Bochum and ThyssenKrupp AG, 10 November 2006.

—— (2006f) Remarks by Dr. Ing. Ekkehard D. Schulz, Chairman of the Executive Board of ThyssenKrupp AG at the Annual Press Conference, Villa Hügel, Essen, 1 December 2006.

—— (2007) *Annual report 2006.*

WAZ (2007) *Umsatz Explodiert*, Westdeutsche Algemeine Zeitung, 24 March 2007.

4 Eindhoven: automotive industry

4.1 Introduction

Economic and social development in Eindhoven have been strongly related to industrial development. The city has seen rapid development since the establishment of the Philips Gloeilampen N.V. in 1891. In that time, the city was a small agricultural town with only 5,000 inhabitants (van den Berg *et al.*, 2005). The city has grown thanks to the growth of the electrical giant Philips and truck manufacturer DAF. Today the city has 209,699 (2007, CBS, n.d.) inhabitants and it is the fifth largest city in the Netherlands after the cities in the Randstad.

The dependency of the city on the two industrial giants implied some risks as performance of the region depended on the performance of the two companies. This became clear in the beginning of the 1990s, when the region went into a recession after problems in the two key companies. Philips did not only lose competition from new competitors from Asia, but also Western firms were more efficient than Philips. This forced Philips to start a large-scale restructuring programme. Production units were either closed or relocated to low-cost countries. It was estimated that Philips cut around 10,000 jobs between 1988 and 1991. In addition, in 1993, DAF collapsed. The bankruptcy cost about 2,500 DAF workers their jobs. Moreover, DAF used an extensive network of regional suppliers and another 2,500 workers of suppliers lost their jobs (van den Berg *et al.*, 1997).

Nowadays, after the Stimulus Programme, a plan to strengthen the socio-economic structure and to reduce unemployment in the region, Eindhoven has a strong knowledge-intensive manufacturing industry. More than half of the employment in the manufacturing sector in the Eindhoven region can be categorised as high- and medium-tech (Commissie Sistermans, 2006a). One of the pillar industries is the automotive industry, with DAF (restarted in 1993), VDL (a bus assembler) and NedCar as key assemblers in the region. In addition, the automotive cluster in the region counts various suppliers of car parts and systems and knowledge institutes that conduct research for the automotive industry. The knowledge base of the cluster was recently strengthened by the relocation of TNO Automotive from Delft to the Eindhoven region.

In this chapter, we focus on the dynamics of the automotive industry in the Eindhoven region and investigate the linkages between the automotive industry and the regional knowledge base. The chapter is organised as follows. Section 4.2 puts the automotive industry into regional economic context. Next, in section 4.3, we introduce the main players of the automotive cluster in Eindhoven and we analyse the geography of the sector. In section 4.4, we discuss the networks in which the cluster actors operate, and we discuss the role of the government in the automotive industry in section 4.5. In the last section (4.6), we draw conclusions.

4.2 Foundations of the Eindhoven region

The Eindhoven region (728,133 inhabitants at 31-12-2006) consists of 21 municipalities in the south-eastern part of the province of Northern Brabant. The core cities are Eindhoven (209,699 inhabitants) and Helmond (86,061) (2007, CBS, n.d.). One of the key sectors in the region is the automotive industry. To understand the current dynamics of this industry, this section puts the industry in the perspective of its regional context. This is done with a description of Eindhoven's regional foundations: the economic base, the knowledge base, accessibility, quality of life and factor conditions.

Economic base

The crisis in the beginning of the 1990s changed the economic structure of the region. Just like in other regions, Eindhoven was undergoing a de-industrialisation process. Companies have changed their strategies from a diverse portfolio of businesses to their core business and are outsourcing production and other functions. A typical example of the development of the manufacturing industry in the region is the copying machine producer Océ. Océ used to produce 25% of the products by itself; nowadays this is only 2% (interview). One of the consequences of the changing economic structure is that the region has become more dependent on foreign companies (10% of the manufacturing companies are owned by foreign bodies), making it more sensitive to international developments. The automotive industry is a typical example of a global industry, and many companies in the Eindhoven cluster are in foreign hands, like DAF (owned by PACCAR, USA), PDE Automotive (part of Bentler Group, Germany) and NedCar (owned by Mitsubishi, Japan).

Although the largest reduction in jobs in the manufacturing sector took place in the 1970–1990s, the region is still in a de-industrialisation process. The employment in Eindhoven's manufacturing sector dropped 16% in the 2001–2005 period, which is above the Dutch average (NV Rede, 2007). Despite this drop in employment, the manufacturing sector in Eindhoven is still important for the Eindhoven region and accounts for 30% of the total manufacturing jobs in the Netherlands and for 40% of the value added. It is worth mentioning that more then 50% of the jobs in the manufacturing industry in Eindhoven can be categorised as high- and medium-tech (Commissie Sistermans, 2006a). Moreover, the manufacturing sector in Eindhoven is still the largest sector; about 18% of the employees work in the manufacturing sector; compared to the Netherlands' average where this is 13% (NV Rede, 2007). Also in terms of value added, the manufacturing sector has a 20% share in the total value added in the Eindhoven region (the Dutch average is 18%).

Manufacturing has a long tradition in the Eindhoven region, but there were large differences within the region. The city of Eindhoven specialised in electronics and automotive (through the presence of Philips and DAF), while Helmond depended heavily on the textile industry. The decline of the textile industry in the 1970s and the crisis in the beginning of the 1990s have changed the economic structure and target industries of the government in the region. These days, the region focuses on business services and knowledge-intensive manufacturing. The automotive industry (5,915 employees in 2005) is (still) one of the major industries in the region. Other pillar industries in the region are: medical technology (5,560 jobs), ICT (18,584 jobs) and mechatronics (36,338 jobs). Moreover, the region has two upcoming industries: (industrial) design and food (NV REDE, 2007). In addition, the Eindhoven region has other strong sectors, such as the machinery industry and the optical industry. Apart from Océ (a market leader in producing copying machines that is located in Venlo), there are also several smaller suppliers in the machinery industry. Often, these companies are highly innovative and are involved in the development of new products.

The major industries offer possibilities for (new) combinations with the automotive industry. For instance, mechatronics and ICT systems are gaining importance in cars, and therefore the presence of these industries are important assets for the automotive industry in the region. Also the growing design industry is important for the automotive industry, since design plays an important role in fitting new car parts and systems into a car.

The south-eastern part of the Netherlands accounts for 14.5% of the Dutch GDP, which is more than the contribution of the region of Amsterdam (11.2%) (Commissie Sistermans, 2006a). Despite this large contribution to the national GDP, the GDP per capita in the Eindhoven region is €29,000, equal to the Dutch average (CBS data, 2003). The unemployment rate in the Eindhoven region varies between 4% and 14.5%, with high percentages in the core cities; 10.1% in Eindhoven and 14.4% in Helmond. The structural unemployment is caused by a mismatch between demand and supply in the labour market (Commissie Sistermans, 2006a). There are no large differences between the unemployment rate in Eindhoven and the Dutch average (NV REDE, 2007).

Knowledge base

A key asset to attract knowledge-intensive parts of the automotive value chain is the presence of R&D facilities. The Eindhoven region has several research facilities from public and private actors. The majority of the research institutes are specialised in technical fields, like ICT and embedded systems, and are often in line with the pillar industries in the region. Major facilities for R&D in Eindhoven include: the Technical University Eindhoven (TU/e), public research institutes (such as TNO, the Netherlands Organisation for Applied Scientific Research), private research institutes and research departments of companies (a major example is Philips Research, 'Nat.Lab') and public–private research centres and 'Triple Helix' research centres (such as the Dutch Polymer Institute, DPI, a partnership between the Dutch national government, industrial partners and universities that focuses on research in the area of polymer materials). Another major R&D facility in the Eindhoven region is High Tech Campus Eindhoven (HTCE) that was founded in 1999 on an initiative from Philips. The campus is a business park of 100 hectares for Philips and non-Philips companies and stimulates open innovation by shared facilities and the physical clustering of various actors that form (potential) research partners.

Other parts of the knowledge base are the labour market and educational facilities which are important to upgrade the labour market. In the Eindhoven region, there are several educational institutes on different levels. Key institutes for academic and higher education are TU/e, the Design Academy Eindhoven (a prestigious higher education institute, which trains students in the field of industrial design) and Fontys University of Applied Sciences (which offers various bachelors and masters degrees in technical and social fields). TU/e offers a complete range of education in many technical disciplines and most of the courses are offered in English. All these institutes are offering courses, or are planning to introduce courses, that are directly or indirectly linked to the automotive industry. Two educational facilities that offer a variety of training programmes on a vocational level are ROC Ter Aa and ROC Eindhoven.

The majority of the educational facilities specialise in technical education (see Figure 4.1). It is worth mentioning that the region offers various possibilities for foreign students and workers, since many courses at the academic and higher educational facilities are offered in English and the region has international primary and secondary schools (NV REDE, 2007).

Companies also play a role in the development of the labour pool. Many companies offer internships to students and/or give specific training courses. For instance, DAF teaches its employees to work according to the PACCAR (the parent company of DAF) standard, based on the just-in-time principle that is used in the production process. Another example is Koninklijke Nedschroef Holding B.V. (a company that producers fasteners for the automotive industry) that teaches the specific skills necessary to deliver high-quality fasteners that are required in the automotive industry.

Despite the good training facilities and the focus on technological training, the manufacturing sector in the region faces a shortage of workers, especially of technical employees. This can be illustrated by a citation from the annual report of VDL Groep: 'We would like to recruit more technical specialists, but unfortunately they are not available. The low intake level amongst young people into technical education is therefore a considerable cause for concern in respect of maintaining and improving our international competitive position' (VDL Groep, 2007a).

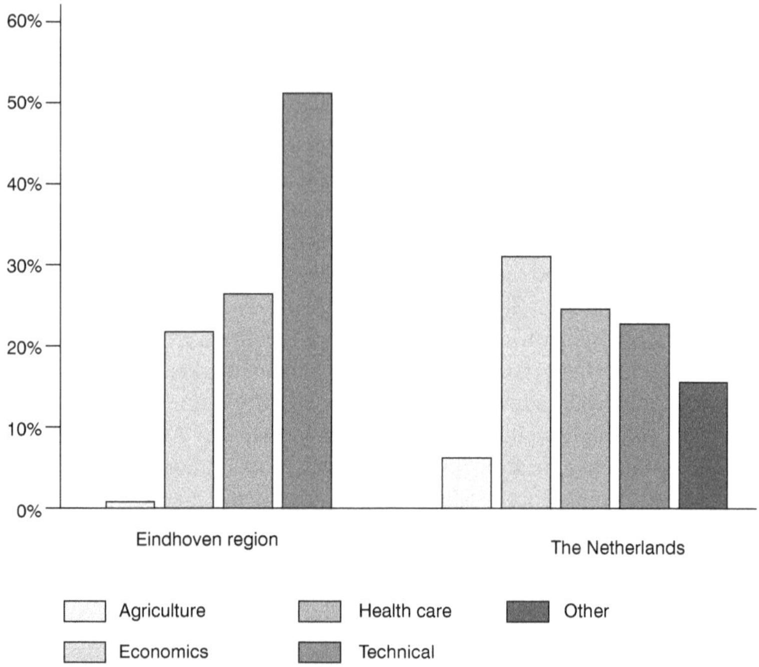

Figure 4.1 Students per major, 2005–2006 (source: NV REDE, 2007)

One reason for the shortage is the image of the industry. Many youths have a negative perception about manufacturing, and therefore they do not want to work in plants. Various (joint) actions are taken by actors in the region such as:

- The *use of foreign employees*. DAF hires about 12,000 foreign employees. The majority of these employees come from Germany and Belgium, the neighbouring countries. But via a specific employment agency it also hires employees from the United Kingdom.
- *Specific promotion programmes* to get youths and pupils in touch with technology. For instance, the Technique Promotion Foundation ('Stichting Techniek Promotie') organises various activities to make youths familiar with technology.

The Eindhoven region is one of the 'technological and innovative hotspots' in Europe. It receives high spending from private R&D, it has the highest patent density per worker in Europe, and it has a strong knowledge infrastructure (Gemeente Eindhoven, 2003; Programme Agency Horizon, 2004). The province of Northern Brabant (where Eindhoven is located), ranks fourth on the European 'scoreboard for innovation power', after Stockholm, Helsinki and Munich. This ranking is set by the European Commission and is based on diverse scientific and technological indicators (Commissie Sistermans, 2006b, DMI, 2004). Eindhoven

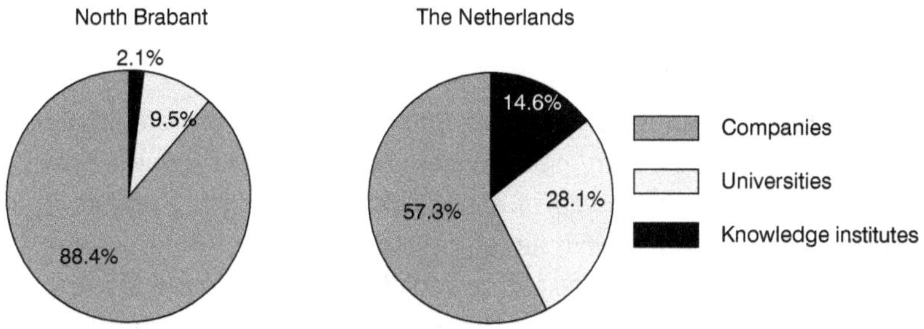

Figure 4.2 R&D expenditure in million euro, 2003 (source: NV REDE, 2007)

also has a high patent density. In 2001, it had a patent density of almost 550 patents per 100,000 active workers in the region, which was more than three times more than the second highest, Oberbayern (the greater Munich region), which had about 150 patents per 100,000 workers (Commissie Sistermans, 2006b).

The Eindhoven region is by far the most important technological region in the Netherlands; 46% of the private R&D expenditures and 43% of the patent requests in the Netherlands are requested in the south-east Netherlands (the Eindhoven region and the northern part of the province Limburg). Even if the number is adapted for the influence of Philips, Eindhoven takes the lead above the other R&D hotspots of Twente and Delft (Gemeente Eindhoven, 2003, 2004). There is a large difference in the share of private R&D spending in the total R&D spending between the province of Northern Brabant and the Netherlands (see Figure 4.2). Therefore, some of our interview partners argue that the region should lobby for more public investment by the state to get more balanced R&D expenditures.

Accessibility

The Eindhoven region is located in the south-eastern part of the Netherlands. It is surrounded by areas with a high population density such as the Dutch Randstad and the German Ruhr area. Although the demand for vehicles is saturated in Western Europe, it is still an important market and it might be an advantage for companies in the automotive industry to be located near these high-density population areas. In addition, Eindhoven's geographical location near Germany, Belgium and France, is of interest for the automotive industry; many plants, research departments and headquarters of OEMs and suppliers are located in these countries, at a relatively short distance from the Eindhoven region. Within the Netherlands, however, Eindhoven is relatively far from The Hague (where the government has its seat), and therefore, Eindhoven often has less lobbying power than the four big cities in the Randstad area.

A major asset for fast international traffic is Eindhoven Airport (946,218 passengers in 2005) and it has direct connections to 18 European destinations, including other technical hotspots, like London and Paris (NV REDE, 2007). The direct connection with Paris might be an asset for automotive research since

Box 4.1 The Phileas project

'The Phileas is an innovative and futuristic transport mode. The good characteristics of a tram/metro system are combined with the flexibility and low exploitation costs of a bus system. Magnetic markers in the road conduct the bus with its 'automatic driver' over the bus lane, the markers serve as reference for the vehicle position. In the first (test) year the Phileas will operate with a driver. As a matter of fact, the Phileas is not limited to the bus lane and, with a driver, the high tech vehicle can also drive over normal roads.

Since the beginning of July 2004, the first Phileas vehicles have driven, with a short break, on the track Eindhoven Central Station – Eindhoven Airport. After finalisation of various tests, the Phileas vehicles have also driven from Eindhoven Central Station to Veldhoven Centre since July 2005.

The Phileas has been developed in close cooperation of the launch customer, the municipality Eindhoven. APTS, a public–private partnership, is now taking care of further marketing of the Phileas. Various cities in France, Germany, Italy, Israel and Korea already have shown interest. By now, the first foreign success is booked; the Northern France city Douai has ordered 12 Phileas vehicles with a value of 16 million euro. In addition, APTS has signed a license and a technology transfer agreement with the Korea Railroad Research Institute.

Source: citation, slightly adapted, from Commissie Sistermans (2006b, p70). Picture from http://www.apts-phileas.com/.

Paris is an important R&D hub for automotive research. Besides the airport in Eindhoven, the region has access to the intercontinental airports of Amsterdam (Schiphol), Brussels and Dusseldorf, which are all located within 150 kilometres by car.

Congestion on the highways and ring roads is a major hindrance for the Eindhoven region. This seems to be a disadvantage for the automotive industry, especially for the production function, since congestion makes Just-In-Time deliveries more difficult. The internal accessibility on roads is not considered a problem by our interview partners, but one interview partner noted that public transport might be improved because of a shortage of exchange points between public transport and other transport modes. Another drawback regarding the public transport is the lack of a direct connection to Schiphol Airport.

In order to improve the accessibility of the region, several infrastructure projects are under development, such as an increase in the number of driving lanes on the highways, and the development of a 'high-grade public transport' network in the region. A first, and major, part of this high-grade public transport network is the Phileas project, an innovative public transport mode that combines bus and metro systems (see Box 4.1).

Quality of life

Eindhoven is one of the oldest cities in the Netherlands, but this is not visible because the city was bombed heavily during World War II. The industrial past of the city played a large role in its development. The industrial heritage is still present in the city, such as the first Philips factory and the DAF Museum. The city also has modern attractions, such as the Van Abbe Museum (a museum for modern arts) and Muziekcentrum Frits Philips (a concert hall). Moreover, the city is directly surrounded by rural areas, consisting of woods and wetlands. These attractive natural surroundings might make the city a pleasant place to live for technical knowledge workers, like engineers from the automotive industry, who appreciate a green and quiet living environment.

Nevertheless, there are also drawbacks. One is that the city lacks a real metropolitan image. Although much is tried, there are too few social facilities, like restaurants and pubs. This makes the city less attractive for more commercial parts of the value chain, like headquarters and marketing departments, that require a vibrant urban surrounding for their workers and for receiving their business customers. Illustrative of this is the movement of the Philips' headquarters to Amsterdam.

Factor conditions

Factor costs (prices for production factors) and 'soft' factors play key roles in the development of industries. The Netherlands has relatively low labour costs in the automotive industry compared to the European average and its neighbouring countries of Belgium and France, which gives it a competitive advantage for car production (see Figure 4.3). It should be noted that the figure depicts the wages in the automotive industry which is relatively small in the Netherlands. Many suppliers in the automotive industry are categorised into other sectors (like metal, plastic or electronics). In these sectors, the differences between the Netherlands and other countries might be less favourable. In addition, it is not only the labour costs that should be considered when comparing production locations, but also labour productivity. High labour costs are not a problem when workers are very

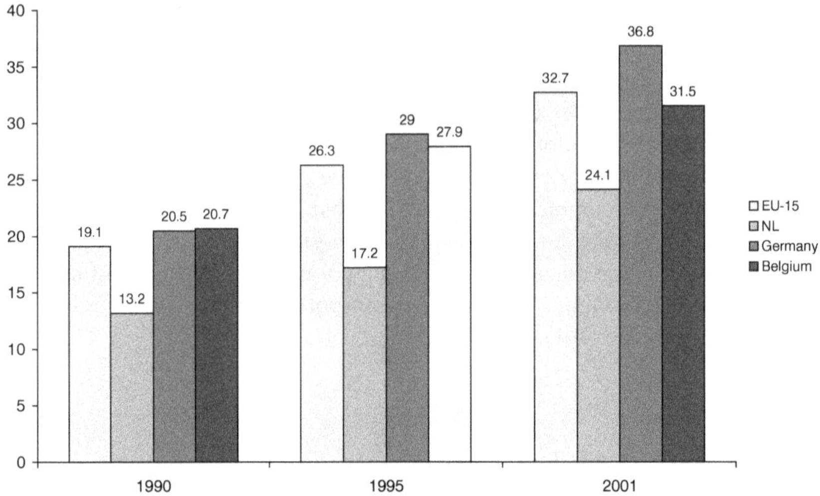

Figure 4.3 Labour costs in the auto industry (conversion to USD, based on PPP) (source: Heneric *et al.*, 2005)

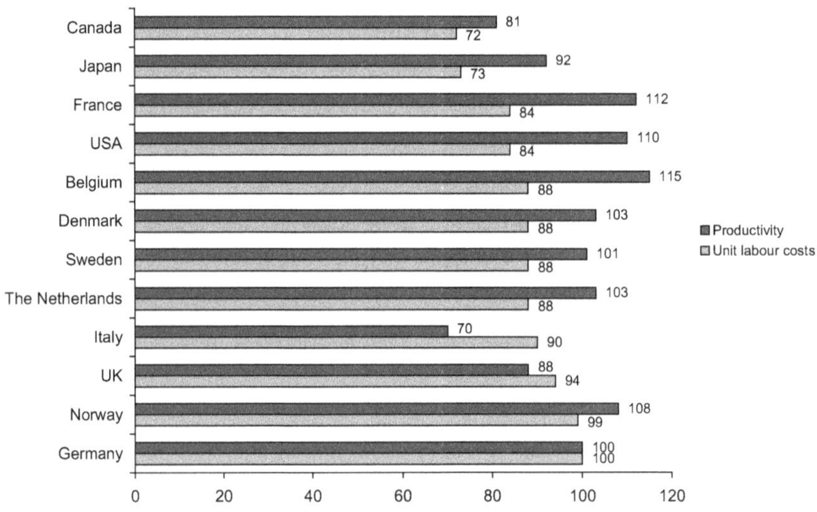

Figure 4.4 Unit labour costs and productivity in the manufacturing sector in 2003 (Germany = 100) (source: iwd, 2004, in Becker, 2006)

productive. In Figure 4.4, the Netherlands ranks a shared fourth place in terms of labour costs, after Germany, Norway and the United Kingdom. In terms of labour productivity, the Netherlands has more or less the same position; it ranks fifth (equal with Denmark) after Belgium, France, the USA and Norway.

Other factors which should be taken into account to compare production locations are the prices of land and real estate. Figure 4.5 shows that the average price to rent a business location in south-eastern Brabant is lower than the Dutch

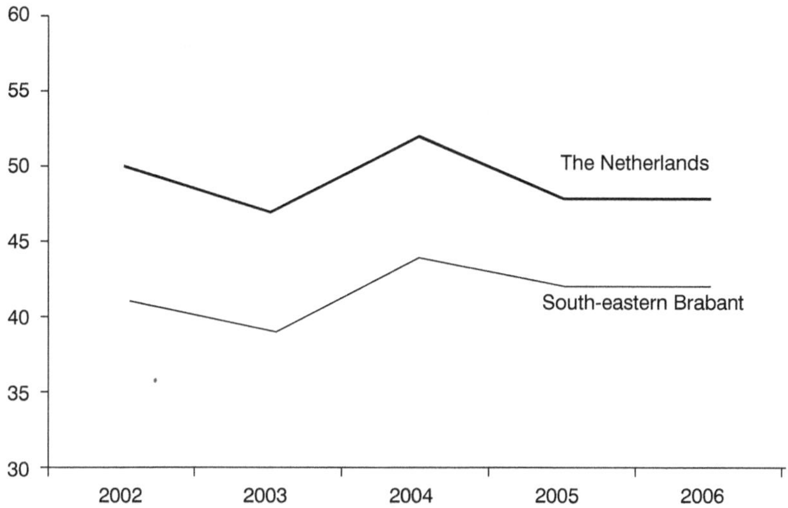

Figure 4.5 Average annual rent for business locations in €/m² (source: Vastgoedmonitor, 2007)

average. This would suggest that the Eindhoven region is more attractive for business than the cities in the Randstad; but it is not only price that matters, but also the availability of facilities and the presence of other businesses.

An important soft factor that distinguishes the Eindhoven region from other regions in the Netherlands (especially compared to Randstad) is a strong willingness to cooperate and strong entrepreneurship which is seen as part of the regional culture. There are direct linkages between the actors and there are many informal networks in which the actors meet each other regularly. Since the Stimulus Programme, firms, the government and knowledge institutes cooperate closely to be competitive and 'to get things done', like lobbying for resources. A clear example is the HTAS (High-Tech Automotive Systems) programme, a joint vision and innovation programme of the Dutch automotive industry (see section 4.5), which is largely concentrated in the Eindhoven region. The strong cooperation in the region is an important asset for the automotive industry where competition increasingly takes place in networks between OEMs and partners.

4.3 The automotive cluster in the Eindhoven region

The south-eastern part of the Netherlands forms the core of the automotive industry in the Netherlands. Over 50% of the automotive-related companies are located in the provinces of Limburg and Brabant (FHA 2006a). Within this area, a major part of automotive firms and knowledge institutes are concentrated in the Eindhoven region (see Table 4.1). It is worth mentioning that this table only includes the direct jobs and companies involved in the assemblage of cars and car parts, and not services like engineering. In addition, it excludes companies which

Table 4.1 Number of companies and jobs in the automotive industry

Categories (ISEC Rev. 3)	*Eindhoven region*		*The Netherlands*	
	Companies	*Jobs*	*Companies*	*Jobs*
Assemblage of motor vehicles (341)	11	4,579	91	10,494
Coachwork and making of followers carriages and semi–trailers (342)	25	1,236	481	8,052
Production of car parts and accessories (343)	12	100	133	6,800
Total automotive	48	5,915	705	25,346

Source: NV REDE (2007)

are only producing a part of their production for the automotive industry. If these companies would be included, the total number of participants in Eindhoven's automotive industry would be much larger. An estimation of the total activities in the automotive industry in the Eindhoven region is given in Box 4.2.

Figure 4.6 shows that automotive industry in the Netherlands has declined in terms of employment. In Eindhoven, the automotive industry has declined, but not as strongly as in the Netherlands in general.

This section discusses the actors in the automotive industry in the Eindhoven region in more detail: OEMs, suppliers and knowledge institutes. Moreover, we give an overview of the geography of the automotive industry in the Eindhoven region and the surrounding areas, because the automotive industry does not stop at the borders. It stretches across south-east Netherlands (the provinces of Northern Brabant and Limburg) and across the border regions into Belgium and Germany. Major nodes of automotive activities include: Eindhoven and Helmond in the Eindhoven region, Lommel and Genk in Belgium, Born in the province of Limburg in the Netherlands (Gemeente Eindhoven, 2003) and Aachen in Germany.

Box 4.2 Key figures in the automotive industry in the Eindhoven region (estimation)

- 120 companies with focus in the automotive industry
- Many knowledge-intensive companies
- Concentrations in Eindhoven and Helmond
- 15,000 employees

Source: Liebrand (2007)

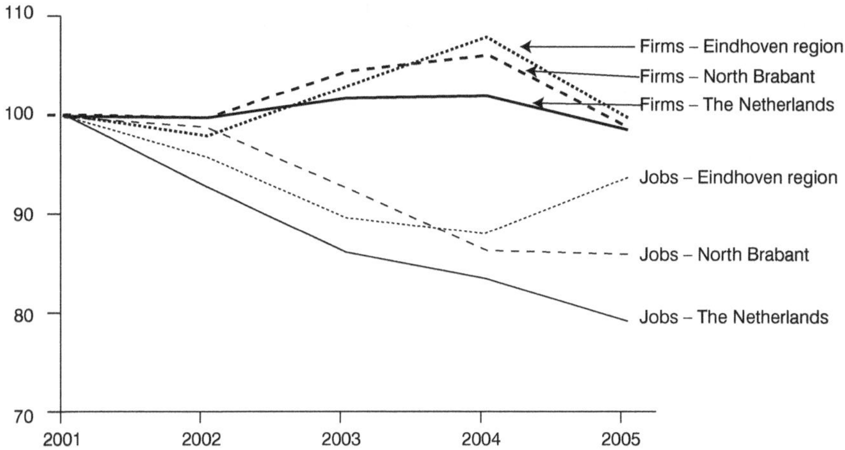

Figure 4.6 Development of the automotive industry (source: NV REDE, 2007)

OEMs

The Netherlands plays only a minor role in the world car production and it has only a few assembly plants for complete vehicles. The Dutch world market share (in 2004) was 1.04% for passenger cars, 1.99% for trucks and 0.1% for buses (Heneric *et al.*, 2005). The assembly of motor vehicles has decreased in the 2001–2006 period, although there are differences between the segments (see Table 4.2). Production of passenger cars has declined steeply, which is caused by problems with NedCar, the only large producer of passenger cars in the Netherlands. The production of trucks (and tracking units) and buses, on the contrary, is increasing. Two major producers of trucks and buses are DAF and VDL Groep (which includes VDL Bova and VDL Bus International), which both have their headquarters and a number of assembly plants in the Eindhoven region. This section describes these two OEMs and NedCar in more detail, including other OEMs in other parts of the Netherlands and in the border regions.

DAF[1] was established in Eindhoven in 1928 as a small welding plant. Today, DAF is one of the leading truck producers in Europe. It has three production plants: one in the city of Eindhoven (5,947 employees at 31-12-2007), one in Westerlo (Belgium, about 70 km from Eindhoven; 2,585 employees) and one in Leyland (508 employees). On its site in Eindhoven, in addition to its production facilities, it has offices and several R&D facilities. The R&D department in Eindhoven has about 500 employees. Recently, it opened a new Engine Test Centre (investment of $60 million) which will play an important role in the development of more efficient and cleaner engines (DAF, 2007b), one of the main research subjects of the truck manufacturer. The engine centre will also develop engines for Kenworth and Peterbilt, the other companies in the PACCAR Group. A related research theme for DAF is the development of

Table 4.2 Assembly and production of motor vehicles in the Netherlands

Brand/ category	*2001*	*2002*	*2003*	*2004*	*2005*	*2006*
Volvo	122,109	108,016	89,191	27,988	0	0
Mitsubishi	67,152	74,352	73,889	87,047	68,145	73,206
Smart	0	0	0	72,565	46,976	14,126
Total passenger cars*	189,261	182,368	163,080	187,600	115,121	87,332
DAF	8,907	8,533	8,329	9,058	11,501	11,881
Scania	n.a	n.a.	21,011	22,700	22,000	25,700
Local	359	411	186	186	245	277
Total truck	9,266	8,944	25,526	31,944	33,746	37,858
DAF	20,331	18,635	20,678	25,904	29,027	31,539
Terberg	475	479	450	594	690	805
Total tractive units	20,806	19,114	21,128	26,498	29,717	32,344
VDL–Bova	588	603	591	660	744	810
VDL–Bus International	666	850	956	801	1,240	1,110
Total buses and coaches	1,254	1,453	1,547	1,461	1,984	1,920
Total vehicles	220,578	211,879	211,281	247,503	180,568	159,454

Source: Bovag–Rai (2007)

Note: * The total passenger cars produced is equal to the production of NedCar and it excludes the production of Dutch luxury sport car producers like Spyker and Donkervoort. These companies have a small output and are therefore not included in the total.

environmentally friendly trucks by using recyclable materials. An important topic in the production process is optimising production by introducing the Just-In-Time (JIT) principle. This has already led to a cost reduction in man-hours of 40% and DAF expects to realise another 40% saving. It is notable that there is a large difference between the production of passenger cars and trucks; trucks are produced in smaller series and have a wider variety. This makes the production process more labour intensive and it requires more flexibility which can be realised by a JIT-based production system.

Western Europe is currently the largest market for DAF, but Central and Eastern Europe (CEE) is the fastest growing market. It is expected that the CEE market will expand to 100,000 trucks in the next five years (DAF, 2007a). The lorry constructor aims to grow by extending its sales and dealer network. DAF does not plan to open production facilities in Eastern Europe yet because it needs to optimise its current capacity first. In a later stage, it aims to extend into Eastern Europe, despite the fact that wages are rising there. The extension into CEE will take time because DAF has to develop a supplier network. This can be done using local suppliers or via follow-sourcing. PACCAR has already opened a parts and distribution centre in Hungary, which supports DAF's growth in CEE (DAF, 2007a).

VDL Bus and Coach[2] is the bus division of VDL Groep (VDL). VDL Groep is a large company, consisting of different subsidiaries which are active in various industries. Next to buses, VDL also delivers various plastic and metal parts, various modules and other finished products. The automotive industry is one of the major customers of the parts and modules produced by VDL. The total group comprises about 75 companies. The companies learn from each other and there are several internal deliveries (about 20% of the combined turnover are internal supplies). The group has 7,321 employees (at 31-12-2006), of which 5,596 are in the Netherlands, 950 in Belgium and 775 in the rest of the world. The company explicitly aims to keep production in the Netherlands as part of its strategy (VDL Groep, 2007a):

> The VDL Groep attaches considerable importance to continuing manufacturing at competitive prices in the Netherlands and the Flanders region of Belgium. We aim to retain and reinforce our competitive position in the international market by continually investing in reliable expertise, robotisation and automation. In addition, our industrial activities in Eastern Europe and Asia enable us to meet the specific wishes of our customers in terms of production in these regions.

To be competitive with companies in low-cost countries, the group focuses on technology in the production process. An illustrative example is given in Box 4.3 which describes the investment in a new (largely) automatic welding robot by VDL Technics. With this new welding robot, VDL Technics is able to compete with companies in low-cost countries that are using manual welding techniques. To keep production in the Netherlands, the majority of the investments are done in the Netherlands. Nevertheless, the group also invests abroad. For example, recently, it constructed a new production plant in Romania, and it extended a production site in China. In addition, the group has become more international through takeovers. For instance, in 2006 it bought Philips ETG, which has two plants in the Netherlands and two plants that produce modules in Singapore and China (VDL Groep, 2007a).

VDL Bus and Coach also conducts research and it has an R&D research centre on its site in Valkenswaard. Just like for DAF, the biggest challenge for VDL is to develop cleaner engines, to meet the stricter environmental rules of the European Union (the Euro 4 and Euro 5 norms). Another ambitious project in which VDL is involved is the Phileas project (see Box 4.1). VDL has a 70% share in APTS, the company that is developing, producing and marketing the Phileas. In addition, VDL is involved in developing infotainment systems for drivers.

Box 4.3 VDL Technics: an example of competing based on technology

New welding installation for VDL Technics.

VDL Technics, located in Boxtel, is specialised in large welded modules. With a new welding robot, VDL Technics is able to economically produce large modules in a largely automated welding process. And this against lower prices than a low wage country like the Czech Republic. The frames are supplied to an establishment of an American stock market listed company that produces lifting systems and telehandlers. ...

Until recently, this work was sourced out by the American company to a Czech supplier which produced the frames by manual welding. However, against the advantage of the low hourly (labour) tariffs, there were also disadvantages: communication problems, inconsistent quality, transport costs and long travel distances for employees. Through automation of the process, it is not only possible to produce quicker, but also at lower costs and a higher quality. Once again, one example that investing in automation of production processes is competitive with low wage countries.

Source: citation from VDL Groep (2007b), translated from Dutch, slightly adapted

NedCar, the only large producer of passenger cars in the Netherlands, is owned by Mitsubishi Motors Corporation. The NedCar plant (capacity of 300,000 vehicles) is Mitsubishi's only production plant in Europe, and it is used as a 'regional production hub' for the European market. Currently, the Japanese assembler is investigating the feasibility of a production plant in Russia which could serve the Eastern European market (Mitsubishi, 2007). It is not clear to what extent a second European plant will have consequences for the production in NedCar plant. The NedCar plant does not have an R&D department; Mitsubishi has a European R&D centre in Germany in the Frankfurt am Main region. Besides its European R&D centre, Mitsubishi has several R&D centres in its home market Japan and two in the USA (in Michigan and California).

Also, in other parts of the Netherlands and in the cross-border regions there are various other OEMs with production plants. In the Netherlands, truck manufacturer Scania has a large production plant (in Zwolle) and a painting plant (in Meppel). Furthermore, some smaller, local truck and passenger car assemblers have plants in the Netherlands, like the truck producers Ginaf and Terberg, and luxury sports car producers Spyker and Donkervoort. In Belgium and in the German state of North-Rhine Westphalia, there are even more OEMs with production plants, like Ford in Genk and Opel in Bochum. In total, there are 14 assembly plants in the Netherlands, Belgium and the state of North-Rhine Westphalia which are producing about 2 million vehicles a year (Liebrand, 2007). These production plants are relevant for the suppliers in the Eindhoven region.

Suppliers

The majority of the value added in the automotive industry is created by suppliers. This might be an advantage for the Dutch automotive industry, where 95% of the companies are suppliers. Figure 4.7 makes clear that the entire value chain is present in the Netherlands; there are suppliers of all tiers, from raw material suppliers to tier-one suppliers (FHA, 2006a). Whereas the OEMs are dominated by foreign-owned companies, the suppliers have a better mix between foreign owned and local companies, like Philips and Koninklijke Nedschroef Holding B.V. ('Nedschroef'). The suppliers vary also in size; there are large multinationals like DSM, but also several (specialised) SMEs. It is important to note that some of the suppliers in the Dutch automotive industry are among the European or even global leaders in a certain niche (see Table 4.3). Most of these firms are raw material suppliers or focus on electronic systems.

The majority of the suppliers are located in the south-eastern part of the Netherlands ('the greater Eindhoven region'), close to Germany, Belgium and France, where the majority of the European automotive industry is located. Paris and Stuttgart, two of the major European automotive clusters, are located less than 500 km from the Eindhoven region. In the interviews we found that there are especially strong links with Germany (and less with France) because there are many similarities in the Dutch and German culture and there is barely a language barrier. The suppliers also benefit from the good knowledge infrastructure consisting of a highly educated workforce and various research institutes. Therefore, various suppliers have opened research departments in the Eindhoven region, like Bentler (PDE Automotive) and Philips. Furthermore, an important asset is the presence of other industries in the region (like the ICT and the mechatronics industry) or near the region (like the chemical clusters in Sittard-Geleen and in the ports of Rotterdam and Antwerp, and the steel industry in the German Ruhr area). These industries form major inputs for the automotive industry.

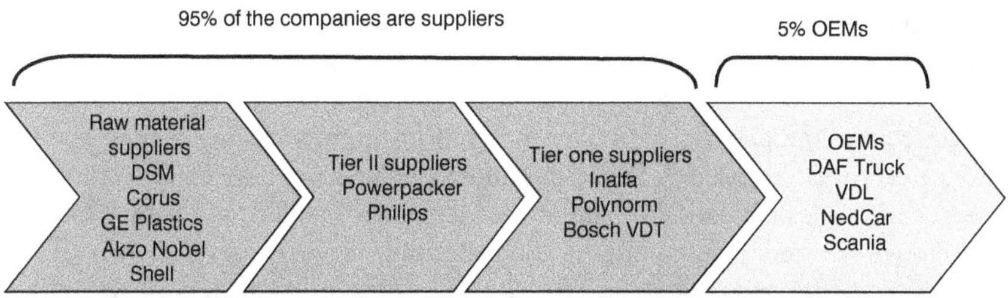

Figure 4.7 Simplified value chain of the Dutch automotive industry (source: own elaboration based on FHA, 2006a)

Table 4.3 Suppliers in the Dutch automotive industry are among the market leaders

Company	*Description*	*Market position*
Akzo Nobel Coatings	Coatings	No. 1 worldwide
Bosch VDT	CVT belts and technology	No. 1 worldwide
Corus	Steel (IHSS) and aluminium	No. 3 in Europe
DSM	Engineering plastics: airbags housing and timing systems	No. 1 worldwide
GE Plastics	Exterior body panels, automotive lighting and glazing	Top 2 worldwide
Groeneveld	Greasing systems	No. 1 worldwide
Inalfa	Roof systems	No. 2 worldwide
Navteq	Digital map data	No. 2 worldwide
Nedschroef	Automotive fasteners	Top 3 in Europe
Philips	Lighting, car radio ICs, in-vehicle networking and car access ICs	No. 1 worldwide
Powerpacker	Hydraulic systems for convertible roofs	No. 1 worldwide
SKF	Bearings	No. 1 worldwide
TomTom	Navigation systems (after market)	No. 1 worldwide

Source: FHA (2006a)

Knowledge institutes

The automotive industry in the Netherlands is one of the major contributors to the total private R&D expenses. In total, about 10% (€ 500 million) of the total private R&D spending in the Netherlands is done by the automotive industry. Furthermore, 7 of the 15 largest R&D spenders derive a large part of their business from the automotive industry. Moreover, several important inventions in the automotive industry have been done in the Netherlands, like the CVT which is a key part in today's automotive transmissions (by Van Doorne in the 1950s), the Philips Carin system (the first 'in-car' navigation system in the 1980s) and the first automotive people-movers (by Frog Navigation systems in 1997). Currently, TomTom is leading in the development of navigation software (FHA, 2006a).

A large part of the research infrastructure in the automotive industry in the Netherlands is concentrated in the Eindhoven region for its highly educated workforce and the presence of many research institutes. The region has several actors that conduct research in the automotive industry. First, TU/e carries out research, provides education in automotive technology and it will start a special Automotive Masters programme in 2008. Second, there are several specialised knowledge institutes that conduct research in fields that are important for the automotive industry. Examples include: the Embedded Systems Institute (ESI, which focuses on embedded systems that are increasingly important in cars) and the Dutch Polymer Institute (DPI, which develops new materials that can be used in car parts). Finally, there are many companies that have their research departments in the Eindhoven region. These concern all types of companies, from OEMs (only NedCar has no research activities), to suppliers, and from large companies (such as Philips) to smaller specialised engineering companies. There are also start-up companies and/or spin-offs of TU/e (for instance, DTI, an innovative knowledge and concepting centre that is located on-site at TU/e).

The region's strong knowledge base acts as a magnet, and it attracts other research institutes. One example is TNO Automotive, the automotive department of the public research institute TNO. In 2005, TNO decided to transfer the automotive division from Delft in the Randstad area to Helmond in the Eindhoven region. TNO had various reasons to do this, including:

- *The labour pool*: there are many automotive engineers in the Eindhoven region. Especially of interest is the presence of automotive studies at TU/e and Fontys Hogescholen, which can guarantee a supply of future researchers.
- *Presence of the automotive industry*: more than 50% of the companies in the automotive industry are located in the south-east Netherlands, as are several research institutes. The companies might be potential clients of TNO.
- A *strong willingness to cooperate* between the various actors in the region.
- *Strong lobbying* from the industry and local governments to the Dutch government to transfer TNO Automotive to the Eindhoven region.
- *Geographical location*: Eindhoven is located next to the German border, where the majority of its customers is located.

The transfer of TNO Automotive to the Eindhoven region has various benefits for the region. First, it is a stimulus for the employment; in total about 250 jobs will move from Delft to the Eindhoven region. Maybe even more important is the symbolic value of the movement. The presence of the key player TNO makes the value chain in the region more complete (ED, 2005) and it gives the region a stronger perception to the national government and as an automotive region. This could increase the lobbying power to the national government.

Another example is TomTom, one of the leading companies in the development of car navigation devices and software. In 2007, the company decided to open a research centre in the Eindhoven region. TomTom's main reason to do this was the labour pool. Eindhoven is one of the largest pools of talent for engineers and software developers. Another reason why TomTom has opted for the Eindhoven region is the presence of other technological companies, software developers, other companies (like design companies) and knowledge institutes. These companies might act as research partners or as developers of new products.

There are also various important research institutes and companies in the cross-border regions around Eindhoven and in other parts of the Netherlands that conduct research in the fields of automotive technology. To name a few: Technical University Aachen (various fields of automotive research), IKA/FKA (R&D institute linked to the Technical University Aachen with expertise in power trains, vehicle safety, vehicle dynamics and acoustics), FEV (technical centre, as a spin-off of the Technical University Aachen, specialises in engines), Ghent University (Automotive R&D in fuels and in-vehicle networks). Another example is Philips, which has research departments for automotive R&D in Eindhoven, Aachen, Nijmegen and Leuven (FHA, 2006a).

Geography of the sector and region

The Eindhoven region forms the core of the Dutch automotive industry. Within the Eindhoven region, the automotive industry is scattered, but there are two major concentrations of automotive activities: the cities of Eindhoven and Helmond (see Figure 4.8). Educational facilities and research centres of firms and specific research institutes are based in Eindhoven or Helmond, like TU/e, the research centres of DAF and TomTom, and TNO automotive. In addition, several companies have production plants or even have their main sites in one of the two cities, like DAF and Nedschroef. There are also important physical clusters. One example is the ‘Automotive Campus Helmond’, a planned business park for firms, knowledge and educational institutes, and network organisations that are active in the automotive industry. The campus is planned at a business site where TNO Automotive and PDE Automotive are located. The core of the campus will consist of the Automotive Research, Development & Engineering Centre Helmond Eindhoven (this centre will supply several production and system test

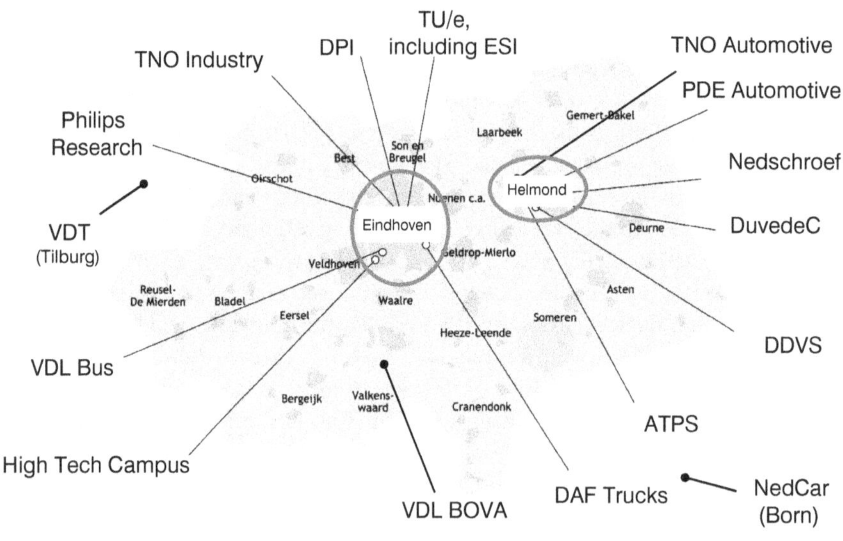

Figure 4.8 Eindhoven and Helmond: two major concentrations of automotive activities (source: own elaboration, based on Liebrand, 2007). Note: this figure is not comprehensive; it does not include all firms and institutes in the automotive industry in Eindhoven and Helmond.

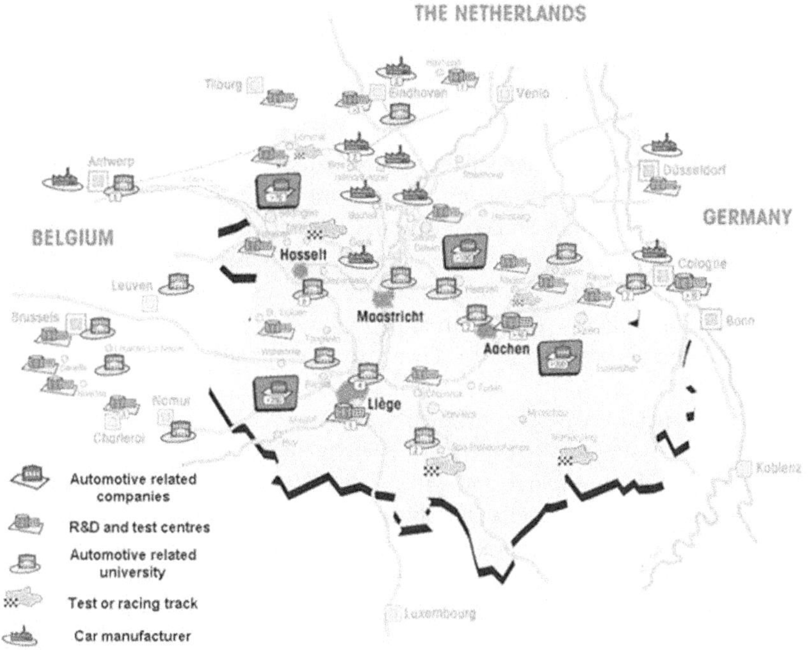

Figure 4.9 The automotive industry in the 'Meuse Rhine Triangle' (source: http://automotive.medienartig.com/index.php?id=33)

facilities for product development), a concentration of automotive studies, and multi-functional research facilities. The aim is to develop an open campus that forms an expertise centre for the Dutch automotive industry.[3]

Finally, as we have seen in this section, there are various important firms and knowledge institutes in other parts of the Netherlands and in cross-border regions in Belgium and Germany. Figure 4.9 shows the major automotive activities in the so-called 'Meuse-Rhine Triangle', the cross-border area just below the Eindhoven region. Moreover, the firms and knowledge institutes in the Eindhoven region are involved in wider production and knowledge networks with foreign partners. These networks are discussed in more detail in the next section.

4.4 Local and global networks

OEMs do not operate independently, but increasingly work in partnerships with other OEMs, suppliers, universities and other research institutes. Competition increasingly takes place between networks. This section discusses these networks in more detail. For this, we make use of the two types of networks that we have distinguished in our theoretical frame: production networks and knowledge networks. Moreover, we deal with the question to what extent do manufacturing and R&D facilities need to be co-located? Finally, we investigate the relations between the automotive industry and research institutes in detail.

Production networks

Production networks have changed through globalisation and other developments. A general trend is that production shifts from Triad[4] countries to emerging countries, like China, India and countries in CEE, for cost reasons or to serve the local markets. Nevertheless, our interview partners believe that it is possible to keep production in Western Europe. There are companies, also in the automotive industry, which explicitly do not aim to go to low-cost countries and continue to keep production in Western Europe. There are several reasons for this:

- Countries in CEE have become members of the EU, and therefore, employees are able to move from countries in CEE to countries in Western Europe. This makes labour more scarce and will increase the wages in CEE, leading to a declining difference in factor costs between CEE and Western Europe. For instance, currently it is not attractive for firms in the automotive industry to move to Slovakia, because of a shortage of employees in the sector. Also in other areas, like Shanghai, labour becomes scarcer and wages are rising.
- In general, the automotive assembly follows the market to minimise the transport costs. This will keep car assembly in Western Europe. Also, image plays a role in this, especially for the German luxury car assemblers. Nevertheless, new investments by OEMs are increasingly done in CEE to serve this market. Also suppliers are following OEMs to CEE in later stages ('follow-sourcing').
- High quality and reliability are required in the automotive industry. For instance, Eaton, a parts supplier in the automotive industry, uses a standard of 1 of 20 million PPM, which means that it aims to only send back to its suppliers one piece in 20 million (interview). Such high quality standards can not always be guaranteed in areas outside Triad countries. Suppliers in Western Europe focus on quality and innovation, and are able to make profits on new products during the first period of a product lifecycle. Each year, OEMs force suppliers to lower prices, and in later stages, the products are not profitable anymore, and companies outside the Triad are able to reproduce ('copy') the product. Via innovation, the suppliers in the Triad try to stay competitive.
- Culture in companies and historical roots. For instance, VDL Groep has strong regional feelings and explicitly aims to keep production in the region (see section 4.3).
- The production process. Products with a high specificity that are produced in small batches require a highly automated production process. For this type of production, it does not make sense to shift production to low-cost countries. An example of this is ASML, which is producing its variety of products in small batches.

Recent investments by firms in production facilities in the Eindhoven region illustrate the willingness to keep production in the region. Some examples are: DAF optimises its production process, VDL invests in various facilities like VDL Parts (a joint distribution centre in Valkenswaard for various VDL companies), and DSM constructs a new chemical plant on its site in Geleen and not on one of its sites in China. For DSM, cost benefits raised by shared facilities on its site in Geleen (for instance the new plant will make use of one central control centre that is also used by another plant) are higher than cost benefits from lower wages in China.

Although there is willingness to stay, the competition in the automotive industry might be a threat for the companies in the Eindhoven region. For example, VDL Bus and Coach is the only bus producer of the 'European Top 6' which does not have production facilities in Eastern Europe. The other five major bus assemblers in Europe have invested in production facilities in CEE to save costs by benefiting from lower wages. To stay competitive, VDL Bus and Coach has changed the production process and increasingly outsources parts of the production process. It is notable that outsourcing of activities is not necessary given to suppliers in low-cost countries; it can also be given to suppliers with an efficient production process in countries with relatively high wages. For instance, VDL Bus and Coach sources roofs for buses from a company in Switzerland. Nevertheless, a threat for VDL Bus and Coach is that suppliers also increasingly invest in production facilities in Turkey to supply the bus assemblers that went there ('follow-sourcing'). This not only increases the geographical distance between the suppliers and the VDL plants in the Eindhoven region (which increases transport costs), but it also leads to barriers through differences in language and culture. We have also seen that various companies investigate possibilities of opening plants in developing countries to serve local markets. For instance, as we have noted before, DAF plans to invest in production facilities in CEE after it has optimised its production in its plants in Western Europe. Suppliers are also doing this. For example, Nedschroef follows the developments in emerging markets, and if it is needed, the company will enter these markets by taking over local firms.

The role of proximity between the car assemblers and their suppliers differs per product. On the one hand JIT deliveries require that certain suppliers are located close to the OEMs. DAF is currently introducing the JIT principle, and therefore, it makes use of certain regional suppliers that can deliver products quickly. NedCar has three suppliers on its site in Born (co-siting) that can directly deliver the parts to the NedCar plant. Also important are internal deliveries of the OEMs from one site to another. For instance, DAF produces cabins and axle assemblies in Westerlo which are transported to Eindhoven where final assembly takes place. Also VDL Bus and Coach sources parts from companies within the VDL Groep. It should be mentioned that this relevant region for production networks is wider than the Eindhoven region; it also includes other parts of the Netherlands and Western Europe, in particular Germany. Suppliers in the Eindhoven region export the majority of their products, and only a limited share of the output is sold to local OEMs. Some of them even export all their products to OEMs abroad. Suppliers of smaller parts supply their products to OEMs on a large distance. For instance, Nedschroef supplies fasteners to various OEMs in Western Europe, which also export the fasteners to their plants outside Europe.

There are also deliveries between OEMs. These deliveries can be between local OEMs (like VDL Bus and Coach that uses DAF diesel engines in certain buses) or between OEMs that are located farther from each other. For example, a certain DAF model uses a cabin from Renault and an engine from another OEM. It should be noted, that DAF does not develop new products with its competitors.

Knowledge networks and the link between R&D and production

The companies in the automotive industry often have their basic research in the home region, with strong regional partnerships, while applied research (or 'development') takes place with more global partners. This is also the case in the Eindhoven region. For instance, DAF, focusing on development of cleaner and more sustainable trucks (use of environmentally friendly materials, increasing the life of trucks), has several regional partners: TNO Automotive, Department of Road Transport ('Rijksdienst Wegverkeer', RDW) and TU/e. It also cooperates with FEV in Aachen, which can develop new engines quickly. The proximity of production and R&D is important for DAF; it has a research centre on the production site and in general it uses research partners located within 100 km. The link between production and manufacturing is important for DAF to produce test vehicles. In addition, research partners need to walk around the truck in the production plant to 'feel' the production process. Proximity of research partners is also important to develop new inventions quicker, and to prevent language and culture barriers. Also Nedschroef conducts its basic research in the Eindhoven region, and is partnered with the Fraunhofer Institutes in Germany. Another example is VDL Groep, which focuses on the development of infotainment systems in buses. So, for new research topics, proximity to production is important, and therefore, as noted by our interview partners, it is important for companies in Western Europe to focus on basic research to keep the competitive advantage, and to keep (a part of) the production in Western Europe.

Nevertheless, there are also examples of companies that can split research and production geographically. One is TomTom which focuses on R&D and marketing of car navigation devices and software, which largely takes place in the Netherlands. It has outsourced all production to suppliers in low-cost countries where product engineering also takes place. This product engineering is jointly done by TomTom and its suppliers, where TomTom gives the requirements for the adaptations to the products. Another example is Mitsubishi, which has split up its European activities. As we have noted before, it produces in the NedCar plant in Born in the Netherlands, while research takes place in Germany. Furthermore, there are also companies which focus mainly on marketing and outsources (parts of) the R&D and production functions. One example is Philips, which was already outsourcing production in the 1970s and 1980s, and is now increasingly outsourcing R&D activities and is thus buying the knowledge. For example, the electrical giant has sold R&D-intensive departments like NXP Semiconductors, and buys R&D services from various specialised R&D centres.

Relations between research institutes and the automotive industry

The TU/e is an important research partner for the automotive industry in the Eindhoven region. For instance, with DAF it develops hybrid vehicles in 'green' zones. For this, TU/e has a special DAF test truck in a laboratory. TNO is also involved in the project and it finances a PhD student in the research project. Another important research partner is Bosch VDT, a supplier of CVT belts and technology. TU/e also has PhD contracts with research partners outside the region, like a PhD programme on the use of active control for passive safety systems (DCT, 2005). Often, personal contacts between professors and industrial partners are important in the research projects. There is also strong cooperation between TU/e and its industrial partners for education. Several companies (often located in the region, like DAF and VDT) offer internships for students, and industrial partners are involved in study programmes. Via the HTAS programme (see section 4.5), various companies support the development of an automotive master at the TU/e. Representatives of companies, like TomTom, have a chair in a special 'result testing group' (in Dutch: 'klankbord groep'). Besides involvement of companies in educational programmes of TU/e, the industrial partners are also involved in educational programmes and internship projects of other educational facilities like Fontys Hogescholen (polytechnic level) and various ROCs (lower and middle level).

Another important research partner in the Eindhoven region is TNO Automotive. Although, the national share has increased, the majority of TNO Automotive's customers are foreign OEMs. Physical distance is often not an issue for TNO Automotive's research projects; vehicles are flown over from the Far East to TNO Automotive in the Eindhoven region for a crash test. Sometimes, projects are done by TNO Automotive with tier-one suppliers or with OEMs and tier-one suppliers. Negotiations with the customers are often long, and the fear of losing knowledge is large. Secrecy, international property rights (IPRs) and liability are the main issues for these negotiations. TNO Automotive has a relatively strong position towards its customers, because it focuses on basic research in certain niches (in particular the integration of active and passive safety) in which TNO Automotive has created inventions that can be sold to its customers. This can be done beforehand (OEMs finance research projects and receive the results afterwards) or afterwards (OEMs buy the results via IPRs or via privatisation of research departments which are sold). In addition, sometimes OEMs use TNO Automotive (and comparable research institutes) as a 'block' against tier-one suppliers which have increased in size and power. TNO Automotive is an independent research institute that competes with these large suppliers. Therefore, the relations between TNO Automotive and tier-one suppliers are sometimes delicate when these actors have joint research projects with an OEM.

TU/e and TNO Automotive are initiative takers of the Competence Centre for Automotive Research (CCAR),[5] a partnership between industrial partners, TNO Automotive and TU/e. Goal of the partnership is to form a platform for market-driven research in various areas in the automotive industry. Projects of CCAR are done by PhD students who use the facilities of the university and project partners. It is expected that the competence centre will create about 100 new jobs (ED, 2005).

There are also some points of criticism in the relations between the industry and research institutes. We found that contact between the industry and educational institutes might be improved for internship projects. Teachers at educational facilities who are supervising students do not always visit production plants to see what their students are doing in practice. Another point of criticism that we found in the interviews is the relatively small scale of TU/e compared to competing universities, like the Technical University in Munich. To even this out, TU/e is cooperating with other technical universities in the Netherlands (the Technical Universities of Delft and Twente) in the 3TU Federation. One of the competence centres in 3TU is High-Tech Systems, which is active in the automotive industry. In addition to the three Dutch universities, TU/e also cooperates intensively with the Technical University in Leuven, Belgium. There is less cooperation with the Technical University Aachen, due to culture differences. With other universities, cooperation is increasing (including for automotive research), and in general, physical distance is becoming less important because of better technology, although physical networks are still relevant. A final point of criticism for the relations between the industry and TU/e is a lack of growth incentives at universities. Professors are judged on publications instead of growth. This might lead to differences in interests between the industrial partners and TU/e.

4.5 Policy and cluster organisations

The government and cluster organisations play an important role in the development of automotive clusters. Often, various governments finance (research) projects, while a main task for cluster organisations is to stimulate cooperation and to link various actors.

The role of the government

The Dutch state recognises the importance of the Dutch automotive industry and the concentration of the industry in the Eindhoven region. This can be illustrated by the decision to move TNO Automotive from Delft to Helmond in the Eindhoven region (see section 4.3). Furthermore, the Dutch government financially supports various projects and institutes in the automotive industry. One example is a financial contribution to the Automotive Technology Centre, a cluster organisation for the Dutch automotive industry. The Ministry of Economic Affairs contributed a maximum of €645,000 to the total project costs of €1,1910,000.[6] Additionally, the state contributes to the first stage of the 'AutomotiveNET.eu' project (an INTERREG project which aims to strengthen the innovative capacity of the automotive industry in the 'Meuse-Rhine Triangle').

Also the provinces in south-east Netherlands, Northern Brabant and Limburg, support various projects in the automotive industry through financial contributions. For instance, the province of Northern Brabant contributed €2.5 million for the relocation of TNO Automotive to the Eindhoven region and it gave subsidies of €215,000 to ATC for the 2006–2007 period.[7] The province of Limburg also contributed €215,000 to ATC and €57,000 to the AutomotiveNET.eu project.[8] In addition, the province development agencies, 'NV Brabantse Ontwikkelings Maatschappij' (Brabant Development Organisation, BOM) and 'NV Industriebank LIOF' (Industry bank LIOF), are 'partners' in various projects like ATC and the AutomotiveNET.eu.

The automotive industry is a key industry in the municipalities of Eindhoven and Helmond and also in the wider region. Various municipalities, private actors and knowledge institutes cooperate in the Brainport Development to initiate programmes to develop and promote the region. One project is the Brainport Programme for Automotive Innovation which aims to strengthen the automotive value chain and industry in the region. Helmond is seen as the core of the automotive industry in the region. One of the main topics in the programme is to improve the labour force through education. Currently, the automotive industry contributes to the regional employment, but the largest share is formed by low-skilled workers in the 'traditional automotive industry' (like parts production and metal treatment). Future growth needs to take place on the university and applied sciences level (contract research, development, testing and engineering).[9]

Business associations and cluster organisations

Business associations and cluster organisations play key roles in developing networks between the industry and other actors, promote regional clusters and have a lobby function for the industry to the government. There are several of these organisations in the Netherlands, including Federatie Holland Automotive (FHA, an organisation that represents the interests of the Dutch automotive industry) and NEVAT Holland Automotive (NEVAT is the business organisation for the suppliers in the Dutch industry).

The Automotive Technology Centre (ATC) was jointly established in 2002 by various companies in the automotive industry, TNO Automotive, TU/e, other network organisations and the development agencies of the provinces of Northern Brabant and Limburg. At that time, the Dutch automotive industry had problems because of the decline in production by NedCar, a shift of production to Eastern Europe and lack of political interest in the sector. ATC was established with the aim to improve the innovation power of the Dutch automotive industry. The cluster organisation has grown quickly from about 10 partners and participants to over 20 partners and over 110 participants in 2007. ATC is organised as a foundation with a financial contribution by the Dutch state (via the Ministry of Economic Affairs), the provinces of Northern Brabant and Limburg and the development agency for

Box 4.4 The HTAS programme

The High-Tech Automotive Systems (HTAS) programme is a market driven innovation programme that is set-up and steered by the Dutch automotive industry. The initiative taker for the programme is FHA. Several companies, knowledge institutes and other stakeholders have been involved in the creation of the programme. Companies from different industries are involved: OEMs and suppliers of all tiers from the automotive industry, software and ICT companies, transport and logistical companies, etc. It is notable that foreign OEMs that are not based in the Netherlands are also involved in the programme; BMW, Fiat, Volvo and Daimler-Chrysler have all contributed to the programme.

Important for the development of the programme is support of the Dutch Ministry of Economic Affairs (EZ). In May 2007, EZ assigned the HTAS programme. The total contribution of the Dutch state to the programme is €42 million for the 2007–2011 period. The total budget costs for the programme are estimated at €158 million.

The main characteristics of the HTAS programme can be summarised as follows:

- The ambition is to increase the turnover in the Dutch automotive industry from €12 to €20 billion from 2006 to 2015 and to raise employment from 26,000 to 50,000 employees (FTE). It is important to notice that growth not only takes place in the automotive industry it serves (direct effects), but also in other sectors that support the automotive industry, like electronics and embedded systems (indirect effects).
- The programme focuses on two key areas for innovation: driving guidance (including the connected car, vehicle control dynamics and integrated human machine interaction) and vehicle efficiency (including efficient powertrain and light constructions). These areas are based on the strength of the Dutch automotive industry and general trends in the automotive industry. For example, driving guidance is largely based on software and electronics, which are increasingly used in cars. The Eindhoven region has a strong electronics and software cluster and can play a key role in innovations in the area of driving guidance.
- The programme deals with technical as well as with non-technical issues. Non-technical issues are described as 'enablers' and include education, knowledge transfer and business. These enablers are an

important input for innovation. The programme has various projects to improve these enablers. For instance, to improve the labour pole, TU/e starts with a specific automotive master.

- The programme is open and demand driven. Via research proposals, actors can receive funding for research projects. Also foreign knowledge partners can be involved in the research projects and can benefit from subsidies.

Source: http://www.pps-automotive.nl/ (accessed 11 March 2008), FHA (2006a, 2006b) and interviews

the Eindhoven region. The target group of ATC is the automotive industry and it includes OEMs and suppliers of all tiers as well as knowledge institutes. In addition, both SMEs and MNOs can participate in ATC. Although ATC is open for companies in the whole of the Netherlands and it has the entire Netherlands as its relevant region, the focus of the cluster organisation is the south-eastern part of the Netherlands.

ATC focuses on technology and it has a major networking function, therefore it organises various activities, like company visits and an annual congress. In addition, it stimulates cooperation between the industry and knowledge and educational institutes, promotes the sector, and it is involved with other networks. ATC also plays a major role in the development of the HTAS programme (see Box 4.4) and it aims to increase the cooperation in the cross-border area between the Netherlands, Belgium and the German state of North-Rhine Westphalia. One of the cross-border projects in which ATC is involved is the 'AutomotiveNET.eu' project and, therefore, it stimulates cross-border cooperation between the automotive industry and knowledge and educational institutes. Although the first task has already been finished (the creation of an Automotive Atlas), our interview partners agree that there is still room for improvement regarding the cross-border cooperation. In particular, the cooperation with German partners is still limited. This is probably related to cultural differences. Another factor might be the funding structure; most of the innovation programmes are nationalistic, which means that potential foreign research partners cannot benefit from subsidies. Only the Dutch Peaks in the Delta and the HTAS programmes are also open to foreign actors. In Germany, in contrast, there are (yet) no comparable programmes in which Dutch or Belgium actors can make use of research funding provided by the German government. This might hamper cross-border cooperation.

4.6 Conclusion

The manufacturing sector has a long and dynamic history in the Eindhoven region. Philips and DAF trucks have played a key role in economic development. The companies were (and still are) among the largest employers in the region, had a strong local supplier network and played an important role in the society through social programmes. Nevertheless, the region was hit hard in the beginning of the 1990s when the two industrial giants got into financial trouble, resulting in large regional unemployment. Meanwhile, the region has been restructured, focusing on R&D and technology. Nowadays, it is among the top technology regions in Europe. Nevertheless, the manufacturing sector still plays a significant role in the regional economy as it is still the largest employer, with about 18% of the regional employment.

One of the lead industries in the region is the automotive industry. In this chapter, we have analysed the automotive industry in the region in more detail. Among other things, we investigated the linkages the automotive industry has with other industries, the relations with the strong knowledge base in the region, the production and knowledge networks of the automotive industry, and the governance of the sector.

In the remainder of this section, we summarise the main findings of our research and draw some conclusions. First we analyse the link between the automotive industry and the regional foundations and second we will look at the structure of the cluster and production and knowledge networks.

Conclusions on foundations

The manufacturing sector still plays a major role in Eindhoven's *economic base*. A strong point in the economic base is the presence of (high-tech) industries, such as mechatronics, ICT and design which form major inputs for the changing automotive industry where electronic systems and software are increasingly used next to or instead of hydraulic systems. The HTAS programme is catering to this trend and the strength of the region (the favourable sector mix) by focusing on driving guidance and efficient vehicles. Another strong point of the economic base is the balanced mix between SMEs and multinationals, also in the automotive industry. A point of concern for the economic base is its sensitivity for economic cycles and international developments. This is especially the case for the automotive industry which is very international, and many companies in the automotive industry in the Eindhoven region are in foreign hands.

The *knowledge base* of Eindhoven is very strong. It has several research institutes, a highly skilled labour pool, and it has more than 40% of the total Dutch private R&D expenditures. It is among the technological hotspots in Europe. There are also several research facilities for the automotive industry, consisting of universities (TU/e), public research institutes (like TNO Automotive), public–private research centres (such as DPI) and research departments of companies (TomTom, DAF and VDL Groep have research departments in the region). In addition, automotive education has improved (like the development of a special automotive masters programme at TU/e) in order to upgrade the labour pool, which now consists of many low-skilled workers, while there is an increasing demand for workers with a university or an applied sciences degree.

Also the *accessibility* of the Eindhoven region is improving. The regional airport has grown, and it has direct connections to some other technical hotspots in Europe, which is favourable for creating 'temporary proximity' relevant for research projects with foreign partners. In addition, the government invests in infrastructure projects to reduce congestion. Nevertheless, currently, congestion forms the major bottleneck for the region which may negatively influence production as well as knowledge networks. Congestion may hinder JIT deliveries (relevant for production networks) and increase the relative distance to large intercontinental airports (relevant for knowledge networks).

The results for the *quality of life* are mixed. On the one hand, Eindhoven's natural and green surroundings make the city attractive for technical knowledge workers. On the other hand, although much is changing, the limited metropolitan facilities and image makes the city less attractive for high-skilled workers in higher services, such as marketing and finance.

For the *factor conditions*, we have seen that the factor prices and labour productivity in the Netherlands are comparable with most of the other Triad countries. Within the Netherlands, Eindhoven generally has lower real estate prices than in the Randstad region. An important asset for the Eindhoven region is the presence of many informal networks and a strong willingness to cooperate and a 'Triple Helix approach'. The HTAS programme, where the industry, knowledge institutes and the government cooperate, is a clear example of this.

Cluster structure and networks

Despite the relatively small presence of OEMs, the entire value chain is present in the Eindhoven region, with a strong emphasis on the R&D function. The majority of firms in the Dutch automotive industry are suppliers. There are suppliers from all tiers. The OEMs in the Eindhoven are a truck manufacturer (DAF), a bus assembler (VDL Bus and Coach) and a relatively small plant for passenger cars (NedCar). Many firms have research facilities in the Eindhoven region. Additionally, there are various other research and educational institutes of which TU/e and TNO Automotive are two key players. The major concentrations of the automotive industry in the region are the cities of Eindhoven and Helmond. In Helmond, the Automotive Campus Helmond is a planned business site where several automotive facilities will be concentrated. The automotive industry is not limited to the Eindhoven region; also in the border area with Germany and Belgium are many firms in the automotive industry and research institutes that conduct research in fields of the automotive industry. Although there are joint initiatives (like AutomotiveNET.eu) and foreign partners are stimulated to participate in joint research projects (foreign companies can benefit from subsidies in the HTAS programme), the cross-border cooperation can be improved. The cooperation with partners in the Aachen region is still seen as low, something that might be explained by culture differences.

From analysing the dynamics in the cluster, we have distinguished two types of networks in our frame of analysis: production networks and knowledge networks.

We have analysed the linkage between these networks, i.e. what is the relation between R&D and production? Furthermore, we have analysed the role of proximity; i.e. we have investigated to what extent the actors in the cluster make use of regional or global networks.

The role of proximity in *production networks* differs per case. Proximity of suppliers is required for JIT deliveries of large modules or systems. For instance, NedCar has three suppliers on its site (co-siting). Nevertheless, we have also seen that suppliers of smaller parts (like Nedschroef) deliver to all global OEMs.

An important development for production networks is a shift of production facilities to CEE and other markets for cost reasons and for the rapid development of these new markets. The influence of this development on the production networks in Western Europe is mixed. We see a strong willingness of companies to keep production in Western Europe. There are various examples of investments in production facilities in the Eindhoven region, like an investment of DSM in a new plant and investments by DAF to optimise production in its plants. The remaining production in Western Europe concerns, in particular, high-tech production, focuses on innovation and high quality. The automotive industry is an industry with high quality standards, and various parts with high quality requirements can be produced in Western Europe. However, we have also seen that OEMs increasingly invest in countries in CEE and other new markets to serve these new markets. In the European bus market, five of the six largest bus producers have opened production facilities in Turkey. Also suppliers have followed the bus assemblers to Turkey to serve the OEMs. This 'follow-sourcing' might be a threat for bus producers which do not have production facilities in Turkey (like VDL Bus and Coach). The shift of suppliers increases transport distances (and thus costs) to plants in Western Europe and culture differences might hinder trade.

For *knowledge networks* it is important to analyse where R&D takes place and who the research partners are. The location for R&D activities depends on the type of R&D. Basic research is often done in the home country, because of the fear of losing knowledge, IPR tensions and the presence of strong research partners in the home base. For instance, DAF, DSM, VDL Bus and Coach, Nedschroef and TomTom all have their basic research centres in their home base in or near the Eindhoven region. Research projects are often done in close cooperation with regional research partners, like TU/e, TNO Automotive, FEV Aachen or other specialised research institutes. Proximity of research partners is important to develop new products quicker and to prevent communication problems through language and/or culture differences. Also the link between R&D and production is important. For DAF, the presence of production for research is required to build test vehicles and the engineers can 'feel' the production process. Nevertheless, we have also found examples where R&D and production are split geographically, like Mitsubishi and its European activities and TomTom, which completely outsources production to suppliers in low-cost countries. Adaptive research and development often take place outside the home base and often at the site of the supplier (or customer) and in cooperation with the supplier. For instance, TomTom adapts its products jointly with its suppliers in low-cost countries.

Finally, we have indications that regions in developing countries and regions in Triad countries are converging and that in both places clusters emerge where (high-tech) production and R&D take place. On the one hand, we see that factor costs in developing countries have risen. This makes producing in these countries less attractive, and production might come back to the Triad countries, especially for high-tech production that is strongly linked with R&D. On the other hand, we have analysed that certain regions (like Shanghai) are upgrading rapidly and are changing from adaptive research to more basic research. Companies are increasingly investing in (basic) research facilities in these countries. Also production in these regions has been upgraded from low-tech (often labour-intensive) production to more medium- and high-tech production. Furthermore, companies in Triad countries focus more and more on marketing and other commercial functions, while R&D is outsourced and bought from specialised regions.

Notes

1 http://www.daf.com/index.asp (accessed 13 June 2006) and interviews.
2 http://www.vdlbus.nl/UK/index.htm (accessed 14 June 2006), corporate brochure VDL Groep 'Kracht door samenwerking' and interviews.
3 http://www.helmondautomotive.nl/index2.php (accessed 29 January 2008).
4 The Triad includes Western Europe, North America and Japan.
5 http://www.ccar.nl/index.php?option = com_content&task = view&id = 25&Itemid = 38 (accessed 2 February 2008).
6 Source: news bulletin of the Ministry of Economic Affairs, 6 January 2006, '8 miljoen euro voor gebiedsgerichte innovatieprojecten'. Available at: http://www.kennispoort.net/nieuws/detail/?FBE64946CA794E7B857A0BA725B84AE4E8C2272D9834483B984019E69CDCEF3E (accessed 15 June 2006).
7 http://www.brabant.nl/Applicaties/Search.aspx?qa = automotive (accessed 19 June 2006).
8 http://www.senternovem.nl/kennis_bij_de_buren/de_vorderingen/autonet.eu.asp (accessed 5 July 2006).
9 http://www.brainport.nl/ (accessed 5 March 2007).

Bibliography

Becker, H. (2006) *High noon in the automotive industry*, Berlin: Springer-Verlag.

van den Berg, L., P.M.J. Pol, W. van Winden and P. Woets (2005) *European cities in the knowledge economy – the cases of Amsterdam, Dortmund, Eindhoven, Helsinki, Manchester, Munich, Münster, Rotterdam and Zaragoza*, Aldershot: Ashgate.

—— E. Braun and J. van der Meer (1997) *Metropolitan organising capacity – experiences with organising major projects in European cities*, Aldershot: Ashgate.

BOVAG-RAI (2007) *Mobility in figures – cars 2007*, Amsterdam: Stichting Bovag Rai Mobiliteit.

CBS (n.d.) Dutch Statistical Bureau, available online at: www.cbs.nl/nl-NL/default.htm (accessed 13 June 2006).

Comissie Sistermans (2006a) *Brainport Eindhoven – Brainport navigator 2013, Lissabon voorbij!*, Eindhoven: Drukkerij Jémé.

—— (2006b) *Bijlagen Brainport navigator 2013, Crossing borders, moving frontiers*, Eindhoven: Drukkerij Jémé.

DAF (2007a) 'PACCAR begins construction of Parts Distribution Centre in Budapest, Hungary', press release, 25 September 2007.

—— (2007b) 'PACCAR earns record annual profit and revenue; capital investments drive global growth', press release, 30 January 2007.

DCEr (Design Connection Eindhoven region) (2006) 'Design Academy Eindhoven', *Capital Guide*, 1.

DCT (Dynamics and Control Technology) (2005) *Annual report 2004*, available online at: www.dct.tue.nl/.

DMI (Dutch Manufacturing Institute) (2004) *Samenvatting businessplan*, Werkendam: Drukkerij Damen.

ED (Eindhovens Dagblad (2005) 'Autotechniek-centrum levert honderd banen op', 28 September 2005.

—— (2005) *Regio motor van de auto-industrie*, 4 October 2005.

EZ (Ministerie van Economische Zaken, Minstry of Economic Affairs) (2004) *Pieken in de delta – Gebiedsgerichte economische perspectieven*, Den Haag: EZ, directie Ruimtelijk Economisch Beleid.

FHA (2006a) *PPS automotive – vision for the Dutch automotive sector*, available online at: www.pps-automotive.nl/.

—— (2006b) *High tech automotive systems – innovation programme*, available online at: www.pps-automotive.nl/.

Gemeente Eindhoven (2003) *Samenwerking: basis voor economisch succes Regio Eindhoven – De kracht van de regionale Triple Helix in vergelijking met Leuven, Helsinki en München.*

—— (2004) *Stadsvisie 2010 – Stedelijk ontwikkelingsprogramma 2005–2000, Projectgroep GSB/SOP 2005–2009.*

Heneric, O., G. Licht and W. Sofka (2005) 'Europe's automotive industry on the move: competitiveness in a changing world', *ZEW Economic studies*, 32.

Iwd (2004) Schröder Ch., 'Productivität und Lohnstückkosten im internationalen Vergleich', *IW-Trends*, 3.

Liebrand, N. (2007) 'Brainport Automotive Innovatie', presentation at the Brainport Automotive meeting, Helmond, 1 July 2007.

Ministries of VROM, LNV, V&W and EZ (2006) *Nota Rruimte – Ruimte voor ontwikkeling*, Den Haag: Ministries of VROM, LNV, V&W and EZ.

Mitsubishi (2007) 'Mitsubishi Motors signed an agreement for the granting of incentives with Russian Ministry', press release, 28 December 2007.

NV REDE (2007) *Feiten en cijfers regio Eindhoven 2007*, Eindhoven.

Programmabureau Horizon (2005) *Design in Brainport Eindhoven Programma 2005 – 2010*, Eindhoven: Programmabureau Horizon.

Programme Agency Horizon (2004) *Eindhoven region, south-east Netherlands – top technology – crossing borders, moving frontiers*, Eindhoven: Kempen Druk.

SRE (Samenwerkingsverband Regio Eindhoven) (2003) *Actie reactie – Sociaal-economich beleid regio Eindhoven 2003–2007.*

Vastgoedmonitor (2007) *Zuidoost-Brabant – Cijfers en trends in Vastgoed*, available online at: www.vgm-zo-brabant.nl/inlei.html (accessed 29 January 2008).

VDL Groep (2007a) *Annual report 2006*, available online at: www.vdl.nl/nl/vdlgroep/frstvdlg.htm.

—— (2007b) 'New welding installation for VDL Technics', press release, 20 July 2007.

5 Munich: automotive industry

5.1 Introduction

Germany is the chassis, body and engine of the European automotive industry. It is the world's third largest car producer, after the USA and Japan. Within Europe, Germany is by far the largest producer (5.35 million passenger cars in 2005), as it produces more than twice as many cars as the second producer (France, 3.11 million) and the third (Spain, 2.1 million) together, resulting in a market share of 34%. The country is home to various leading car assemblers (see Figure 5.1), that contributed nearly half (48.8%) of the total gross net value added of all OEMs in the EU in 2002. Germany also houses a number of world-leading suppliers: Robert Bosch, Siemens VDO (recently sold to Continental), Continental, ThyssenKrupp and ZF Group are all ranked on the list of the 15 largest automotive suppliers in the world (based on returns in 2005). In Europe, the dominance of German suppliers is even larger: 9 of the 15 largest automotive suppliers in Europe are German, and German suppliers contributed 43.3% of the total gross net value added of all automotive suppliers in the EU in 2002 (IGM 2006a,b,c). Finally, the German automotive industry is a leading innovator, with over 3,600 patents registered in 2005; the German automotive industry accounted for 35% of all applications worldwide, which is more than any other country (VDA, 2006).

The German automotive industry is concentrated in a number of clusters around the headquarters and production plants of the major OEMs (see Figure 5.1). One of the major clusters is the Munich region, home of two luxury car assemblers Audi and BMW.[1] The automotive industry is one of the leading manufacturing industries of this high-tech cluster. Despite Munich's relatively late industrial development,[2] manufacturing is of great importance for Munich and the Free State of Bavaria;[3] 23.8% of the employees in the city and 27.5% of the employees in Bavaria are active in the manufacturing sector (Landeshauptstadt München, 2006a,b; BMEAITT, 2005).

This chapter analyses the dynamics of the automotive industry in Munich and its wider region. In particular, via analysis of production and knowledge networks, we address the changing role of manufacturing in the automotive value chains, as well as the spatial linkages between production on the one hand, and knowledge-

Figure 5.1 Automotive clusters in Germany around headquarters of OEMs (source: own elaboration)

related activities on the other. In addition, we reflect on the management of this automotive cluster.

The rest of this chapter is organised as follows. Section 5.2 describes the regional economic context of Munich, while section 5.3 introduces the major players in the automotive industry. Section 5.4 investigates the major production and knowledge networks that influence the development of Munich's automotive cluster, and analyses the major platforms and policy initiatives. The last section (5.5) draws conclusions.

5.2 Foundations of the Munich region

Munich (2.5 million inhabitants in the urban region), the capital of Bavaria, has a modern service economy, but also major (high-tech) manufacturing industries (Landeshauptstadt München, 2006a,b). It is one of the leading automotive clusters in the world. To understand its current dynamics, it is important to put it in the perspective of the regional context. Therefore, this section provides a description of Munich's regional economic foundations: the economic base, the knowledge base, accessibility, quality of life and factor conditions.

Economic base

Munich is one of the leading cities in Germany in economic terms. Its GDP per capita (€52,396 in 2004) is twice as high as the German average (€25,900) (see Figure 5.2). Bavaria (€30,995) is also performing better than the national average, and its share in the total German GDP is 18.4% (in 2005), while the share of population is 15.1% (BMEAITT, 2007). In terms of purchasing power per capita (2004), Munich (€24,262) scores well above the German average (€17,422) and even outperforms other prosperous German cities, such as Stuttgart (€20,014) and Frankfurt am Main (€19,232). Also in terms of unemployment, Munich has an unemployment rate of 7.2% (2005) which is better than Frankfurt (8.0%) and Stuttgart (8.4%) (Landeshauptstadt München, 2006a,b). Another sign of economic health is the high start-up rate. Bavaria has the highest number of start-ups in Germany. In 2005, 38,207 companies were founded, which is about one-fifth of the German total (BMEAITT, 2007).

One of the key factors of Munich's economic success is its late industrial development and the lack of heavy industries. Therefore, and unlike industrial areas such as the Ruhr area, Munich has not suffered from the old industrial sectors in the 1970s and 1980s (van den Berg *et al.*, 2005). Today, the manufacturing sector has a share of 23% in value added and 28% in employment, making it the second largest sector after the service sector (71% of the value added and 62% of the workforce) (Landeshauptstadt München, 2006a,b). The manufacturing sector in Bavaria is dominated by the automotive industry, mechanical engineering, electronics engineering and the food industry (see Figures 5.3 and 5.4). It is worth noting that the automotive industry is relatively large compared to the German average. Moreover, car assembler BMW has 32,000 employees and is the second largest employer of the city, after the City of Munich itself (35,000 employees), but above electrical giant Siemens (28,000) (interview).

Munich is known for its highly diverse and balanced economy. The term 'Munich Mix' ('Münchener Mischung') is often used to describe this, referring to the variety in terms of firm size (globally operating multinationals as well as local SMEs), sector mix, and the mix between foreign firms and domestic companies. Moreover, there is a variety of functions. For instance, Munich is an important

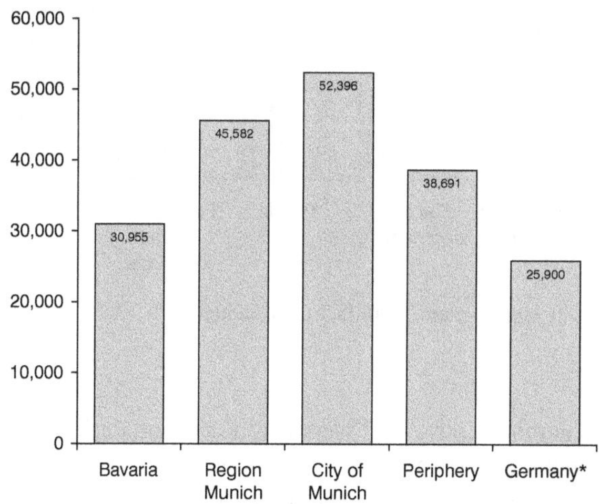

Figure 5.2 GDP per capita (in €, 2004) (source: Landeshauptstadt München – Referat für Arbeit und Wirtschaft, 2006). * Note: data for Germany is from the year 2005 and concerns the real GDP, in contrast to the other areas, which concern the nominal GDP.

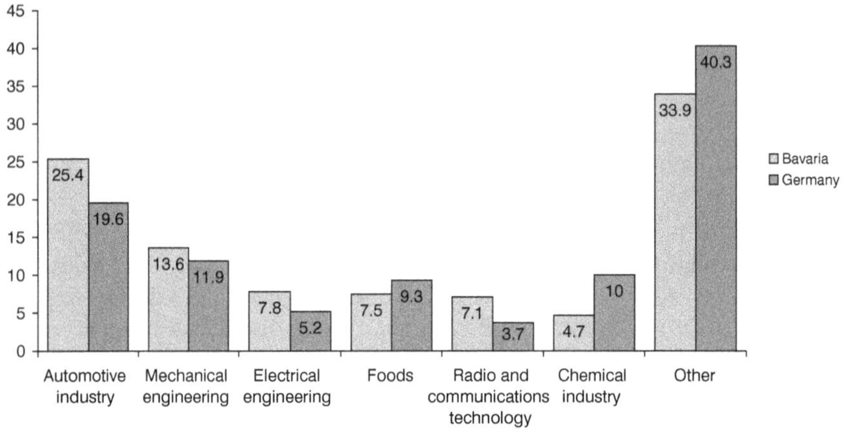

Figure 5.3 Share in manufacturing sales, %, 2004 (source: BMEAITT, 2005)

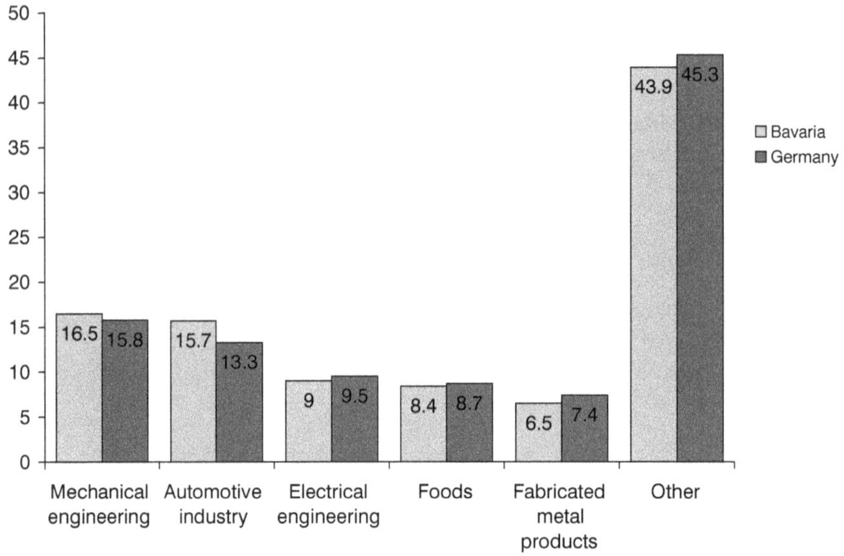

Figure 5.4 Share in employment manufacturing industry, %, 2004 (source: BMEAITT, 2005)

international economic centre,[4] and has many R&D and design centres, but also has major production facilities. The mix has at least two advantages. First, it reduces the city's vulnerability to economic cycles in certain industries. Second, the variety of industries offers possibilities for 'new combinations' or intersectoral cooperation that is the source of innovation and new business opportunities (for instance bio-informatics) (van den Berg *et al.*, 2005).

The highly diverse economic structure of Bavaria and Munich is an important asset for the automotive industry. The automotive sector combines a variety of technologies and uses a range of products from other manufacturing industries (plastics, electronics, to name just two), and multiple services. Thus, the variety of the economic base helps automotive firms to find relevant partners and local knowledge. Moreover, the presence of international headquarters reduces the dependency on decisions taken elsewhere.

Knowledge base

The greater Munich region is a successful technology centre that is on par with other top technology regions in the world, such as Silicon Valley and Greater London.[5] On the European 'scoreboard for innovation power' of 2003, which is set up by the European Commission and is based on several scientific and technological indicators, Upper Bavaria[6] ranks third after Stockholm and Helsinki. In terms of patents per employee, Upper Bavaria ranks second (in 2001), after the Eindhoven region with about 150 European Patent Office (EPO) patents per 100,000 active workers (Commissie Sistermans, 2005). In 2005, more than one-quarter (28.3%) of all applications received by Germany's Office of Patents and Trademarks were submitted by companies in Bavaria. In terms of patent density per inhabitant, Bavaria ranks second (with 110 patents per 100,000 inhabitants), after the state of Baden-Würtenberg (BMEAITT, 2007). The company with the most patents in Germany is Siemens, with 3,892 registered patents in 2004. Just like Siemens, almost all companies in the top 20 'patentees' in Germany are either headquartered in Munich (Infineon and BMW) or have research institutes (GE, Philips, Bosch) or cooperate with one of Munich's research institutes (City of Munich, 2006). In terms of R&D expenditures as share of the GDP (in 2003), Bavaria ranks third after the states of Berlin and Baden-Wurttemberg with a share of 3%. This is in line with the Lisbon Treaty target which says that 3% of the GDP should be spent on R&D. The R&D ratio for Germany as a whole is only 2.5% (BMEAITT, 2007).

The success of Munich in these rankings can be explained by its strong and diverse knowledge infrastructure. A large number of corporate research centres are located in or near Munich. For instance, Siemens and BMW have the largest corporate R&D departments in Munich, each with over 5,000 researchers (City of Munich, 2006), and also the public research infrastructure is world class. The Bavaria region boasts 11 universities, 17 polytechnics, 11 Max Planck institutes and 10 centres maintained by the Fraunhofer Society (BMEAITT, 2007). Within Bavaria, Munich is the dominant centre for R&D. Major universities in Munich are the Ludwigs-Maximilians University (LMU, 44,865 students in 2005) and the Technical University Munich (TUM, 19,627 students). The universities in Bavaria rank among the top in Germany. The Ludwigs-Maximilians University takes the lead in a ranking of 40 German universities based on several indicators during 2002–2004. Three other universities (including the Technical University in Munich) in Bavaria are in the top 10 of the list (RVI, 2006). Besides universities, there are various centres of applied research and other institutes. Together, these institutes have over 85,859 students. Furthermore, there are various other public research institutes in Munich, like the Max Planck Society and the Fraunhofer society, both of which have their headquarters in Munich.

A second, and related, success factor is Munich's highly educated labour force: 18.2% of the employees have a university or applied sciences degree (Landeshauptstadt München, 2006a,b). This is reflected in the manufacturing sector, where high-tech companies dominate, accounting for more than half of the employment (57%) and for nearly two-thirds (65%) of the earnings in the Bavarian manufacturing industry. In Germany, these percentages are, respectively, 51.8% and 55.1%. (BMEAITT, 2007). Also our interview partners perceive the high quality of the labour pool in the region as a key factor for its economic success. The many institutes for higher education provide for a continuing influx of talent into the region. Despite high costs of living in Munich, these institutes attract students from all over Germany, and increasingly also from abroad. According to some of our interviewees, the only problem in the education system might be design education, which has to increase in quality.

Hence, Munich has an excellent knowledge base with strong research competence and a highly educated labour pool. These are key location factors for the automotive industry, especially for luxury car makers such as BMW and Audi that employ relatively many highly educated workers compared to volume car manufacturers.

Accessibility

Accessibility is a key factor for car logistics as well as for the transport of knowledge workers. Munich has good external accessibility. It has a well-developed road and rail infrastructure and it has good connections with Western Europe as well as the upcoming markets of Central and Eastern Europe (CEE). The latter is important for the movement of automotive components produced there to the production plants in Bavaria, and also to distribute spare parts from distribution centres in Bavaria to repair and service points in Europe. In addition, Munich is connected to many destinations by air. Munich airport is the second largest airport in Germany after Frankfurt am Main. In 2005, it handled 28.6 million passengers and offered 250 destinations (22 domestic, 159 European and 69 intercontinental) (City of Munich, 2006). The airport, just like the HST connections of the city, is a key asset for the transport of researchers to other knowledge hubs, as well as for the distribution of spare parts. Especially for luxury car makers such as BMW and Audi, it is important to deliver spare parts quickly in order to prevent image problems.

Munich's external connections are excellent, but within the region, it struggles with some problems. Just like many other cities, Munich suffers from congestion on the roads, especially during peak hours. An important reason for this is a concentration of jobs in the city centre and growth of residential areas in the suburbs of the city. This has resulted in a strong growth of the traffic load in the

Munich region (City of Munich, 1999, in: van den Berg *et al.*, 2005). The metro lines running between the suburbs and the city centre also face capacity problems. A barrier is that eight tracks come together on one main axis. There are plans to solve this problem by adding a second axis. Another drawback in the rail system is the connection between the airport and the city centre, which takes about 45 minutes. For many years there has been a discussion between different actors for the construction of a Maglev link between the airport and the city centre which could reduce the travel time to ten minutes.

Quality of life

Munich is known as 'the largest village in the world' due to a combination of many cultural and natural attractions within a small distance. The city has a large number of urban amenities (culture attractions, but also parks), while the wider region offers mountains and lakes. This makes Munich attractive for various target groups. The city ranks second in the Cushman & Wakefield survey of 30 cities (2008) after Barcelona. It is one of the most visited cities in Germany with 4.15 million arrivals and 8.3 million bed nights in 2005 (Landeshauptstadt München, 2006a,b). The number of inhabitants has grown, and the city is especially attractive for high-skilled people (such as automotive engineers).

However, the attractiveness also has a downside: the city is one of the most expensive cities in Germany. Rent levels and house prices rank among the highest in the country. Currently, a special instrument (SOBON) is being developed by the city of Munich to promote the construction of social housing for lower income groups. The aim is to reach a balanced mix between different social groups. Another critique on the quality of life might be that Munich lacks an 'underground culture' which 'real global cities' such as Tokyo and New York have. For this reason the city might be less attractive for certain groups, such as painters and designers. But it should be noted that this does not seem like a problem for automotive designers, since they define themselves as 'car freaks' rather than 'bohemians' who enjoy an underground culture (interviews).

Factor conditions

Factor costs and 'soft' factors are major indicators for business in their location decisions. Germany has relatively high labour costs. The labour costs in the automotive industry in Germany are the highest in the world (see Figure 5.5). In 2006, the average hourly labour costs in Germany were €40.65, which is three times as much as in Portugal. Several car producers have threatened to relocate to cheaper locations if labour costs remain so high. Recently, deals have been made between labour unions and car manufacturers. In 2006, Volkswagen and IG Metall (the largest union) agreed to have a longer working week in exchange for reaching a certain production level (Financial Times (FT), 2006).

Within Germany, there are significant differences in wage levels. Gross annual wages in the manufacturing industry (in 2005) range from €36,578 in Schleswig-

Holstein to €48,421 in Hamburg. Bavaria is somewhere in the middle with €39,887, which is just above the German average (€39,471). In Bavaria, average gross wages for manufacturing in the automotive industry are relatively high (€39.14 per hour) compared to the average of all manufacturing industries (€26.03). Only the wages for the processing of oil (€39.25) are higher (BMEAITT, 2007).

Labour costs should be weighed against labour productivity. From Figure 5.6 it becomes clear that Germany is the most expensive location in terms of labour costs. In terms of labour productivity, however, Germany ranks eighth place out of the twelve investigated countries. In a study about labour productivity in several European metropolitan areas, the Munich region has the second highest labour productivity (in 2004, in all sectors) after the Paris region. The annual average growth of the labour productivity in the Munich region (>2%) during the 1995–2004 period is above the average of the EU-15 and of the 'top 20' of all investigated cities (Manshanden *et al.*, 2005).

Our interviewees confirmed that the high labour costs do not seem to be a problem for the automotive companies in Bavaria because of the high labour productivity. Luxury vehicle manufactures (like BMW and Audi) can afford to pay higher wages because they have relatively high profit margins compared to volume car makers. Furthermore, vehicles built in Germany can be sold as 'made in Germany', referring to the image of Germany as high quality production country, and therefore, German companies can afford to pay relatively high wages.

The real estate prices in Munich are high compared to other German cities. Munich ranks second in terms of prime office rents (€28.50 per m2), after Frankfurt am Main (€34.00 per m2) (see Figure 5.7). Average office rents are much lower with €13.50 per m2 in 2005. Munich is a relatively expensive city for housing, too. Munich has the highest house rents in Germany (as does Hamburg),

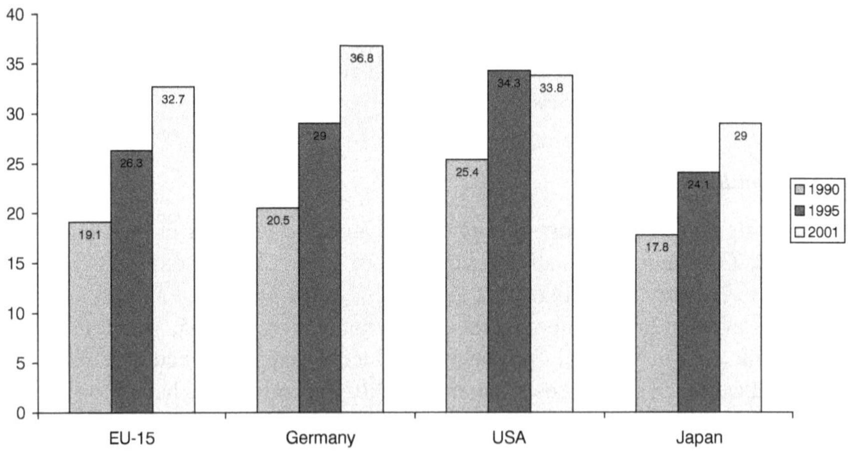

Figure 5.5 Labour costs in the auto industry (conversion to USD, based on PPP) (source: Heneric *et al.*, 2005)

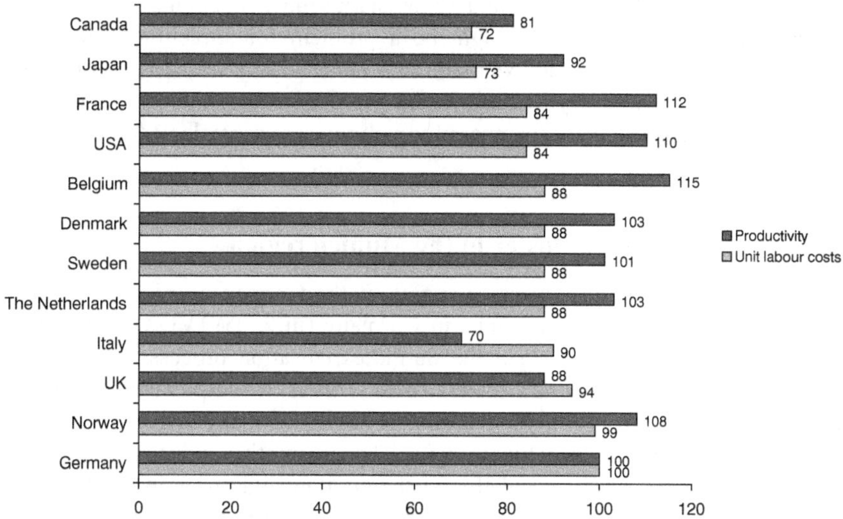

Figure 5.6 Unit labour costs and productivity in the manufacturing sector in 2003 (Germany = 100) (source: iwd, 2004b in Becker 2006)

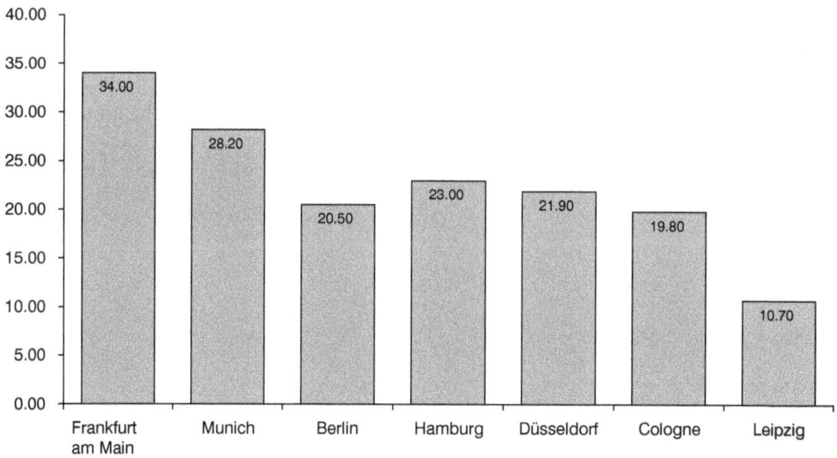

Figure 5.7 Prime office rents, €/m², 2005 (source: Landeshauptstadt München, 2006)

with an amount of €11 per m2 in 2005. However, the average house rent has declined in the last three years (Landeshauptstadt München, 2006a,b).

The high real estate prices might be a problem for attracting lower paid workers, who can not afford to live in the city. This does not seem to be a problem for the automotive industry in the Munich region, because the two luxury car manufacturers use relatively (compared to volume car assemblers) few low paid workers.

5.3 The automotive cluster in the Munich region

The automotive industry is a major engine of the German economy. It counts for 13.2% of all manufacturing jobs in Germany (in 2005) (VDA, 2006). The automotive industry has escaped the de-industrialisation process: it has even created more than 100,000 new jobs (+16%). This is in sharp contrast with the development of the overall manufacturing sector in Germany that lost 1 million jobs (8%) since 1995[7] (see Figure 5.8). The automotive industry is also an important innovator. It is one of the most important industries in Germany in terms of R&D. The total expenditures on automotive R&D more than doubled in ten years, from about €7 million in 1995 to about €16 million in 2005, and made up for more than one-third of the German total expenditures on R&D. In total, 85,000 employees of the sector (11%) are involved in R&D (VDA, 2006).

The Munich region is, after the Stuttgart cluster, the second largest cluster, having more than 100,000 employees (Iwd, 2004a). With Audi and BMW as luxury car manufacturers, it is one of the two automotive clusters in Germany for luxury cars (the other is Stuttgart/Baden-Wurttemberg with Daimler-Chrysler and Porsche). The cluster has a dense concentration of headquarters, production plants and research/testing facilities. In terms of employment, the automotive

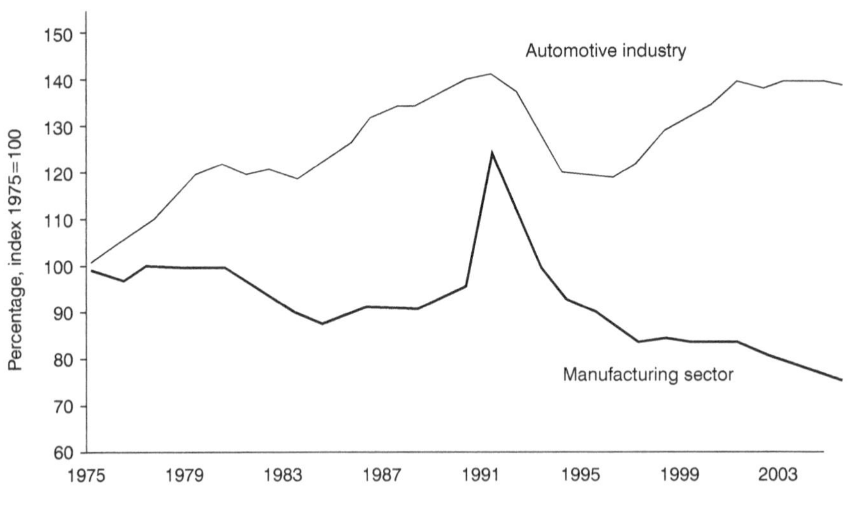

Figure 5.8 Development of manufacturing employment in Germany (source: VDA, 2006)

industry (15.7% share) is the second largest manufacturing industry in Bavaria, after the mechanical engineering industry (16.5% share). The industry's sales and employment have grown sharply since 1997 (see Figures 5.9 and 5.10). The total employment (in 2004) amounts to 182,657 persons, of which 7,772 work in automobile bodies and trailers sector, 61,891 in the automotive parts sector and 113,024 in the vehicles and engines production sector (BMEAITT, 2005). Within Bavaria, the share of automotive employment differs widely; in Munich, which has in absolute terms the largest automotive industry, the share of total

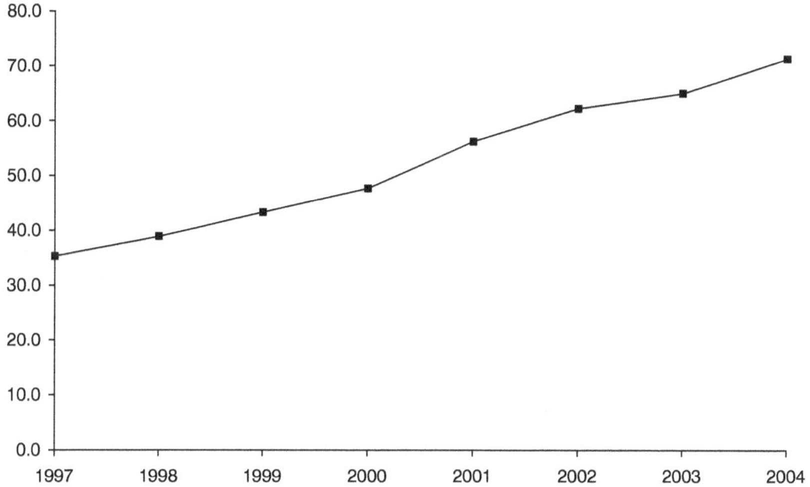

Figure 5.9 Development of sales generated by companies from the automotive industry based in Bavaria (1997–2004; in billion euro) (source: BMEAITT, 2005)

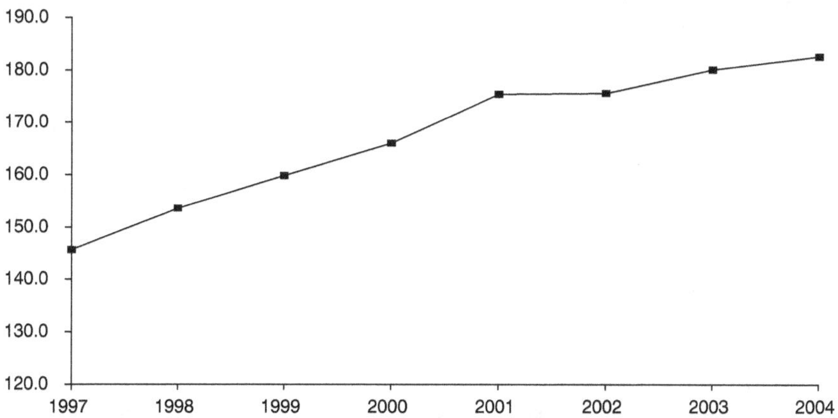

Figure 5.10 Development of employment in the automotive industry in Bavaria (1997–2004, in 1,000 persons (source: BMEAITT, 2005)

Table 5.1 Key data of the automotive industry in Bavaria, 2004

Plants	215
Employees	182,637
Employees per company (average)	849
Sales in million €	71,271
Sales per employee in €	390,231
Export ratio in %	61.8
Wages and salaries per employee in €	47,502

Source: Bavarian Ministry of Economic Affairs, Infrastructure, Transport and Technology (2005)

employment is only 3.8%, compared to 20.7% in Landshut and even 22.2% in Ingolstadt (Iwd, 2004a).

Other key data of the automotive industry in Bavaria (for 2004) are summarised in Table 5.1. It should be noted that most of the employees work in large companies; about 87% of the employees work in companies with more than 500 employees (BMEAITT, 2005).

The remainder of this section discusses the major actors in the automotive industry in the Munich region: OEMs, suppliers, and education and research institutes. It is worth noting that the region is home to many other actors that are active in the automotive industry. For instance, all leading consultancy firms that focus on the automotive industry are based in Munich.

OEMs

BMW (established in 1917 in Munich) is one of the leading automotive companies in the world in the premium segment. All brands of the company are from this segment: BMW, MINI and Rolls-Royce. Besides automobiles, BMW also produces motorcycles. Most of the assembly plants are located in BMW's home country of Germany, and in England (the home country of Mini and Rolls-Royce) (see Figure 5.11). Also, in terms of employment and procurement, Germany is the prime location. The share of employment in Germany in total employment has grown, from 55% in 1996 to 75% in 2004 (see Figure 5.12), and Germany accounts for about half of the volume of purchases (BMW, 2005a). In Bavaria, BMW has three plants for the production of complete vehicles (in Munich, 200,200 cars; Regensburg, 301,200 and Dingolfing, 279,200) and two plants for automotive parts (in Landshut and Wackersdorf). Munich is the company's main base; it contains a production plant, the headquarters, several research centres, the central training centre (in Unterschleißheim) and some leisure facilities (such as the BMW museum).

BMW has several completely knocked down (CKD) plants in new growth markets. There, vehicles are assembled from part sets, which are imported from abroad. This technique is often used if there is no possibility to manufacture

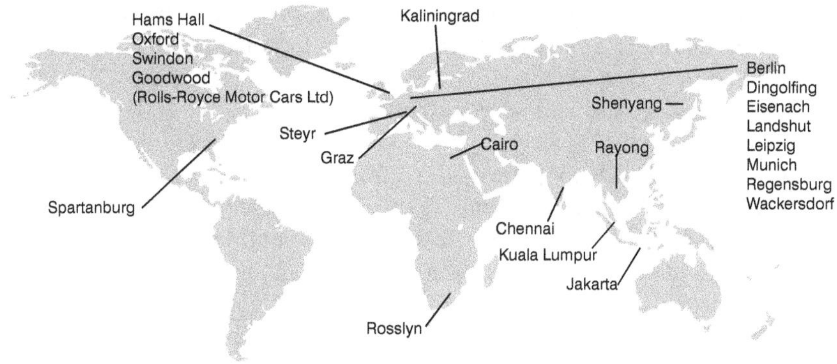

Figure 5.11 Production network of BMW in 2007 (source http://www.bmwgroup.com)

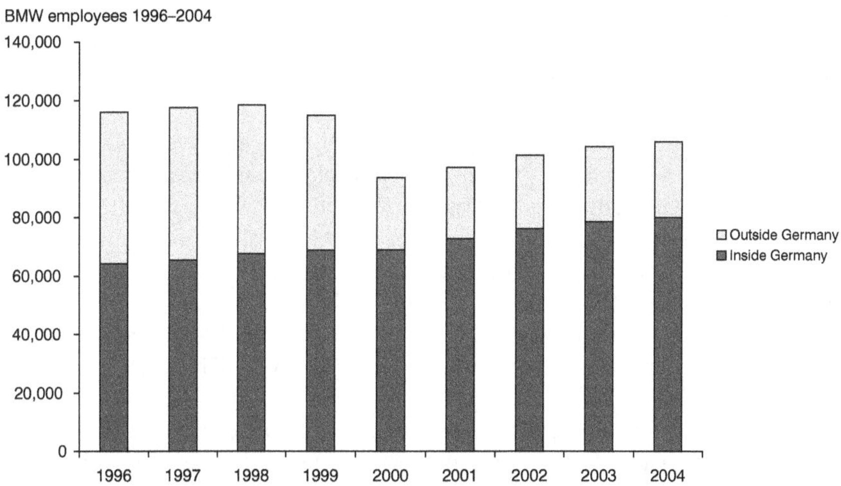

Figure 5.12 Employees of BMW inside and outside Germany, 1996–2004 (source: BMW, 2001and 2005)

the necessary parts in production locations, which is often the case in new production locations. Moreover, CKD helps to overcome high import duties on finished vehicles (Becker, 2006). In CKD plants, which are often set up with local partners, BMW produces vehicles for local markets. In China, BMW has a joint venture plant with the local company Brilliance China Automotive Holdings Ltd. The Chinese government forces foreign OEMs to cooperate with local partners to develop the Chinese automotive industry. Next to production and CKD facilities, BMW has several sales offices, which rose from 24 in 2000 to 36 in 2006.[8]

BMW has invested in almost all global regions since the beginning of this century:

- *Asia*. In Malaysia, BMW opened a new central parts and distribution centre, an ICT centre and a sales joint venture with a local partner in 2003 (BMW, 12–12–2003). One year later, BMW opened a joint venture (JV) plant in Shenyang, in the north-eastern part of China (BMW, 2004). BMW is considering expanding its capacity in China, either via its current JV partner Brilliance or via new JV partners (FT, 2007).
- *Europe*. In 2001, BMW opened a modern engine plant in Hams Hall in England (BMW, 2002a). In 2005, a new central distribution centre for spare parts in Dingolfing was opened (BMW Group Dynamics Centre). A sophisticated logistics system makes it possible to deliver to Germany in 12 hours and to other parts of Europe in 24 hours. The location of Dingolfing was chosen because of the proximity of the largest production plant of the BMW Group (BMW, 2005b). Another important investment in Europe is the Leipzig plant (opened in 2005).
- *USA*. In May 2007, BMW announced that the annual capacity in its Spartanburg plant will be extended from 140,000 to over 200,000 vehicles. The extension will be an important step in lowering the US dollar currency risk, which is an important part of the BMW Group's strategy. (BMW, 2007a)

BMW also invests in its home country, and the number of employees in Germany is increasing. However, in terms of sales, Germany lags behind the sales abroad (see Figure 5.13). In Germany, sales increased to 20% in 2001–2005, while the growth in sales outside Germany was 56%. The share of the total sales in Germany dropped from 27% in 2001 to 23% in 2005. A major growth market is Asia, which grew 82% in the same period. BMW is aware of the shift in global growth regions. This can be illustrated by a quote from BMW's Chairman of the Board of Management during the 87th Annual General meeting of the company (BMW, 2007b, p.11):

> Twenty years from now, Asia will have replaced the United States as the world's largest automotive market. Demand for premium vehicles is still strongest in the G6 industrial nations. But the strongest growth rates are achieved in China and India as well as in markets such as Brazil and Russia. We keep both things in mind – the mature markets and the young, emerging markets. In China, for example, our retail volume rose by 65 per cent between the end of 2003 and 2006. In India, we are just at the beginning of this development. In March, I inaugurated our BMW assembly plant at Chennai in Southern India. For the time being, we can produce up to 1,700 cars a year there.

The shift in global growth regions might be a threat for the automotive industry in Germany (and other parts of Western Europe) in the long run, especially because the automotive industry follows the market. This is also explicitly stated in the sustainable value report of the company (BMW, 2005a, p. 8): 'In line with the motto 'Production follows the market', the BMW Group invests in the development of new, highly promising markets …'. R&D and design are core activities for the BMW Group.[9] BMW's global R&D activities are spread across

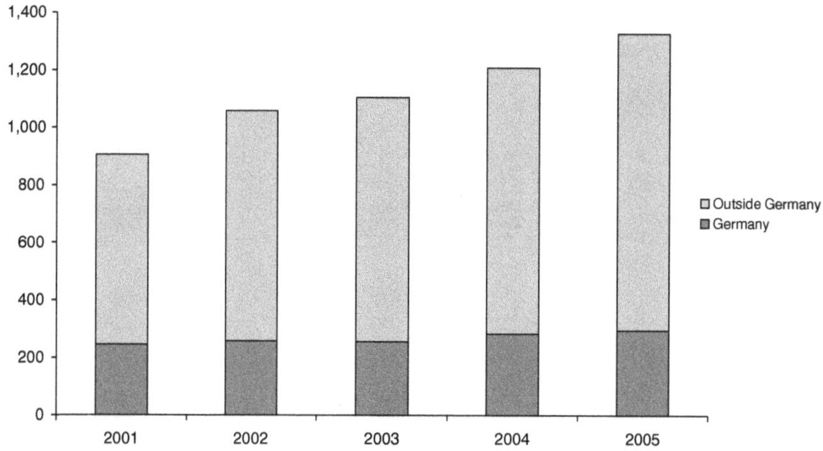

Figure 5.13 Sales of BMW inside and outside Germany, 2001–2005, in 1,000 units (source: BMW, 2006)

eleven locations in five countries (see Figure 5.14), and employ 9,400 employees (end of 2005). Total spending on R&D from BMW was €3.1 billion, which was 6.68% of the revenue (BMW, 2006).

BMW has four research facilities in Bavaria (three in Munich and one in Landshut):

- *BMW Group Research and Innovation Centre* (Forschungs- und Innovationszentrum, FIZ) is BMW's largest research centre, employing 6,000 engineers and 1,500 other specialists (City of Munich, 2005). The centre employs workers from the BMW Group and from suppliers. In the centre, the researchers develop vehicles and technologies for the entire BMW Group.
- *BMW Group Research and Technology* was established in 1985 as an independent think tank (100 employees), physically separated from all other development areas, but the researchers of the centre have close contact with other BMW departments.
- *BMW Group Car IT* (established in 2001) is responsible for the development of vehicle-related software and IT. This is important since the emphasis in the automotive industry has shifted from hardware to software. BMW Group Car IT cooperates closely with FIZ.
- *Landshut Innovation and Technology Center (LITZ)* focuses on the development of new, cost-effective lightweight design technologies, as well as the integration of additional functions and parts into the vehicle components and systems.

In California, BMW has a technology office in Palo Alto, a design centre in Newbury Park, and an engineering and emission test centre in Oxnard.

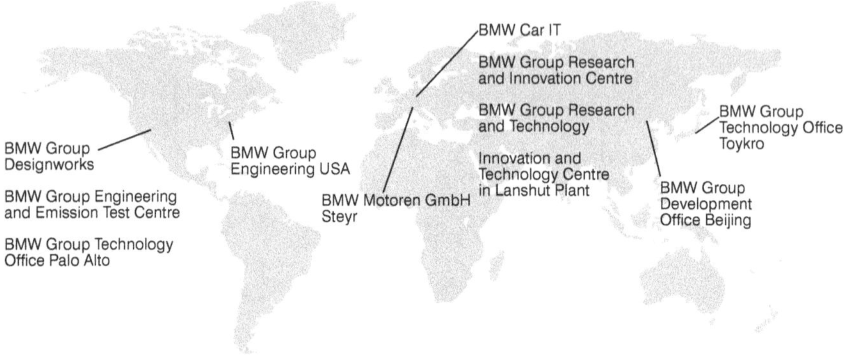

Figure 5.14 The global research network of BMW in 2007 (source http://www.bmwgroup.com)

BMW is involved in numerous research projects on different topics, such as ICT in cars, safety, traffic management/guidance systems and cleaner vehicles/new energy. For its energy strategy BMW concentrates on hybrid vehicles and the optimisation of current engine systems combined with lightweight automotive engineering concepts. The engine has always been one of the core competences of BMW. The energy strategy in the long run focuses on the development of hydrogen vehicles (BMW, 2005a; 2007b[10]).

Audi is the second leading luxury car assembler in the Munich region. Audi Group, part of Volkswagen Group, includes the Audi and Lamborghini brands. Since 1949, Audi has been located in Ingolstadt. A second German production location is Neckarsulm. Besides the locations in Germany, Audi has production locations in South America, Europe and Asia (see Figure 5.15). Ingolstadt is the main location: it employs nearly 60% of the total workforce and nearly 70% of the German Audi workforce. The Ingolstadt location consists of Audi's headquarters (with the core areas of administration, sales and technical development/R&D), a production plant, a design centre, a training and service centre and a centre with leisure facilities ('Audi Forum').

Figure 5.16 shows the importance of Germany as production location for Audi; about 85% of its workforce is in Germany. During 2002–2006, the workforce was relatively stable. Audi indicated that it wants to secure jobs in Germany (Audi, 2007c). The firm signed an agreement with labour union IG Metall in 2005, in which it promises to avoid redundancies for operational reasons before 2011 (Audi, 2006a). Moreover, Audi recently decided to produce three new models in Germany: the Audi Q5 and A5 (in Ingolstadt), and the Audi R8 sports car in Neckarsulm (Audi, 2006b).

Just like for BMW, Audi's growth markets are abroad (see Figure 5.17). In 1997–2006, sales in Germany increased by only 8%, while sales abroad more than doubled (110%). During 2002–2006, China was the main growth market where sales have more than doubled (121%). China is expected to be Audi's second largest export market after the USA in the near future (in 2006, this was the UK).

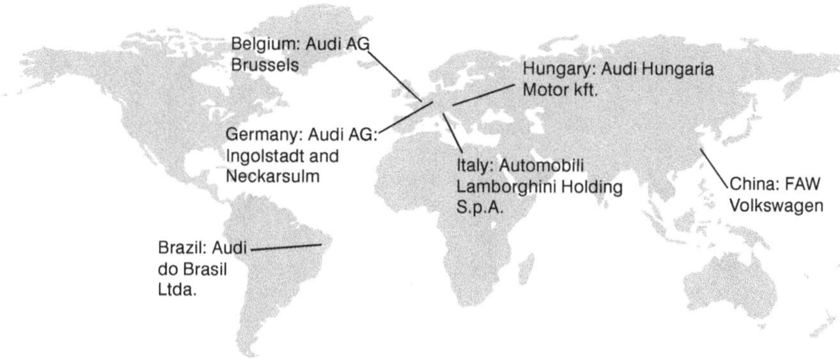

Figure 5.15 Production locations of Audi, 2007 (source: own elaboration)

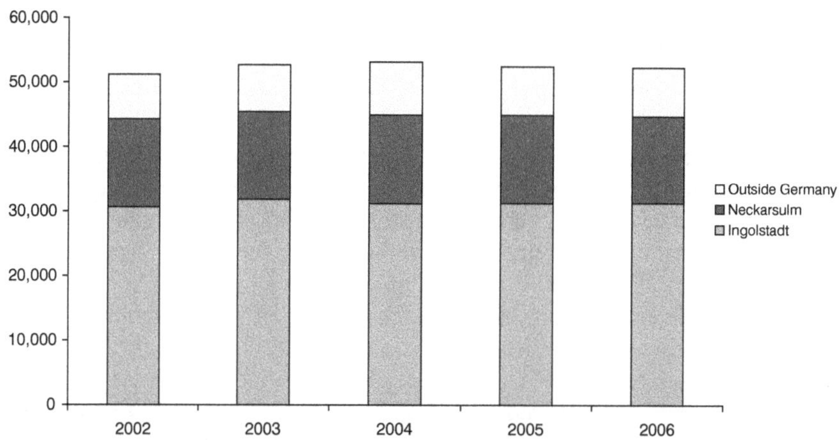

Figure 5.16 Employees of Audi in Ingolstadt, Neckarsulm and abroad, 2002–2006 (source: own elaboration, data from http://www.audi.com/com/brand/en.html)

Also Eastern Europe showed large increases in sales, especially in Russia (Audi, 2007b). Hence, also Audi has seen a shift in global sales regions.

However, until now (2006), Audi is not globalising its production function. The production outside Germany is limited to some models for local markets: the Audi A3 is produced in Brazil, the Audi A6 in China and the Audi TT Coupé and TT Roadstar are assembled in Hungary. The body shells of the models that are assembled in Hungary are produced in Ingolstadt. The plant in Hungary was set up as an engine plant in 1993 and develops and produces engines for Audi, VW and SEAT. In 2006, it had an output of 1.9 million engines and 23,675 vehicles.[11] For the realisation of growth in India, Audi opened seven new dealerships in 2007 and it started assembling locally in autumn 2007.

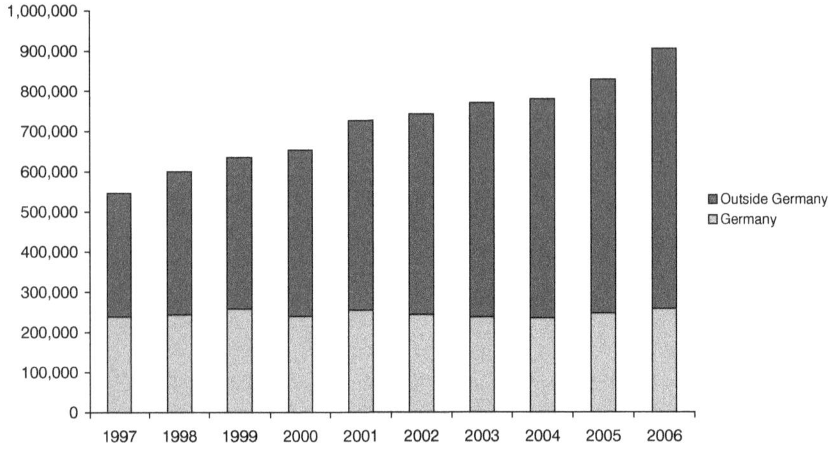

Figure 5.17 Sales of Audi inside and outside Germany, in units, 1997–2006 (source: own elaboration, data from http://www.audi.com/com/brand/en.html)

R&D is a key function for Audi. In 2005, the company spent €1,585 million on R&D[12], which is 5.96% of the revenue. The group's R&D departments employ 5,921 people (2005). Most of them work for Audi AG (5,685 employees). The others are Audi Hungary (98 employees) and Lamborghini (Italy, 138 employees) (Audi, 2006a).

Audi has different research centres at their production sites in Germany. Three major research centres in Ingolstadt are:

- *Audi electronic centre* ('Audi Elektonik Zentrum', more than 750 employees) focuses on electronics in vehicles. This is important given the increasing use of electronics in vehicles. A study by Mercer and two Fraunhofer-Institutes (2004) forecasts that the total share of electrical systems and electronics in the average vehicle in the total value creation will increase from 20% in 2002 to 35% in 2015.
- *Audi Wind Tunnel Centre* ('Audi Windkanal Zentrum', about 50 employees) It focuses on the development and optimisation of Audi models and other brands of the Audi Group (Audi, 12–07–2006).
- *INI.TUM* is an establishment of the Technical University Munich (TUM) in Ingolstadt near the Audi plant (see section 5.4).

Audi also has research institutes in other places, like the Audi Aluminium Centre in Neckarsulm (focusing on development and production of new aluminium body shells for Audi and Lamborghini) and a research centre on its site in Györ in Hungary (a centre that develops engines for various brands of the VW Group).

Design is another key function for luxury car manufacturers. Globally, Audi has about 290 employees involved in the design for the Audi, SEAT and Lamborghini brands, spread across six locations (two in Germany, two in Spain, one in the USA and one in Italy). Two design centres are located in Bavaria, in Ingolstadt and Munich. The design centre in Ingolstadt (Ingolstadt Main Design Centre) is the largest one, and employs about 150 employees. The design centre in Munich, the so-called 'Concept Design Munich',[13] is located in the city centre in order to catch the latest consumer trends. The ambition is to use the studio as 'an interdisciplinary platform for communication and to invite creative, technology and trend experts, opinion leaders and students to Munich so that they can share ideas and work together'.[14] The design centre will have 15 permanent employees. The designers have contacts with all other design centres of Audi. Furthermore, there are plans to develop closer relations with universities all over the world. The design centre is not only active in vehicle design, but it also designs other Audi products (such as racing clothes). Within Audi Group, there is competition between different design teams for new design projects. The team that offers the best design proposal for a certain project gets the order to do the project.

Suppliers

The Munich region has a large supplier base. Five of the 'top 100' automotive suppliers are headquartered in Bavaria: Siemens VDO Automotive, INA, ZF Sachs, Brose and Webasto. Many other German and foreign companies of the global top 100 have subsidiaries in Bavaria, such as Robert Bosch and Delphi. The strong presence of these large tier-one suppliers attracts many local SMEs (Auto World, 2006). Munich, Ingolstadt and Hof are the most important locations for suppliers.

Munich is home to suppliers from all tiers and these have various functions, such as headquarters (like Webasto), plants (Iwis Ketten) and R&D institutes (Magna and Lear). Many suppliers have opted for Munich because of the proximity of Audi (Ingolstadt is about 80km from the city of Munich) and BMW, the highly skilled workforce and the strong knowledge infrastructure, the availability of automotive services (such as consultancy, design and financial services), a good international accessibility and a good quality of life that might attract highly-skilled employees.

Most automotive suppliers in Ingolstadt are clustered around the Audi plant in a dedicated logistics centre and supplier park. The idea for this was jointly developed by the city council and Audi in 1993. Audi's basic aim was to have suppliers near to the production plant. This gives Audi many logistical advantages, stemming from the demands based on sequential JIT deliveries (Larsson, 2002). Currently, more than 100 automotive suppliers are located in the automotive cluster of Ingolstadt. Ingolstadt is an attractive location for suppliers given the presence of Audi and the proximity of BMW factories in Bavaria and automotive supplier clusters in Franken and Schwaben. The Ingolstadt cluster includes several facilities, like a distribution centre (GVZ), a competence centre for automotive engineers (SE Park, housing 17 suppliers) with the focus on (joint) R&D and a railway container terminal (City of Ingolstadt, 2006a,b). The latter is relevant, since half of the incoming goods to the automotive suppliers are delivered by rail (Larsson, 2002).

Hof is a major location for suppliers because of its 156-hectare supplier park. The development of this cluster began when INFUN, a major Spanish supplier of aluminium and magnesium parts, opened a subsidiary in 2003. Since then, many other automotive suppliers have located in the Hof supplier park and other parts of Upper Franconia. Currently, in total about 150 automotive suppliers are active in the region (Auto World, 2006). The major attraction factors are (relatively) low wages, low real estate prices and a favourable geographical position close to the border of the Czech Republic. An important facility on the supplier park is the Automobiltechnikum Bayern, a research centre with several research facilities and a shared platform for joint R&D.

Apart from these locations, suppliers are spread across Bavaria with major concentrations around BMW facilities in Regensburg, Dingolfing and Landshut, as well as the region of Nuremberg.

Research and education

The Munich region is one of the most important regions for automotive research and engineering in the world. By 1892 Rudolf Diesel received a patent for his engine, which he had invented during his study at the Technical University Munich (TUM). Nowadays, more than half of all vehicles in Europe are powered by diesel engines. Bavaria, especially the Munich region, is still one of the leading areas for automotive research and engineering in Germany and abroad. About 25% of German patent applications in automotive engineering and 16.3% of automotive component suppliers are from Bavaria (City of Munich, 2005).

Corporate R&D is highly significant. Audi, BMW and suppliers have large research centres in the Munich region, and there is a multitude of small and medium sized specialised engineering centres delivering engineering services to automotive suppliers and OEMs. Most of these companies are located in the Munich region. Furthermore, there are specialised development companies, such as EDAC, which conduct development and pre-development projects for OEMs and large tier-one suppliers.

In addition, many universities and other institutions conduct automotive research. In total, ten universities are active in the field of automotive engineering, with TUM as the most important one. Nine institutes and chairs of TUM are active in automotive research. Also, 16 universities of applied science (Fachhochschule, FH), and several other research institutes conduct research related to automotive technology. To name a few: University of Applied Research Munich (Fachhochschule München, FHM), the New Materials Competence Centre and the Competence Centre for Vehicle Power Electronics (City of Munich, 2005; BMEAITT;[15] Autoworld, 2006).

The region also has an extensive infrastructure for automotive education. The universities and the universities of applied sciences churn out highly qualified junior personnel. Two universities and ten universities of applied sciences in

Bavaria offer special training for the automotive industry (Auto World, 2006). Furthermore, several firms have special training centres in Bavaria, like BMW that has its central training centre in the Munich region. In 2005, more than 500 training courses were held in this central training centre and about 25,000 participants from about 80 countries participated in these courses (BMW, 2005a). Audi (together with SEAT Germany) has an open training centre for service and diagnosis engineers in Ingolstadt, the Service Training Centre (STC). Currently, about 7,000 engineers from 118 companies are trained in STC, which creates about 150 jobs in the service sector (City of Ingolstadt, 2006a,b). Furthermore, Audi, BMW and the suppliers in the Munich region are playing a major role in employing apprentices from the educational institutes in the region. For instance, Audi employs more than 1,400 apprentices on its site in Ingolstadt, which is about 4.6% of the total workforce in the Ingolstadt location.[16]

5.4 Local and global networks

The actors in Munich's automotive cluster are deeply involved in global and local networks. In this section, we investigate these networks in more detail and focus on two types of networks: production networks (related to operational functions such as production and procurement), and knowledge networks which are linked to R&D and design. Moreover, we analyse the spatial linkages between these types of networks. Finally, this section deals with relevant policy programmes and cluster organisations that influence networks in the automotive industry. Relevant questions are: what are the relations between the actors and the functions in the networks, and what is the role of proximity in these networks?

Production networks

One aspect of the production networks is the purchase of (raw) materials from suppliers. OEMs face pressure to cut costs because of heavy competition in the automotive industry and a strong increase in raw material prices. Audi counteracts the rising prices in particular through intensive partnerships with its suppliers and through long-term agreements. Audi uses three purchasing instruments (Audi, 2006a):

- *Global sourcing* which has been done by 18 local purchasing teams (LPT) that are spread over the world. These LPTs are central contact points for local suppliers and have the task of assessing the potential of local suppliers and to train them.
- *Online negotiations*, which is used alongside 'traditional' negotiations and is aimed at improving the market transparency.
- *Online supplier platform*. A special website (www.vwgroupsupply.com) from VW offers Audi a database of over 24,000 potential suppliers. The platform allows project-relevant data to be exchanged by Audi AG and potential suppliers.

China, India and Russia are gaining importance as procurement markets. For Audi, China is a strategic market. Through its production operations in China, Audi has deep knowledge about local suppliers (Audi, 2006a). Audi's global sourcing strategy might be a threat for the suppliers in Europe in the long run.

Despite the strong growth of sales markets outside the Triad countries, BMW still purchases the bulk of its inputs from Germany (52%, based on 'production material', in 2005) and other parts of Western Europe (20%) (BMW, 2006). This is similar to 2001, when 79% of the production material was purchased from suppliers in Europe (including CEE) (BMW, 2002a).

Geographical proximity of certain suppliers is required for logistical reasons. Complete systems have to be delivered just-in-time or just-in-sequence. For this, certain suppliers have to be located close to production plants of OEMs. In general, when an OEM opens a new plant, some suppliers are forced to open an establishment nearby to supply the production plant (so called 'follow-sourcing'). For instance, when BMW opened its plant in Leipzig in 2005, Faurecia opened a plant that manufactures seats, cockpits and front ends for the BMW 3 Series (Faurecia, 2005). Audi has even developed a supplier park around its plant in Ingolstadt. Similarly, BMW opened a supplier park in Wackersdorf. Other products, which are less time-dependent and with low logistical costs are sourced from suppliers outside the region – from other parts of Germany and from other parts of the world. A solution to overcome the problem of the absence of a local supplier network is the use of CKD plants, in which vehicles are assembled from part sets that are imported from abroad. BMW uses this CKD strategy to enter new growth markets.

An important trend is that OEMs are outsourcing an increasing number of tasks to suppliers. An extreme example is the production of the BMW X3. The production of this model is outsourced completely by BMW. The car is assembled by the large tier-one supplier Magna Steyr in Graz, Austria.

Knowledge networks

Besides production, development work is increasingly outsourced to suppliers. Proximity between OEMs and suppliers is therefore important. BMW integrates suppliers from an early stage in the development process of vehicles or vehicle parts. BMW, among others, is doing this through a new supplier management system and through the involvement of suppliers in development projects in FIZ. BMW is aware that 70–80% of the value added is created outside OEMs (particularly by suppliers). According to the company, close cooperation with suppliers is important for the improvement of products and processes and therefore is a critical success factor (BMW, 2006).

For Audi and BMW, research institutes are important partners in innovation. The TUM is a leading partner, and has joint research projects with Audi and BMW. The research projects of TUM have a time span of about three years, which equals the time of a PhD project. The research projects concern advanced 'pre-competitive' research. An important institute for joint research is INI.TUM, an establishment of TUM in Ingolstadt (see Box 5.1). INI.TUM facilitates direct contact between Audi and TUM. The relation of TUM with BMW is also close, making it possible to start projects fast, without bureaucratic barriers. The relations between TUM and suppliers, however, are limited to some short-term projects. A major reason is that smaller suppliers can not afford the expensive medium-term projects.[17] With large tier-one suppliers such as Siemens, more cooperation can be expected in the future. Apart from joint research, the relation between TUM and companies is also of interest to exchange labour or to train employees. For instance, during their projects, PhD students work temporarily at the OEMs.

After the success of INI.TUM, Audi has set up more university-based partnerships in Bavaria, other parts of Germany and even abroad, such as the University of Sankt Gallen in Switzerland. In Hungary, Audi has signed cooperation agreements with the Technical University Budapest and the Györ István Széchenyi University, and in 2006 it set up a centre comparable to INI. TUM ('Audi Hungary Institute', AHI) focusing on the development of diesel engines. This centre conducts research (mainly through PhD projects) and offers

Box 5.1 INI.TUM

On 11/1/2003 INI.TUM (Institute of the Technical University Munich, Ingolstadt), a cooperation of the partners AUDI AG, Technical University Munich and the city of Ingolstadt started. The task of the new research centre is application-orientated research in the area of vehicle and information technology. The financing of the INI.TUM is project-related via AUDI AG, which stipulates the research and development projects together with the TU-Munich. The city of Ingolstadt finances the initial equipment of the offices. INI.TUM is located in the SE Centrer (Simultaneous Engineering Center) in the north-east of Ingolstadt.

The goal of INI.TUM is to create a regional authority centre, in which science and practice work hand in hand. This is done in close cooperation with the AUDI AG. An emphasis of the Ingolstadt Institute of the Technical University of Munich is the treatment of up-to-date research topics related to automotive engineering. At the same time the qualification of young engineers takes place who see their future operating field in vehicle technology. First research work is taken up to the driving analysis and production process of cars. The projects concentrate on the analysis of the driver, the vehicle and the driver's environment as well as on topics concerning mechatronics in the production process and forming simulation.

Source: citation, slightly adapted, from http://www.ini.tum.de/ENG/index.htm (accessed 15 June 2007)

courses for (young) engineers.[18] In total by the end of 2006, Audi was sponsoring 54 doctoral projects in its partnerships with German universities, seven projects with the University of Applied Research in Ingolstadt and seven projects in Hungary. In addition, Audi has around 30 doctoral projects internationally (Audi, 2006a, d; 2007a).

Similarly, BMW cooperates with universities in Bavaria, other parts of Germany and abroad. Its main partners in Bavaria are TUM and LMU. An example of university-based cooperation by BMW abroad, is a graduate engineering programme at Clemson University in South Carolina (USA). The programme includes masters and doctoral studies in automotive engineering (BMW, 2002b).

Cooperation also takes place between OEMs and specialised research institutes, such as the Max Planck institutes and Fraunhofer Institutes, located in the Munich region or in other parts of Germany. For instance, together with Dortmunder Oberflächen Centrum (DOC, a small research company that is owned by ThyssenKrupp, SMS Demag and the Fraunhofer Society), BMW conducts research on coating techniques of certain car parts.

Thus, the luxury car manufacturers in the Munich region, BMW and Audi, have several relations with university and non-university research institutes in and outside the region. There are also relations with research institutes abroad, but we have no indications for increasing research activities in the new growth markets outside the Triad.

The global automotive industry is characterised by large groups (like VW Group) consisting of a number of OEMs with several brands. Within these groups, but also increasingly between groups, cooperation takes place in operational functions and the field of research and product development. One example is the development of a new four cylinder petrol engine by BMW and PSA Peugeot Citroën, which will be used in future BMW Mini models and in some models of PSA. The aim of the cooperation is to share development costs and to combine expertise and facilities of the partners by applying economies of scale. The engine was developed by BMW in FIZ in Munich, while the production and manufacturing is the responsibility of PSA and will take place in France. Assembly will take place in PSA plants in France and BMW's Hams Hall plant in England (Douvrin, 2005; BMW, 2006). A second example is a joint project between Audi and Jaguar in the field of aluminium body repairs, in which Jaguar hopes to benefit from the expertise that Audi has in this field (Becker, 2005).

An important research theme in the automotive industry is the development of cleaner vehicles and new energy. A main driver is the initiative of the EU to create stricter rules for emissions. This is especially a threat for luxury car manufacturers, since luxury cars, such as Audi and BMW, are generally more polluting than smaller cars. For instance, in a list of the 26 lowest polluting cars, Audi ranks 16th with an average CO_2 emission of 179g/km, and BMW ranks 21st with an average CO_2 emission of 192g/km (Algemeen Dagblad, 2007b). By comparison, the least polluting car is the Smart with an emission of only 116g/km.

For the development of cleaner vehicles and/or new energy, there are many research projects between OEMs and with other actors, such as governments or energy companies. Some examples of such research projects in which BMW is involved include:

- The *German Fuel Cell Alliance*, which consists of 20 leading associations and state fuel cell initiatives, representing more than 300 firms, including Daimler-Chrysler, BMW and Siemens. The goal of the alliance is to obtain policy support, so that Germany can compete internationally in research, development and production of fuel cell technology (Fuel Cells Bulletin, 2005).
- The *development of hydrogen as a transport fuel*. This is one of the main research themes of BMW, which the company is doing with a number of (international) partners, like Magna Steyr, GM and Honda (BMW, 2005a). A major barrier for the use of hydrogen as transport fuel is the need for refuelling infrastructure. This infrastructure is likely to be developed first in cities, since there is a high concentration of vehicles and an anticipated benefit to air quality from a shift to hydrogen as transport fuel. BMW is involved in a research project that is investigating strategies for the development of a hydrogen infrastructure in London. Other partners in this project are the Greater London Authority and BP (Joffe *et al.*, 2004).
- The '*Lighthouse Project*' is a project in which vehicle manufacturers (BMW, Daimler-Chrysler, Ford, GM, MAN Nutzfahrzeuge and Volkswagen) and energy companies (Shell Hydrogen and Total France) have announced a joint approach to push hydrogen as a fuel for road transportation in Europe (Fuel Cells Bulletin, 2006).

Geographical link between R&D and manufacturing

An important question for this research is to what extent are R&D functions separated physically from production functions? In this case study, it is remarkable that both Audi and BMW have clustered all functions on or near the main site, where the headquarters are located. The explanation is that the different units and departments need to interact intensively in the development of new vehicles. At least four departments cooperate from the beginning of the development process: marketing, design, R&D and production. The marketing department makes sure that customers' wishes and preferences are taken into account in the development process. The R&D department develops technical specifications and determines technical feasibilities. The design department provides ideas on shape and design. Finally, the proximity of the production department is required to test whether the new car can be produced efficiently. The different departments need frequent face-to-face interaction: each department gives its input to which others can react to prevent impossible combinations. One interviewee spoke about a 'ping-pong game' between the design department and R&D department, to prevent impossible design/technical product combinations. Physical proximity matters as it shortens communication lines. Video conferencing is not seen as a substitute, since it precludes non-verbal communication and lacks the possibility to show concept cars physically. The latter aspects are especially crucial for design: designers prefer to walk around a prototype, touch it and feel it.

Paradoxically, some R&D and design functions are deliberately put at some distance from the main operations to promote fresh thinking and independence. Audi opened its design studio in Munich for these reasons. Munich's urban ambiance (compared to Ingolstadt) helps designers to develop fresh ideas. Another example is BMW Group Research and Technology, which is set up as an independent think tank; 'a branch office which would also be physically separated from all other development areas'.[19] Despite the more independent characteristics of these departments, communication with other departments is still important, and therefore the departments are located in the same region of the main operations of the OEMs. For instance, the Audi Design Studio in Munich can be reached in one hour by car from the operations in Ingolstadt. BMW Group Research and Technology is located less than a five-minute drive from the headquarters of BMW.

It is important to note that BMW has research centres abroad at locations where it does not have production functions; it has R&D functions in the leading technology clusters of Tokyo and in Silicon Valley. Audi also has research functions in locations where it does not have production facilities.

A key challenge for OEMs is to manage and coordinate the activities of their establishments worldwide. Audi's strategy is based on a 'develop at home, sell globally' model. The basic research and the development of new models takes place in Ingolstadt. Later, if the development of a new model is finished, the same model is produced in plants abroad, like the FAW-VW plant in China. For some foreign markets, Audi modifies its models slightly, for instance to meet specific government regulations.

Despite the fact that most of the research is conducted in Germany, Audi and BMW are moving towards a more 'meta-national' approach by tapping knowledge from other parts of the world. Audi uses the networks of the Volkswagen Group, of which it is part. For instance, the design centre of Audi in Ingolstadt has a forum to exchange ideas with Audi Design in California, SEAT Design in Spain and also with Lamborghini designers in Sant'Agata in Italy.[20] Similarly, within the BMW Group, there is frequent contact between the different departments in the global research network.

Policy and cluster organisations

Government influence in the automotive industry is substantial, but the reverse is also true. The automotive industry is a politically 'sensitive' industry. When leading companies announce job cuts, it is headline news and there are questions in parliament. The automotive industry is considered as a flagship industry for Germany, at the front of technological innovation, and with global visibility. There is a strong and influential political lobby from the automotive industry.

Government support comes in different forms. We identified four types of government support: 1) R&D funding for automotive research (for instance, the development of clean energy is a key objective for the German government, and therefore the federal government is a major contributor to new energy projects, such as the German Fuel Cell Alliance) 2) support to set up pre-competitive research institutes 3) location policy and support supplier parks (for example the Audi supplier park has been set up in cooperation with the City of Ingolstadt), and 4) investment grants to attract automotive companies (for instance the city of Hof and the districts of Hof and Wunsiedel give grants to attract suppliers and other automotive companies to the supplier park in Hof).

The automotive industry in the Munich region is complex and linked to many local and global networks. Evidently, there is not a single 'cluster manager' governing the regional network relations; OEMs and main suppliers manage their own global networks.

Nevertheless, there are some initiatives (often initiated by public entities) to solve collective action problems and to promote local networking and learning. A significant initiative in this respect is Allianz Bayern Innovative,[21] initiated by the Freestate of Bavaria. Among many other things, this programme supports 19 'cluster platforms' related to the major industries in Bavaria. The goal of these cluster platforms is to improve the competitiveness of Bavarian companies by facilitating and supporting innovation networks, and to improve the image of Bavaria as a high-tech hotspot. The main task for the cluster platforms is to create and maintain networks between different actors, such as companies, research centres consultants and other players in the cluster. The cluster platforms have a time horizon of five years from 2006, and Bavaria invests €50 million in the clusters.

One of the platforms (BAIKA) is set up for and by the automotive industry. BAIKA is an 'over-regional' network that promotes the cooperation between OEMs and suppliers. Its members are 12 major international OEMs from Germany and abroad, suppliers (including some of the world's ten largest tier-one suppliers), national and international cluster organisations, branch organisations, universities and universities of applied research. The key tasks of BAIKA include: stimulation of innovation through cooperation, technology transfer, information rationalisation and qualification. BAIKA is an open network, including for actors outside Bavaria, and membership is free. Members only have to pay for the events in which they participate.

Besides the BAIKA network, some smaller (local) networks can be found in Bavaria, such as 'www.OfraCar.de'[22] (a platform for the Upper-Franconian region that is established by the Chamber of Commerce of Bayreuth, and aims to build up a cooperation network between automotive and related companies) and the 'Lightweight Construction Cluster'[23] in Landshut (an interdisciplinary network of companies, universities and other research centres for cooperation on lightweight construction materials, lightweight construction design and manufacturing technologies). The activities of these smaller cluster organisations are similar to that of BAIKA.

5.5 Conclusions

In the last decades, Germany has faced a deep process of de-industrialisation. Employment in manufacturing has decreased by 3 million workers since the mid-1970s. The automotive industry has escaped the trend, however: employment in this sector has even increased substantially.

The automotive industry is a key industry in Germany. It is an important job generator, a major contributor to German exports and it is an important generator of R&D; more than one-third of the German R&D expenditures is generated by the automotive industry. Germany is still an important production location; it has various regional automotive clusters with production plants of OEMs and suppliers. The German market is saturated, just like many other Western markets, and an increasing share of the production in Germany is exported.

In this chapter, we have analysed the development of the automotive industry in the Munich region, one of the leading automotive clusters in Germany. A specific aim of the study is to analyse the evolving functional and spatial relationship between physical production ('manufacturing') and other activities in the value chain. In the remainder of this section, we summarise the main findings of our research and draw some conclusions.

Conclusions on foundations

Munich's *economic base* provides fertile soil for the automotive industry. Its diverse structure is an important asset for the automotive industry. The presence of other manufacturing industries is important since the automotive industry uses many inputs from other manufacturing industries and draws from a wide range of technologies. The 'Munich Mix' offers a wide variety of cooperation partners. Furthermore, the automotive industry benefits from the presence of multiple services in the Munich region, such as marketing, consultancy and financial services. Also the *knowledge base* in Munich is very strong. The region has multiple world-class (public and private) research institutes, many of them active in research areas that are highly relevant for the regional automotive industry. There are dense research networks and a multitude of joint research projects. Furthermore, Munich has a high-skill labour pool with many automotive engineers and designers. This fact alone makes the region attractive for the knowledge-intensive segments of the automotive industry. Despite some relatively minor problems (congestion on the highways around Munich and heavily occupied public transport during peak hours), we can say that the *accessibility* of the Munich region is excellent, both for passengers and for automotive logistics. Munich has a central location in Europe, with good connections to other destinations in Europe, including the increasingly relevant markets in CEE (for sales as well as supplies). The *quality of life* is another strong asset of the Munich region; the city has a lot to offer in terms of high-level amenities and it is very attractive for knowledge workers, including engineers in the automotive industry. In terms of *factor conditions*, Munich has relatively high factor costs; wages in the automotive industry in Germany are among the highest in the world and the city has relatively high real estate prices. However, we have indications that the relatively high factor costs are not a major problem for the OEMs in Bavaria; Audi and BMW are both luxury manufacturers which have relatively high margins and can afford to pay higher wages. Furthermore, they use many high-skill workers.

Cluster structure and performance

Bavaria is one of the leading automotive regions in Germany and in the world. It is the home of two leading car manufacturers in the luxury segment, Audi and BMW, both of which have one or more plants in Bavaria. All parts of the value chain are present in Bavaria: production plants (of OEMs and suppliers), R&D, design, sales, consultancy, etc.

The automotive cluster in Bavaria has experienced a remarkable growth. Employment has grown from 145,700 employees in 1997 to 182,600 employees in 2004, at an average annual growth of 3.2%. Sales have even grown on average by 10.5% per year, from €35.3 billion in 1997 to €71.3 billion in 2004.

Two important sub-clusters can be identified. One is in Munich. There, manufacturing takes place in the plants of BMW, and in the plants of a number of suppliers, such as Webasto and Iwis Ketten. In terms of R&D and design, Munich is a leading centre, with important universities (especially TUM), universities of applied sciences, research and design centres of large automotive OEMs and suppliers (among others, three institutes of BMW, Audi's Concept Design Munich and Magna Engineering Centre), many small and medium sized research and design companies, and other research centres. Munich is also home to the headquarters of BMW and a number of suppliers.

A second important cluster is Ingolstadt, where Audi has all its functions. Furthermore, there is a large supplier park and other facilities, such as SE park (engineering function) and STC (a training centre).

The two dominant OEMs, Audi and BMW, both have all functions (headquarters, R&D, design, production and commercial functions) geographically clustered on or near one site. Both firms have indicated that a geographical separation of these functions is not desirable. All major departments cooperate closely from the beginning of the development process. Having the production plant nearby makes it easier to test the quality and efficiency of the production process of a new model.

Production and knowledge networks

To analyse the production and knowledge networks in the cluster, we distinguish two types of activity: exploration (R&D, design, product development) and exploitation (production and distribution). The exploration stage is divided into two phases: pre-competitive research and product development.

Concerning *production*, we have observed that Germany is still the main location for BMW and Audi, and this will remain the case. Both firms have a stable or increasing labour force in Germany, and show explicit commitment to

Germany as a production location, verbally and through new investments. On the other hand, we have seen that the OEMs also open plants outside Germany, sometimes in the form of joint ventures with local partners. Thus, the 'holy unity' of production and R&D does not always hold.

In some of its foreign production plants, BMW adopts a complete knocked down strategy to overcome problems of a weak local supplier base, or to evade high import duties on finished vehicles. With ongoing growth in new markets, both for BMW and Audi, the importance of Germany as a sales market is likely to decrease further. Another ongoing trend is global sourcing: an increased share of supplies is purchased from low-cost countries, thus hollowing out the manufacturing base for suppliers in Germany. Note that some types of suppliers are required to locate near production plants for logistical reasons.

For *pre-competitive research*, we find that Munich companies traditionally have their strategic partners in the region. There are fruitful cooperations with Bavarian universities (INI.TUM is the best example of close collaboration between Audi and the Technical University Munich). However, the leading companies increasingly look beyond the region for partners. Audi has recently entered into a number of partnerships with universities outside Bavaria and abroad. Its partnership with Györ University in Hungary is worth noting in this respect.

For *product development* networks (including design), the scope of Munich's companies is more global. The OEMs are involved in global research networks, especially for the development of cleaner vehicles and new energy. They are parts of global automotive groups and cooperate with research departments of sister brands. We observe that the global orientation is on the increase. For design skills, BMW taps Californian expertise, and the company has R&D sites in Japan and Palo Alto. Audi works with design competence within the global Volkswagen group. We may conclude that Audi and BMW are still firmly based in the Bavarian knowledge economy, but they increasingly adopt 'meta-national' innovation strategies, and tap specific pockets of knowledge in the world.

Notes

1 Besides Audi and BMW, the truck and bus manufacturer MAN Nutzenfahrzeuge AG (MAN) is also located in the Munich region. However, this case study mainly concentrates on passenger cars and therefore MAN has not been investigated in more detail.

2 Industrial development of Munich started only after the Second World War (Van den Berg *et al.*, 2005).

3 In the reminder of this chapter, we simply speak about 'Bavaria' instead of the Free State of Bavaria.

4 It is home of the international headquarters of German firms such as BMW, MAN Nutzfahrzeuge and Siemens, as well as the German headquarters of foreign firms like Microsoft.

5 http://www.invest-in-bavaria.com/?PHPSESSID = xxx (accessed 26 January 2007).

6 Upper Bavaria is one of the seven administrative regions (so-called 'Regierungsbezirke') in Bavaria. Munich and Ingolstadt are located in this region.

7 Since the mid-1970s, the total loss of jobs in the German manufacturing industry has been 3 million (VDA, 2006).

8 http://www.bmwgroup.com/bmwgroup_prod/e/0_0_www_bmwgroup_com/investor_relations/corporate_events/hauptversammlung/2007/_pdf/rede_presseversion.pdf (accessed 13 February 2007).

9 Information on the research facilities of the BMW Group is obtained from: Annual report BMW 2005, BMW Group Sustainable Value Report 2005/2006 and the BMW website http://www.bmwgroup.com/ (accessed 14 February 2007).

10 http://www.bmwgroup.com/bmwgroup_prod/e/0_0_www_bmwgroup_com/investor_relations/corporate_events/hauptversammlung/2007/_pdf/rede_presseversion.pdf (accessed 14 February 2007).

11 http://www.audi.hu/deu/index.php (accessed 17 July 2007).

12 This concerns both R&D expenditures not recognised as an intangible asset (€999 million) as well as R&D expenditures that can be recognised as an intangible asset (€586 million Euro) (Audi Annual report 2005).

13 www.audiworld.com/news/03/121203/content.shtml and http://media.seat.com/en/view_object.html?obj = 1200,1610,1202,c,10643 (accessed 14 February 2007).

14 http://media.seat.com/en/view_object.html?obj = 1200,1610,1202,c,10643 (accessed 14 February 2007).

15 http://www.invest-in-bavaria.de/en/bavarias-clusters/mobility/automotive-engineering/ (accessed 11 March 2010).

16 http://www.audi.com/audi/com/en2.html (accessed 11 June 2007).

17 For instance a three year project costs on average about €40 million (interview).

18 http://www.audi.hu/index.php (accessed 18 July 2007).

19 http://www.bmwgroup.com/e/nav/index.html?/0_0_www_bmwgroup_com/home/home.html&source = overview (accessed 26 April 2007).

20 www.audiworld.com/news/03/121203/content.shtml (accessed 14 February 2007).

21 http://www.invest-in-bavaria.de/BavariansCluster/index.html?PHPSESSID = 4faaa74393067479d1a93645c3ae2bd8 (accessed 7 January 2007).

22 http://www.ofracar.de/Kompetenzatlas/netzwerk.htm (accessed 27 February 2007).

23 http://www.invest-in-Bavaria.com/News/?status = regions&mode = detail&monatID = 99&PHPSESSID (accessed 27 February 2007).

Bibliography

Algemeen Dagblad (2007a) *De superzuinige Duitse auto komt eraan*, 23 April 2007.

—— (2007b) *Het mes op de keel*, 21 April 2007.

Audi (2006a) *Annual report 2005*, Ingolstadt: Audi AG.

—— (2006b) ‘New Audi Q5 to be built in Germany’, press release.

—— (2006c) ‘Topping out ceremony for new Audi climatic wind tunnel’, press release.

—— (2006d) ‘Audi Lab for Market Research’ inaugurated at the University of St. Gallen’, press release.

—— (2007a) *Annual report 2006*, Ingolstadt: Audi AG.

—— (2007b) ‘Annual Press Conference: Audi remains on course for success’, press release.

—— (2007c) ‘Audi acquires extra production capacity’, press release.

—— (2007d) ‘Record results for vehicle sales, revenue and profit’, press release.

—— (2007e) ‘Audi takes over management responsibility at Brussels plant’, press release.

Auto World (2006) ‘Made in Bavaria’, *Auto World* 2, available online: www.localglobal.de/sixcms/list.php?page_id=1089&_t=Search&vt=made%20in%20bavaria (accessed 15 March 2010).

Becker, H. (2006) *High noon in the automotive industry*, Berlin: Springer-Verlag.

van den Berg, L., P.M.J. Pol, W. van Winden and P. Woets (2005) *European cities in the knowledge economy – the cases of Amsterdam, Dortmund, Eindhoven, Helsinki, Manchester, Munich, Münster, Rotterdam and Zaragoza*, Aldershot: Ashgate.

BMEAITT (Bavarian Ministry of Economic Affairs, Infrastructure, Transport and Technology) (2005) *Setting the page – Bavaria's industry 2004/2005*, Munich: Bayern International GmbH.

—— (2006) *Allianz Bayern Innovativ: Eckpunkte bayerischer Clusterpolitik*, Munich: Bayerisches Staatsministrium für Wirtsschaft, Infrastruktur, Verkehr un Technologie.

—— (2007) *Invest in Bavaria facts and figures*, Munich: Bayern International GmbH.

BMW (2001) *Sustainable value report 2001/2002*, Munich: BMW.

—— (2002) *Annual report 2001*, Munich: BMW.

—— (2002a) 'BMW Plant Hams Hall successfully completes first year of production', press release, 7 February 2002.

—— (2002b) '10 Years Later: BMW Manufacturing Corp.to invest $400 million and add 400 jobs', press release, 26 September 2002.

—— (2003) 'Ground-breaking ceremony for BMW Group Parts Distribution Centre in Malaysia', press release, 12 December 2003.

—— (2004) 'BMW Group and Brilliance officially open Shenyang plant', press release, 20 May 2004.

—— (2005a) *Sustainable value report 2005/2006*, Munich: BMW.

—— (2005b) 'BMW Group Dynamics Centre in Dingolfing opens', press release, 15 June 2005.

—— (2006) *Annual report 2005*, Munich: BMW.

—— (2007a) '*BMW Group expands US plant in Spartanburg*', press release, 15 May 2007.

—— (2007b) Statement by Dr Norbert Reithofer, Chairman of the Board of Management of BMW AG, 87th Annual General Meeting of BMW AG, Munich, 15 May 2007, available online: www.bmwgroup.com/bmwgroup_prod/e/0_0_www_bmwgroup_com/investor_relations/corporate_events/hauptversammlung/2007/_pdf/rede_presseversion.pdf (accessed 15 March 2010).

City of Ingolstadt (2006a) 'Ingolstadt an der Donau, attraktiv und innovativ, Wirtschaftsstandort', available online: www2.ingolstadt.de/media/custom/465_2265_1.PDF?loadDocument&ObjSvrID = 465&ObjID = 2265&ObjLa = 1&Ext = PDF (accessed 26 February 2007).

—— (2006b) 'Ingolstadt an der Donau, automotive and logistik, GVZ Ingolstadt', available online: www2.ingolstadt.de/media/custom/465_2379_1.PDF?loadDocument&ObjSvrID = 465&ObjID = 2379&ObjLa = 1&Ext = PDF (accessed 26 February 2007).

City of Munich (2005) *Sector information – Munich's automotive sector*, Munich: Department of Labor and Economic Development.

—— (2006) *Munich because* ..., Munich: Department of Labor and Economic Development.

Comissie Sistermans (2005) *Brainport navigator 2013, Crossing borders, moving frontiers – Bijlagen*, Eindhoven: Drukkerij Jémé.

Cushman & Wakefield (2008) *European cities monitor 2008*, London: Cushman & Wakefield.

Dannenberg, J. and C. Kleinhans (2004) 'The coming age of collaboration in the automotive industry for automakers, it's all about brand management', *Mercer Management Journal*, 17, 88–94.

Die Welt (2007) 'Erfolg um jeden Preis', 13 October 2007.

Douvrin (2005) '*Cooperation between PSA Peugeot Citroën and BMW Group*', Industrial Cooperation Day, 21 June 2005.

Faurecia (2005) '*Press kit Faurecia in Germany*', November 2005.

Financial Times (2006) 'Volkswagen choose to swim against the current', 7 November 2006.

—— (2007) 'BMW looks to expand in China', 15 March 2007.

Fuel Cells Bulletin (2005) 'German fuel cell alliance calls for political support', *Fuel Cells Bulletin*, 2, 7–8.

—— (2006) 'Euro industry outlines hydrogen infrastructure', *Fuel Cells Bulletin*, 11:6.

Heneric, O., G. Licht, G.W. Sofka (2005) 'Europe's automotive industry on the move: competitiveness in a changing world', *ZEW Economic studies*, 32.

IGM (2006a) *Branchenanalyse Automobilindustrie 2006 Teil 1: Überblick*, Frankurt am Main: IGM.

—— (2006b) *Branchenanalyse Automobilindustrie 2006 Teil 3: Entwicklung des internationalen Umfeldes – Deutsche Automobilindustrie in Europa and der Welt*, Frankurt am Main: IGM.

—— (2006c) *Branchenanalyse Automobilindustrie 2006 Teil 4: Automobielzulieferindustrie*, Frankurt am Main: IGM.

Iwd (2004a) 'Automobilindustrie Vom Netzwerk profitieren', *Iwd*, 47, 18 November 2004.

—— (2004b) Schröder Ch., 'Productivität und Lohnstückkosten im internationalen Vergleich', *IW-Trends*, 3.

Joffe, D., D. Hart and A. Bauen (2004) 'Modelling of hydrogen infrastructure for vehicle refuelling in London', *Journal of Power Sources*, 131:1/2, 13–22.

Landeshauptstadt München (2005) 'Brancheninfo Automobiltechnik München', available online: www.muenchen.de/Wirtschaft/publications/181545/index.html (accessed 15 March 2010).

—— (2006a) *München. Der Wirtschaftsstandort. Facten und Zahlen 2006*, Munich: Department of Labor and Economic Development.

—— (2006b) *Münchener Jahreswirtschaftsbericht für 2005*, Munich: Department of Labor and Economic Development.

Larsson, A. (2002) 'The development and regional significance of the automotive industry: supplier parks in Western Europe', *International Journal of Urban and Regional Research*, 26:4, 767–84.

Manshanden, W.J.J., W. Jonkhoff, P.J.M. de Bruijn, S.P. Newrly and R. en Koops (2005) *De top-20 van Europese stedelijke regio's 1995–2004; Randstad in internationaal perspectief*, Delft: TNO.

Mercer and Fraunhofer Institutes IPA and IML (2004) *Future Automotive industry Structure (FAST 2015) – die neue Arbeitsteilung in der Automobielindustrie*, Munich: Mercer.

OICA (2006) Production statistics, available online: http://oica.net/category/production-statistics/2006-statistics (accessed 15 March 2010).

RVI (2006) *Innovationsbericht 2006*, Essen: Rheinisch-Westfälisches Institut für Wirtschaftsforschung.

VDA (2006) *Auto annual report 2006*, Frankfurt am Main: VDA.

6 Ostrava: automotive industry

6.1 Introduction

The Moravian-Silesian region (MSR) is known as the 'industrial heart of the Czech Republic', for its large concentration of mining, steel and other (heavy) manufacturing industries. In particular, in the capital Ostrava and the city of Karviná, there are several large steel plants and mines. Industrial development started with the discovery of coal in Mähhrisch-Ostrau (the former name of Ostrava) in 1763; and the first ironworks, Rudolfhütte (later Witkowitzer Eisenwerke), was founded in 1828. In the beginning of the 1980s, the total hard coal production was 25 million tons, which was four-fifths of the total production in Czechoslovakia. At that time, Czechoslovakia was the largest producer of hard coal per inhabitant in the world (Eckart *et al.*, 2003).

Since the 1990s, however, all the major industries in the region (coal mining, coke making, metallurgy and heavy engineering) have declined; mines are closed and large steel plants are involved in restructuring programmes. The restructuring of the region has had a large impact on the labour market and employment. The region has a high unemployment rate of 13.9% (in 2005), which is far above the Czech Republic's average of 7.9% (Eurostat data).[1] The government, together with various stakeholders, including the largest steel and mining firms, takes various measures to guide the restructuring process to improve the economic and social situation.

Today, the situation has improved again, and the region receives an increasing number of foreign direct investments (FDI). MSR benefits in particular from the endeavours of CzechInvest, a business and development agency of the state, which provides direct incentives to foreign investors. Companies from the automotive industry especially are making use of the incentives from CzechInvest and are investing in MSR due to its favourable geographical position, next to other countries in CEE, and for the region's tradition and capabilities in the automotive industry. The automotive industry in MSR started with Tatra, which produced the first passenger car in Central Europe in 1897. Currently, the automotive industry in MSR revolves around a promising cluster of old firms (like Tatra) and newly established foreign automotive companies and suppliers, like Siemens. The decision in 2006 of the Korean automaker Hyundai to open a production plant in MSR has been seen as a boon for the region.

In this chapter, we focus on the dynamics of the automotive industry in MSR. The rest of this chapter is structured as follows. Section 6.2 describes the regional economic conditions of the region. Next, in section 6.3, we analyse the geography of the sector, and we introduce the main players of the automotive cluster in MSR. In section 6.4, we discuss the networks in where the cluster actors operate and, furthermore, we discuss the role of the government in the automotive industry (in section 6.5). In the last section, section 6.6, we draw conclusions.

6.2 Foundations of MSR

MSR is the largest region in the Czech Republic with 1.25 million inhabitants (in 2005) (CzechInvest, 2007). The core of the region, consisting of the region's capital Ostrava, Karviná and Fýdek-Místek districts, forms the largest residential, demographic and industrial region after the Prague area (Eckart *et al.*, 2003). The region is undergoing a restructuring process from a one-sided economic structure towards a more diverse economic structure with a combination of some remaining traditional industries with future potential and new industries. One of the traditional industries is the automotive industry. To understand the current dynamics of the automotive industry, this section puts the industry into the perspective of its regional context. This is done using a description of MSR's regional foundations: the economic base, the knowledge base, accessibility, quality of life and factor conditions.

Economic base

The largest manufacturing industries in MSR – coal mining, coke making, metallurgy and heavy engineering – were defined by the socialist regime. These industries were dominated by large state-owned enterprises with a large number of low-skilled workers (CZSO, 2005). The three key firms, Ostrava Coal Mines (OKD), Vítkovice (steel and engineering) and Nová Hut (mainly steel) had large plants in the region and played an important role in the society. The enterprises provided the workers a wide range of services, like housing of workers in enterprise flats, and the companies contributed to the region's economy and society by subsiding road construction and other technical infrastructure, or by sponsoring cultural or sporting events (Nesporova, 1998).

Since the beginning of the 1990s, all the major industries had been declining and restructuring of the economic structure started. Low profitability of the coal mines, with many loss-making pits, overcapacity of the steel mills, which by far exceeded the national needs, and problems with exports of the coal mines were problems faced by the region, making it clear that restructuring was unavoidable (Nesporova, 1998). Suchácek (2005) notes that the de-industrialisation of the region started relatively late. The region had not undergone large changes since World War II, because the socialist policy had stalled its economic and social structure. At the beginning of the restructuring process, MSR provided 86% of the Czechoslovak coal mining, 82% of the coke production and 70% of the steel production. About half of the employees worked in the manufacturing sector.

One of the consequences of the restructuring is a huge decline in employment. For instance, OKD reduced its workforce from 118,000 in 1989 to less than 50,000 in 1995 (Nesporova, 1998). MSR performs worse than the Czech average and most other regions in the country. MSR has an unemployment rate of 14.2% which is above the Czech average of 8.88% (see Figure 6.1). Within MSR the unemployment rate varies from 10.9% in Nový Jicín to 18.6% in Karviná. The city of Ostrava has an unemployment rate of 14.82%, which is comparable with the average in MSR (CZSO, 2006). Eurostat data show that the unemployment rate of the Czech Republic is lower than the EU-15 average, but MSR is above it. The unemployment rate of MSR is comparable with the average of the ten new member states of the European Union. The Eurostat data also show that the unemployment rate in MSR has been fluctuating around 14% since 1999. It is worth mentioning that the highest unemployment was reached in 2004 when there were 120,000 job seekers. Since that time, unemployment has declined, and it has halved to just over 60,000 job seekers in 2007 (ARR, 2007a).

In terms of GDP per capita, MSR is comparable to other regions in the Czech Republic. Figure 6.2 depicts that apart from Praha (the larger Prague region), which is performing far above the national average, all regions have a GDP per capita between €7,569 (in Hradec Králové) and €9,055 (in Central Bohemia). Eurostat data show that the Czech Republic (and thus also MSR) is performing worse than the EU-27 average; the GDP per capita in the Czech Republic is more than 2.5 times lower than in the EU-27 average.

Data from CzechInvest suggest that MSR is one of the favourable regions for foreign investors. The region received US$3.19 million from foreign investors, creating 21,020 new jobs. Only Ústí and Central Bohemia received more FDI (see Figure 6.3 and 6.4). To date, 170 foreign investors from 28 countries decided

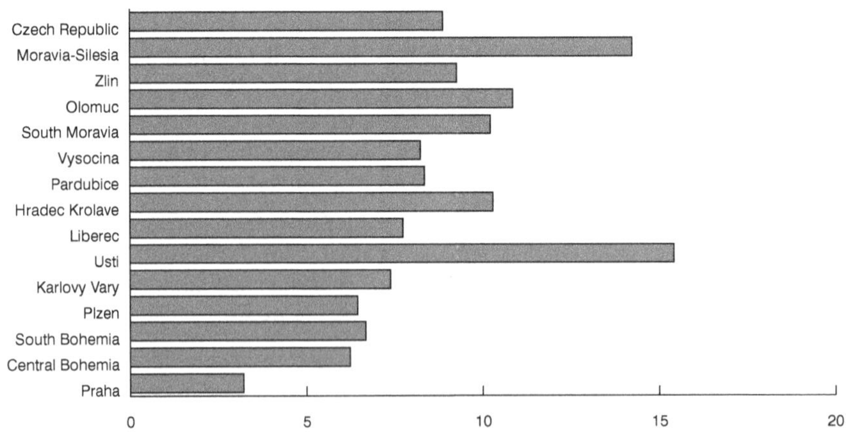

Figure 6.1 Unemployment rate in Czech regions (31/12/2005) (source: CZSO, 2006)

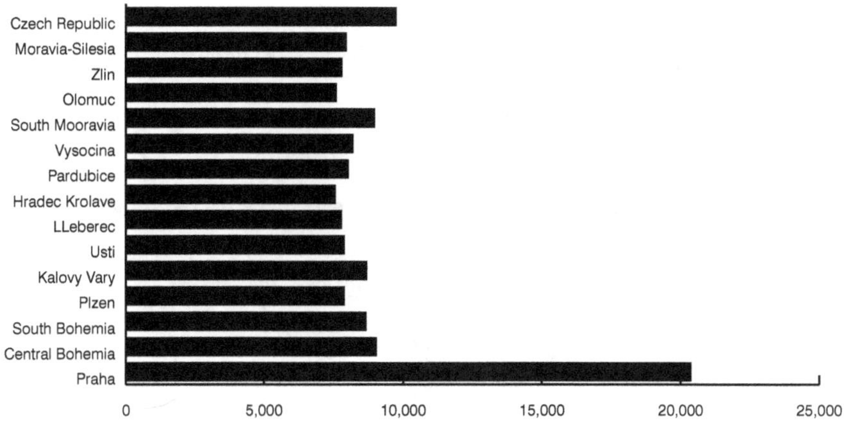

Figure 6.2 GDP per capita, euros, 2005 (source: CZSO, 2006)

to set up businesses in the region or are extending their current business in the region. Most investments are done by the automotive industry (17% of the total FDI) and heavy engineering (12%), but increasingly, foreign companies are also investing in services, like tourism and real estate (ARR, 2007b).

The economic structure in MSR shows that the manufacturing sector still plays an important role in the economy of the region (see Figure 6.5); 27% of the employees are active in the manufacturing sector. In Ostrava, this share is even higher, where 35% of the employees work in the manufacturing sector (City of Ostrava, 2006). The service sector in MSR is diverse; there is no dominant service industry. On the contrary, the manufacturing sector is dominated by the steel industry (31% of the total manufacturing employment), followed by mining (17%), machinery (12%) and the construction of transport equipment (11%) (see Figure 6.6). The automotive industry is included in the latter. This figure suggests that the economic structure is still one-sided, and heavily dependent on these four traditional industries. Nevertheless, it has to be noted that the figure only includes large firms with over 100 employees, headquartered in MSR. If SMEs are also included, the dominance of the four traditional industries would probably be less, since these industries are characterised by large firms, while SMEs play a larger role in modern industries, like biotechnology.

In Ostrava, the dominance of the steel industry ('metallurgy') is even larger; 61% of the employees in the manufacturing sector are active in this industry. The largest manufacturing industries in Ostrava are machinery, 'transportation vehicles and equipment' (including the automotive industry), and 'food, drink and tobacco industries' with shares of 10%, 7% and 6% respectively in the total manufacturing employment. All other manufacturing industries have a share of 5% or less. It should (again) be noted that these data only include firms with 25 employees or more (City of Ostrava, 2006). Most of the largest firms in Ostrava and MSR are from the dominant industries (in particular steel and machinery), and most of

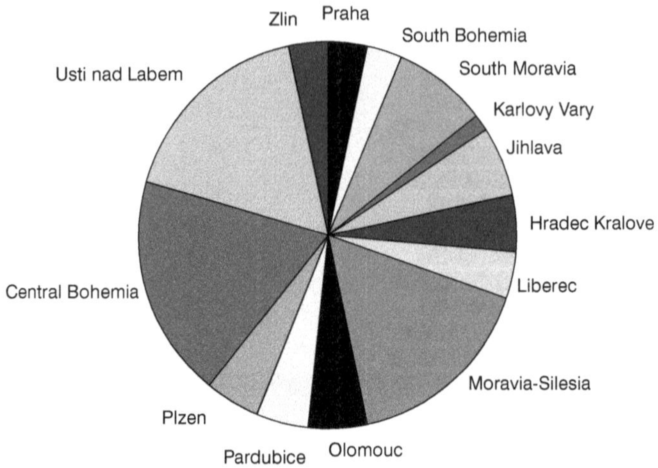

Figure 6.3 Regional shares in FDI through CzechInvestment projects 1993–31/03/2007, in USD; total = $19,455 (source: own elaboration, data from CzechInvest)

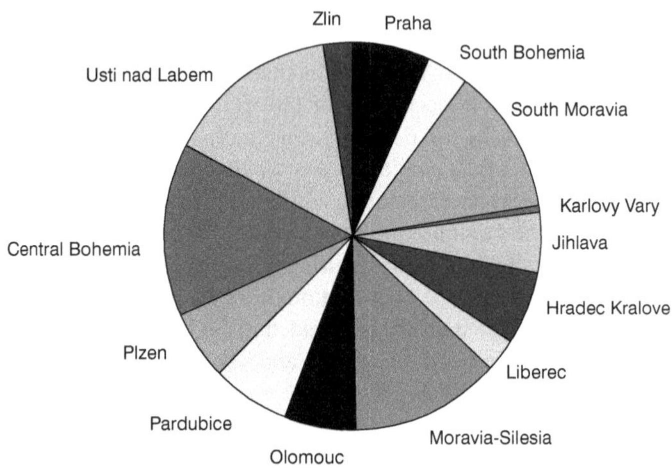

Figure 6.4 Regional shares in total number of jobs created through CzechInvest projects, 1993–2007, total = 161,957 (source: own elaboration, data from CzechInvest)

them are remaining establishments of the key firms Vítkovice and Nová Hut. The latter was taken over by the steel giant Archelor Mittal steel in 2003. The service sector in Ostrava (and to a lesser extent in MSR) is underdeveloped and largely consists of state and mainly non-commercial services such as health services and education. There are few advanced services like banking. Nevertheless, this is changing through the growth of the manufacturing sector, levering the service sector. Recently, the massive investments of foreign manufacturing firms gave

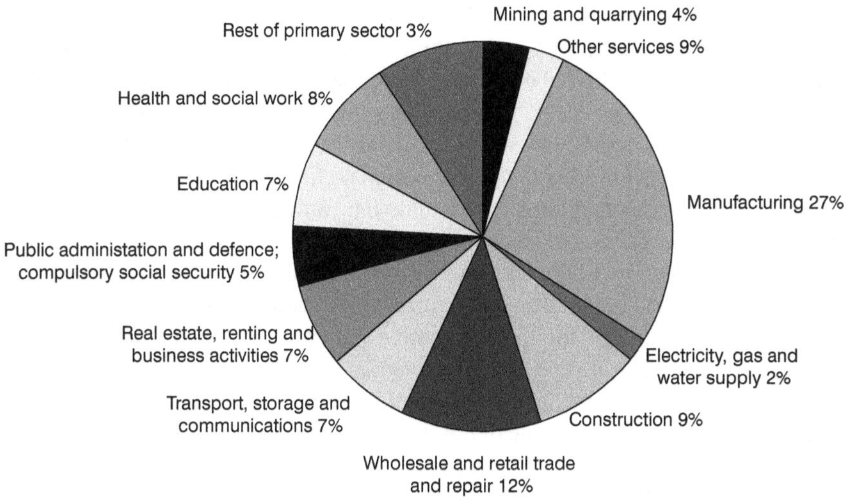

Figure 6.5 Economic structure MSR, in employees, 2005, total = 535,000 (source: own elaboration, data from CZSO Regional Office Ostrava)

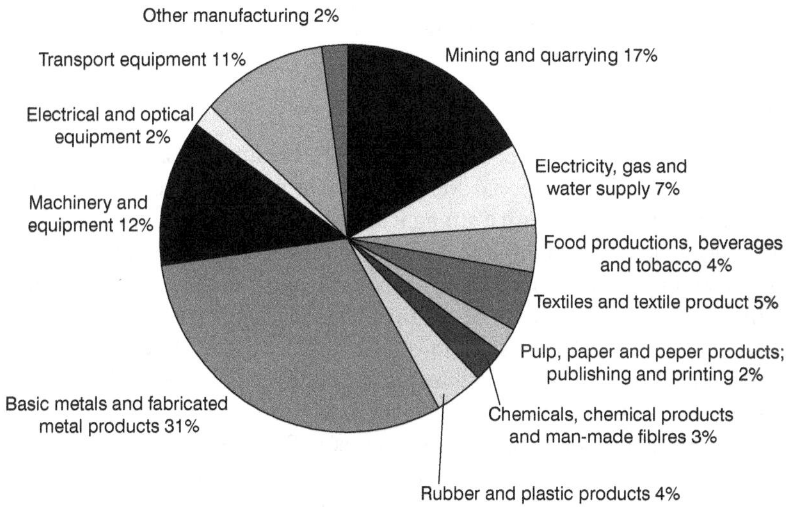

Figure 6.6 Structure of manufacturing sector in MSR, employment, 2005, total = 90,600 (source: CZSO Regional Office Ostrava)

a boost to the regional economy and the service sector is now growing fast. These investments are provoking a boom in the construction of new hotels and apartments, and many business service companies locate to Ostrava in the slipstream of industrial firms. This is important for the restructuring process, supporting the development of a more diverse economic structure.

Hence, the economic base of MSR is relatively weak, but is improving quickly through the large FDI with a quick decline in unemployment as a consequence. The region is in the middle of a restructuring process from a one-sided economic structure (dominated by steel, mining and some related industries) to a more diverse economic structure with wood, engineering, automotive, IT, hydrogen and renewable energy as key clusters. These regional clusters have been selected in a national cluster policy that aims to stimulate innovation. The region suffers from the restructuring process and has a relatively high unemployment rate, but many new jobs are created by FDI, and growth in the manufacturing sectors is ahead of growth in the service sector. The current one-sided economic structure of Ostrava might be a disadvantage for the automotive industry, since it is dependent on many other manufacturing and service industries. In particular, the small presence of business services might lack the attraction of the higher parts of the value chain. Nevertheless, the presence of the steel industry can be an asset to attract the automotive industry, since steel is an important input for the automotive industry. Also some of the key clusters might be a promising asset for the automotive industry; hydrogen and renewable energy are important fields for the automotive industry, especially in relation to the pressure on the automotive industry to develop cleaner vehicles. Also, engineering and IT are industries used by the automotive industry, especially the increasing use of electronics and software in vehicles that links the IT sector and the automotive industry.

Knowledge base

Just as in many other regions, policymakers in MSR realise that a good knowledge base is important for international competition. The region has set the ambition to become 'a growing region with a strong R&D base', also necessary to upgrade its manufacturing activities and to reach a higher value added of the manufacturing sector. According to Brusková (2007), important actors in the current knowledge infrastructure are: *four universities* (VŠB Technical University Ostrava, 20,216 students in 2005/2006; University of Ostrava, 8,523 students; Silesian University in Opava and Karviná, 4,947 students; and School of Business Administration in Karviná) (CZSO, 2006); the *Business Innovation Centre (BIC) Ostrava-Vítkovice* (this centre was originally set up as a Technology-Innovation Centre for Vítkovice a.s., but nowadays, it acts more as a cluster organisation aimed to support SMEs); *Ostrava Science and Technology Park (Ostrava S&TP)*, and the *cluster organisations of the key clusters*. The major aim of the Ostrava S&TP is to support SMEs and to stimulate start-ups, and therefore, incubators pay a lower rent in the beginning. Most of the companies on Ostrava S&TP are active in the IT or the bio-technology industry. There is space also for supporting services like advertising companies. So far, there are no firms from the automotive industry in the park.

Just like the economic structure, the knowledge structure is in a transformation process. The VŠB Technical University has been transforming from a traditional mining school towards a polytechnic university with cooperation projects with other universities. For instance, the Technical University supported the establishment of the other universities in the region in the 1990s and researchers of the Technical University are teaching at the other universities (Eckart *et al.*, 2003). It should be noted that some of the faculties of the Technical University are still focused on research in the traditional industries. Moreover, the Technical University is mainly local-oriented (90% of the students come from the region) and all bachelors and masters courses are offered in Czech, thus hindering foreign students from studying at the university.

The knowledge infrastructure is not only public-sector oriented. Key firms, like Vítkovice and Mittal Steel Ostrava, as well as other (foreign) firms have R&D centres in the region (Siemens and Autopal are good examples).

Another aspect of the knowledge base is the labour pool. Over the last decade, many workers were made redundant by the job shedding of the traditional industries. Retraining is a key challenge. The region has many secondary and apprentice schools, private training facilities in companies (in particular Vítkovice, OKD and Mittal Steel) and universities that can play a role in the further upgrading and adjustment of regional human resources, especially youths, to the changing demands of the labour market (Nesporova, 2003). The key enterprises play an important role in retraining workers as these firms retrain workers for new requirements in their own or other organisations. The retraining of workers takes place in special training centres that are owned by the key firms (Eckart *et al.*, 2003).

Despite the presence of training and educational facilities, and the willingness of employees to be retrained, the skills and methods or habits of work of the employees are relatively dated, and do not match with the requirements in modern production technologies. The mismatch is also caused by the one-sided economic structure (in particular heavy industries) (Nesporava, 2003). It is notable that the educational level of the population in MSR is only slightly below the Czech average (see Table 6.1).

A positive development in the local labour market is the increasing level of English language skills. Through many foreign investments, an increasing number of workers are speaking English or are prepared to learn English. The first international school in the region, which was opened in Ostrava in 2005, is an important facility to teach English to foreign and Czech students. Nevertheless,

Table 6.1 Educational level of the population, %, 2006

	MSR	*Czech Republic*
Basic, pre–primary and without education	19.36	21.28
Secondary education without GCE	37.21	39.54
Secondary education with GCE	32.55	30.11
Tertiary education	10.88	9.08

Source: CZSO http://www.czso.cz/
Note: GCE = General Certificate of Education

there is still (much) room for improvement. For instance, the Technical University in Ostrava does not offer bachelors and masters courses in English. This hinders potential foreign students from taking a course in the region.

A problem facing MSR, just like many other European cities, is a shortage of technical workforce. This is, among others reasons (like educational and labour policies), caused by the negative image of manufacturing, especially in the mechanical engineering industry. Just as in many other European cities, youths in the Ostrava region do not want to work in production plants since they associate work in plants with heavy and dirty work. In reality, work in plants has changed; through automation, work is not heavy and plants are clean. One interview partner even noted that 'assembly plants are like hospitals'. The negative image is strengthened by negative media attention of the mechanical engineering industry. It is also strengthened by the restructuring in the manufacturing sector in the region. Many youths see the high unemployment in the manufacturing sector and opt for non-technical studies, like business and economics, to increase their chances on the labour market. The problem of the shortage of technical workers is identified in all key clusters and is increasing due to new requirements to the workforce. Therefore, the Regional Development Agency Ostrava started ForTech (in 2006), a programme to enhance the interest of young people in technical branches, like engineering and electronics. Furthermore, a regional employment strategy is in development (Brusková, 2007).

R&D and patent applications are important indicators to measure the quality of the knowledge base. MSR spends 2.18 million CZK on R&D (in 2006), which is 3,508 CZK per employee. This is much lower than the Czech average (see Figure 6.7), although the 'average value' is influenced by the high values invested in the

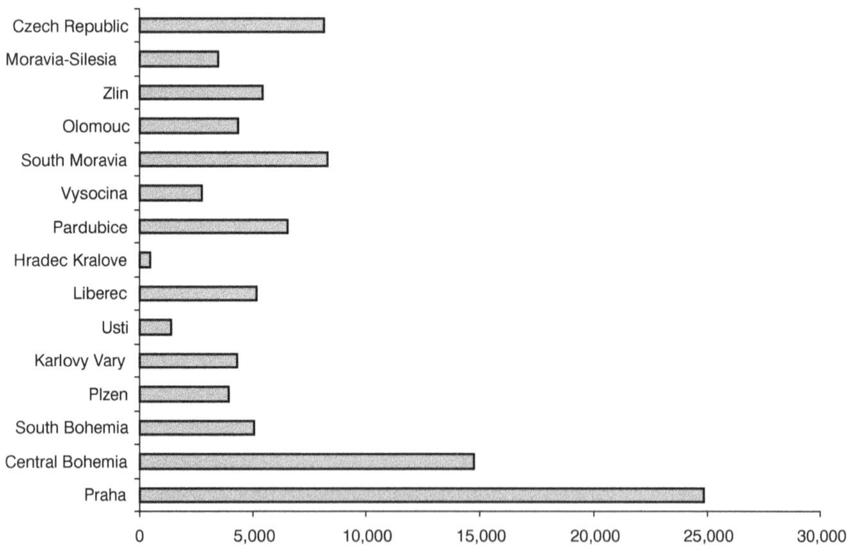

Figure 6.7 R&D expenditures per worker, 2006, in CZK (source: CZSO Regional Office Ostrava, 2006)

Prague region. Nevertheless, also compared to other regions, the investments in R&D per worker in MSR are relatively low; only Vysocina, Hradec Králové and Ústí spend less on R&D per worker. Also the share of R&D employees in the total workforce is with 0.62% relatively low in MSR. Again, only Vysocina (0.33%), Hradec Králové (0.09%) and Ústí (0.23%) have relatively fewer R&D workers. The share in Czech average (1.26%) is again largely influenced by Praha, where more than 4% of the workers are classified as R&D workers (CZSO Regional Office Ostrava, 2006). In terms of patent applications, MSR performs much better. In 2006, MSR had 53 patent applications, which is nearly 10% of the total patent applications in the Czech Republic. Only South Moravia and Praha had more patent applications (CZSO Regional Office Ostrava, 2006).

Accessibility

MSR is located in the south-eastern part of the Czech Republic, bordering Poland and Slovakia. This is a favourable location for the automotive industry for at least three reasons. First, just across the border is the Žilina region (Slovakia) and Upper Silesia (Poland), where other large automotive clusters can be found. OEMs in MSR can use the supplier base in these regions and vice versa. Second, Poland and Slovakia are important upcoming markets with increasing motorisation rates and vehicle sales, just like other countries in CEE. Third, the region is relatively far from the capital, Prague, which makes it possible to keep the factor prices low (regions closer to Prague have rising prises).

The infrastructure in MSR and the Czech Republic in general is behind competing countries in CEE, and according to some of our interview partners development of the infrastructure connecting MSR with other regions is too slow, although progress has been made. For transport by train, MSR is well located at the crossing of two major railways, the Czech arterial railway 'Praha-Bohumín' which goes to Poland and Slovakia and the 'Ostrova-Prerov' line, which continues to Vienna and further to the Balkans (CZSO, 2005). The first line is being modernised and forms an important corridor for freight transport (Nesporova, 1998; Suchácek, 2005; Brusková *et al.*, 2006), and it might act as an asset to transport car parts and entire vehicles to other automotive markets and clusters in CEE. The road network has also improved. The region has been missing a connection to the state motorway network for a long time, and the construction of a motorway that connects Ostrava with Brno (with further connections to Prague, Vienna and Bratislava) on the one side and the Polish Upper Silesian region on the other was postponed many times (Nesporova, 1998; CZSO, 2005; Suchácek, 2005). Currently, this motorway, the D47, is under construction and is expected to be opened in the near future.[2]

Besides road and rail connections, MSR is also accessible by air. Ostrava Airport is the second largest airport in the Czech Republic. It is a fast growing airport with scheduled connections to Prague and Vienna. The number of passengers has grown quickly from 115,363 to 300,736 in 2006. Also the transported cargo has nearly doubled, from 1,023,937 in 2000 to 2,042,768 in 2006.[3] The airport can receive all types of planes, which is important for future growth of the airport. According to our interview partners direct linkages with more destinations should be a priority for further development of the airport.

Regarding the accessibility of Ostrava, we can conclude that the region has potential for lower parts of the automotive value chain. The favourable geographical location and the improving infrastructure are important assets for automotive production and logistics. In contrast, the facilities for passenger transport are relatively weak (no HST connection and regular flights from the airport to only two destinations). This makes MSR less attractive for the higher parts of the value chain, like R&D. These functions are increasingly done in international project teams, which require quick connections for knowledge workers to other regions.

Quality of life

The (heavy) industry has a significant influence on the quality of life in MSR. Mining and industrial development have destroyed many natural landscapes in the region (Eckart *et al.*, 2003). In addition, MSR has an image problem; the region is still seen as the steel heart of the country. This negative image is often kept alive by the national media, which makes it difficult to change the perception of people outside the region (Suchácek, 2005). Finally, the presence of the heavy industries has a negative effect on the quality of life from environmental problems. The steel mills and coking plants of Vítkovice are for the largest part located in the city of Ostrava, which makes living in the city less attractive (Nesporova, 1998).

The government and the companies have taken various measurements to improve the quality of life in the MSR. The government tries to improve the natural environment by reforesting the landscape and by protecting some natural areas (CZSO, 2005). Air pollution declined strongly, especially in the beginning of the 1990s. According to Eckart *et al.* (2003), the situation in the 1990–1993 period mainly improved by the decline of engineering production and restructuring of the manufacturing sector. Nowadays, it is hard to improve the situation because of rising costs for environmental protection. Especially in Ostrava and Karivá, the core of the mining and steel industry, environmental protection is hardly maintained (Eckart *et al.*, 2003). As a consequence, air pollution remains stagnant, despite the large difference compared to the 1990s (see Table 6.2). Also companies take measures to reduce pollution. For instance, Mittal Steel Ostrava invested heavily in various environmental programmes including modernisation of its facilities for production of coke, sinter, and protection of surface and underground water. The company is at one-tenth of its pollution in 1990 and it plans to invest over 5.5 million CZK in the next three years to reduce pollution even further (Mittal Steel Ostrava, 2007). Nevertheless, the government puts even more pressure on companies to reduce pollution. Mittal Steel Ostrava and other companies in the region suffer from this pressure, since companies in other countries in CEE (like Ukraine) have less strict environmental regulations. This gives the companies in the Czech Republic a competitive disadvantage.

Table 6.2 Air pollution in MSR, specific emission in tonnes/km²

	1995	*2000*	*2001*	*2002*	*2003*	*2004*
Specific emissions (REZZO 1–3)						
Sulphur dioxide (SO_2)	14.2	5.2	5.3	5.2	5.4	5.2
Nitrogen oxides (NO_X)	6.9	4.1	4.4	4.3	4.4	4.3
Carbon monoxide (CO)	49.6	25.0	24.7	23.5	25.8	26.8

Source: CZSO http://www.czso.cz/xt/edicniplan.nsf/t/3E004997E6/$File/81010102.xls

Despite the presence of natural attractions (hills and mountains) and relatively low costs of living (like housing, food and other consumer products), which might attract employees from the automotive industry from other regions, the region seems to be less attractive for certain types of high-skilled knowledge workers as the cultural offering is limited. The central city in the region, Ostrava, is relatively small (about 300,000 inhabitants) and therefore it lacks real metropolitan facilities and atmosphere.

Factor conditions

Compared to other Czech regions, MSR has relatively high wages in the manufacturing sector. In 2004, MSR paid 18,061 CZK as average monthly gross wages of employees, which is above the Czech average (17,014 CZK). Only Prague and Central Bohemia paid higher wages of 22,529 CZK and 18,336 CZK respectively. There are differences within the manufacturing sector. The wages in the Czech automotive industry are about 12% (in 2004) higher than the average of the manufacturing sector, making it the most expensive industry after the chemical industry and the 'coke, refined petroleum, and nuclear fuel industry' (CZSO, 2006). If we compare the wages of the Czech Republic with other countries, we see that the wages in the Czech Republic (€6.03 per hour in 2006; Eurostat data) and other countries in CEE (€5.68 for the ten new member states) are still lower than the European average (€21.00 for EU-27), which can be a competitive advantage for these countries for labour-intensive production, like certain car parts.

Nevertheless, low wages are not sufficient to reach comparative advantage with the labour factor. The productivity of labour is also crucial. Within the Czech Republic, MSR, with €16,216 (in 2004), has the fourth highest annual labour productivity of all eight districts in the Czech Republic on the NUTS level 2 after the districts of Praha (€22,446), Ústí (€17,166) and Central Bohemia (€17,095) (Eurostat data).

Other indicators to compare the attractiveness of MSR as a production location are real estate prices. The average price of dwellings (in 2002–2004) in MSR is 6,605 CZK/m2, which is nearly 40% lower than the Czech average. Only Ústí has lower real estate prices (4,382 CZK/m2). The difference between the average prices for other real estate types (family houses and multi-dwelling buildings) in

MSR (1,175 CZK/m2 and 922 CZK/m2) and the Czech average (1,313 CZK/m2 and 1,118 CZK/m2) are lower and are comparable with most other regions in the Czech Republic. Only Prague, the economic centre of the Czech Republic, has much higher real estate prices (CZSO, 2006).

An important asset for MSR is the availability of free space. Due to the closure of manufacturing plants in the traditional industries, production sites have become available for new production firms or other functions. Especially in the competition for FDI, these brownfields might be an advantage. A drawback of the brownfields, however, are the costs for cleaning the soil, since many brownfields are polluted. This may hinder new investors. Another problem for brownfield development is that the brownfields are owned by many different owners, like local governments and private companies, which all have different interests and resources. For instance, Tatra, a former state-owned truck company owns many plots in Koprivnice. This company is currently owned by a foreign consortium and would like to solve its internal problems before starting to think what to do with unused plots. Other plots are owned by small municipalities which are willing to redevelop brownfield sites, but do not have enough financial resources to do so.

6.3 The automotive cluster in MSR

The automotive industry in MSR is one of the key clusters in MSR. Companies from the automotive industry top the ranking of foreign investments by sector. Firms in the automotive industry are responsible for 17% of the total FDI in MSR (March 2007) (see Figure 6.8).

In this section we first describe the relevant region for the automotive industry, which is much wider than MSR, and the major concentrations of automotive activities in MSR. Second, we investigate two key players, OEMs and suppliers, in the automotive cluster in more detail. Another group of key players, knowledge

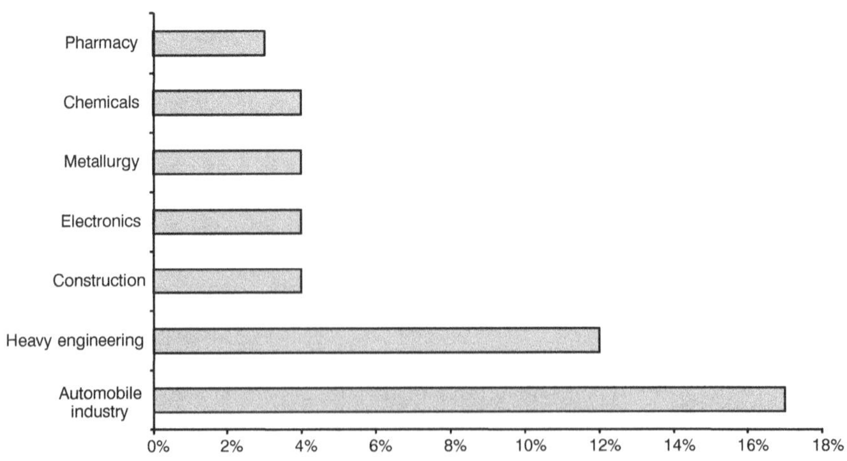

Figure 6.8 FDI in MSR by sector (source: ARR, 09/03/2007)

institutes such as universities, have been described in more detail in section 6.4. The last part of this section describes the major consequences of FDI in the automotive industry on the region.

Relevant region for the automotive industry and concentrations of the sector within MSR

The relevant region for the companies in the automotive industry is not limited to the borders of MSR; many suppliers in MSR also supply to OEMs other than Tatra and Hyundai, which are located across the borders of MSR. GM/Opel has a production plant just across the Polish border in Gliwice; Fiat has production facilities in Tychy and Bielsko-Biala. In addition, Kia has its first European plant in Žilina, just across the Slovakian border. All these production plants are located in less than 100km from Ostrava and can be supplied by suppliers from the region. According to our interview partners, the relevant region is even larger, and encompasses a circle with a radius of about 300km around Ostrava. Seven different assemblers of passenger cars have various production facilities in this circle, with a total annual production capacity of about 3 million cars (interview), see Figure 6.9. The presence of these OEMs offers a potential for all suppliers. The OEMs in MSR (Hyundai and Tatra) can also benefit from the suppliers in the larger relevant region. One interview partner even noted that Ostrava has the potential to become 'the central place in the automotive industry' in CEE.

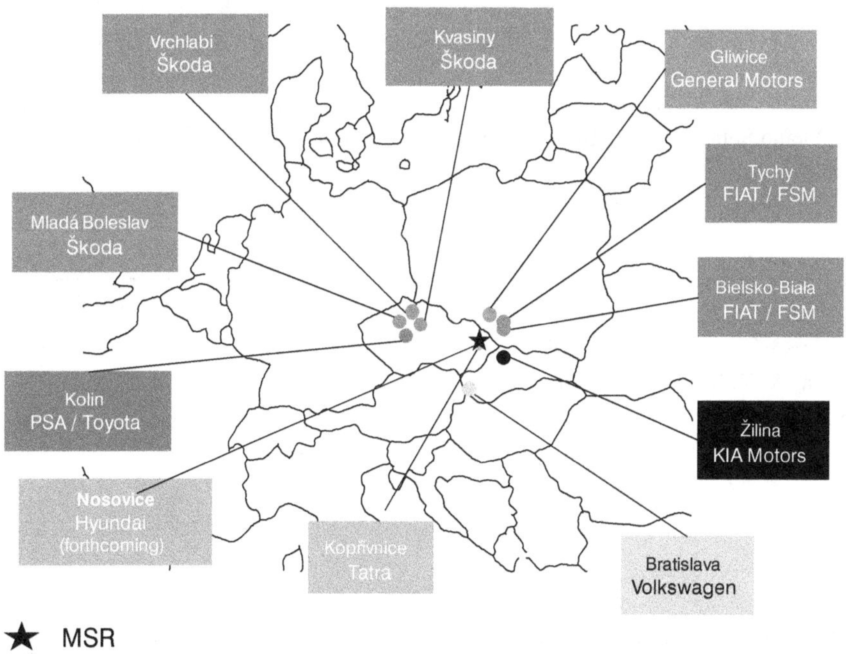

Figure 6.9 The relevant region for the automotive industry (source: own elaboration)

Within MSR, there are some important concentrations of automotive firms. In the first place, this is the location of Hyundai in Nošovice and Tatra in Koprivnice. Hyundai has attracted various suppliers to the region. Three of them are located near the Hyundai plant in Nošovice for logistical reasons.

Koprivnice has a long tradition in the automotive industry because of Tatra. In proximity of the Tatra plant, the town has an industrial zone, Koprivnice Industrial Park, which focuses on the automotive industry. Of the seven companies in the park, only Bang & Olufsen does not have its core business in the automotive industry (see Table 6.3). The land of the industrial park is owned by the town of Koprivnice and the development has been supported by various state institutes (like the Ministry of Trade and Industry) and development agencies. For instance, the production hall of Cirex CZ was constructed by the town of Koprivnice and co-financed by the EU Phare 2000 programme. The total grants received by the park amount to €4.7 million. The companies in the park benefit from each others presence trough cluster advantages, like the presence of customers or suppliers. For instance, one of the reasons for Erich Jaeger to come to the park is the presence of Brose, which is an important customer of Erich Jaeger.

Another important location for the automotive industry in MSR is the capital Ostrava. Ostrava has also developed various industrial zones. One new industrial zone is Ostrava Mosnov. The three companies which have invested in the zone up to now, Plakor, Behr and Cromodora Wheels, are all companies in the automotive industry. The zone is located next to the airport, which might be an advantage for products which require a quick delivery to customers. Spare parts in the automotive industry are a good example of this. Another industrial zone in Ostrava relevant for the automotive industry is Ostrava–Hrabova. Tenants of this zone which are active in the automotive industry are GDX Automotive, Goodyear and Blackner. In addition, various service providers, like logistical companies as UPS and Schenker can offer services to the automotive firms in the park and other locations in MSR.

Table 6.3 Companies in Koprivnice Industrial Park

Company	*Country of origin*	*Opened*	*Investment*	*Employees*
Brose CZ	Germany	2005	€45 m	1,000
Bang & Olufsen	Denmark	2006	€8 m	130
Cirex CZ, s.r.o.	The Netherlands	2003	n.a.	110
Dura Automotive Systems CZ	USA	2002	n.a.	400
Erich Jaeger, s.r.o.	Germany	2005	€2.6 m	200
Roechling Automotive CZ s.r.o.	Germany	2007	€6.5 m	n.a.
Rieger Automotive, s.r.o.	Germany	2007	n.a.	n.a.

Source: own elaboration

Apart from the companies in the Ostrava Mosnov zone and Ostrava, there are various automotive companies spread across Ostrava and other parts of MSR (like Siemens with various plants in Frýdek-Místek and Ford and Visteon/Autopal in Nový Jicín).

OEMs

MSR has been the home location of one traditional Czech OEM (Tatra) for a long time, and since 2008 the Korean car maker Hyundai has produced vehicles in the region for the European market. These OEMs will be discussed in more detail in this section.

Tatra[4] is an old car assembler that was established in 1850. Nowadays, it produces only trucks. The company covers the entire value chain, from development and design of trucks to after-sales and repair services. It has the largest testing track in the Czech Republic, which is designed for testing trucks and special purpose vehicles. The testing facilities (roads and tracks) are also used by many other institutes, research centres and automotive companies. In relation to the testing functions, Tatra is also active in engineering trucks and truck parts. All the companies and the testing facilities are clustered on one site in Koprivnice. Tatra also offers after-sales and repair services, and therefore, the company has a special training school that trains employees to repair Tatra cars. The training school is open to Tatra workers as well as to workers from other companies. The truck company is also active abroad. For instance, since 1997, Tatra has a joint venture plant, 'Tatra Vectra Motors Limited' (formerly 'Tatra Trucks India Limited'), together with Vectra Group, a British industrial consultancy company. Tatra has also various sales offices abroad in China, India, Russia and Slovakia. These sales offices are important since these countries are important growth markets for vehicles.

Tatra, as a former state-owned company, still has a large influence in its home town, Koprvnice, and the development of both is related. Tatra is the largest employer of the city and owns nearly half of the plots in the city. The company more or less created the city. After the Velvet Revolution (1989), Tatra declined, resulting in vast unemployment. Since that time, unemployment has been fluctuating depending on the performance of Tatra, which has had different owners over time. Since 2006, Tatra is owned by an international consortium Vectra Limited (a British company) and a public actor, and is growing again. One of the largest problems the company presently faces is to find skilled labour, the same problem faced by many other companies in the region.

Hyundai invested €800 million to €1 billion in a production plant in Nosovice, in the southern part of MSR. This is the largest (foreign) investment in Czech modern history. The plant will have an annual production capacity of 200,000 vehicles and will build two models. In 2011, in the second phase of production, the annual production will be extended to 300,000 vehicles by a third model (Hyundai, 2007b; Eurofound, 2006). The investment is limited to production facilities; the company does not consider investing in R&D and design facilities and the majority of the employees will tend to be manual workers. Nevertheless, the firm needs some process engineers to control the production facilities and employees in services functions. Currently, Hyundai has an office in Ostrava for basic services.

One of the reasons why Hyundai has chosen MSR as a production location, is the strong supplier base in the wider region and the proximity of a Kia plant in the Slovakian Žilina region, about 90km from Nosovice. Kia, part of Hyundai Group, opened its site in Žilina in 2006. The site includes the Kia factory, a 3km test track, a training centre, two parts plants operated by Kia's suppliers, and road and rail linkages to other parts in Europe. The total investment costs of the site are €1.3 billion (BBC News, 2007). In addition to the two suppliers on the Kia site, other (Korean) suppliers followed Kia to or near to the Žilina region. In total, 11 key Korean-based suppliers to the Hyundai-Kia Automotive Group serve the Kia plant, and another three have announced they will set-up operations in the wider cross-border region (Hyundai, 2007b). For instance, the Korean body supplier Sungwoo Hitech in Ostrava delivers 23 different body parts to the Kia plant in Žilina, which produces 700 cars per day (interview). The Korean suppliers which supply Kia can also be used by Hyundai, and Hyundai can also share other facilities of Kia.

Hyundai also benefits from the improving transport connections in MSR, and it receives incentives of 15% of the total investments, the maximum permissible under EU regulation. The state provides Hyundai direct subsidies to acquire cheap land and to create jobs and it offers the company tax relief. The government also finances the infrastructure for the new plant, including the costs related to the construction of roads and railways demanded by the Korean car maker. For instance, the state will construct an expressway connection from Nosovice to the Slovakia border (which is important for the cooperation with the Kia plant in Žilina) and plans to connect the Hyundai site with the Czech motorway network (Eurofound, 2006).

The wider Ostrava region is only one node in the global network of Hyundai Group that aims to have all parts of the value chain in each regional market (which means per continent). It has its global headquarters in Korea, just like a number of production plants and major R&D and design centres. Outside Korea, Hyundai has regional headquarters in each continent (like in Dubai for Africa and the Middle East), production plants spread across the world, and several R&D and design centres in Europe, the USA and Japan (Hyundai, 2007a).

The production plants of Hyundai Group in Žilina (Kia) and Nosovice (Hyundai) are seen as the fourth 'pillar' of the group in Europe next to Hyundai Motor Europe GmbH (European headquarters, marketing, sales, after-sales services, and technical training), the European Design Centre and Hyundai-Kia R&D Centre. It is notable that production follows R&D and design (which were in Europe before the production plants), and that the production function can be separated geographically from the other functions. All functions other than production and sales (management, marketing, R&D, design, etc.) are located in the greater Frankfurt region.

Suppliers

MSR has a long tradition in the automotive industry through Tatra and currently it has a strong supplier base, including suppliers from all tiers. The suppliers can be categorised in three groups based on the origin:

- *Traditional Czech suppliers.* These suppliers started to produce parts or systems to Škoda and Tatra, the Czech car makers, but are increasingly sourcing to other OEMs and tier-one suppliers. The most successful companies, like Brano, have grown and are integrated in large groups with several establishments in the Czech Republic, and increasingly also abroad. The companies have production, but also development tasks. It is worth mentioning that the two major steel companies in the region, Vítkovice Steel and Mittal Steel Ostrava, do not supply to the automotive industry.
- *Korean suppliers that follow Kia and Hyundai to Europe.* Hyundai Group prefers to have its own Korean suppliers, since it has had negative experiences with local suppliers in Canada. The suppliers start with production facilities in a first stage of the investment in order to supply parts to Kia, but in later

Box 6.1 Sunwhoo Hitech: an example of follow sourcing

Sungwoo Hitech (Sungwoo) had no intention of coming to Europe, but it was asked by Kia to come to Europe, because Kia was not able to find good local body suppliers. Before, Kia had negative experience with local suppliers in Canada and therefore it changed its strategy which includes the use of key Korean suppliers abroad. This strategy works well in India (where 13 Korean suppliers followed Kia) and therefore Kia uses the same strategy in Europe. In the beginning, Sungwoo refused to come, nevertheless later the body supplier decided to come. This relative late entrance gives the company a trail to other Korean suppliers which came some years earlier, and therefore Sungwoo has to set up the plant as quickly as possible and start production. Later Sungwoo will also open an R&D centre on its site in order to develop parts for the Kia model which Kia develops for the European market.

Sungwoo investigated various locations for the production plant in Poland, Slovakia and Czech, with a maximum distance of 100 km from the Kia plant. Ultimately, it has chosen Ostrava as production location, because of the labour pool (a high unemployment rate) and incentives offered by CzechInvest. Another advantage is the favourable geographical location of Ostrava; with the presence of seven OEMs within a radius of 300 km.

stages, the companies intend to open R&D centres to develop specific parts for the European car models. An example of a Korean supplier that followed Kia to Europe is Sungwoo Hitech (Box 6.1).

- *Other foreign suppliers*, especially from the USA and Germany, which come to MSR to supply to the OEMs in the larger region. The suppliers benefit from relatively low wages and often receive governmental incentives to open production facilities in the region. Just like the Korean suppliers, the other foreign suppliers start with production in the Czech Republic in the first stage, and can open R&D centres at a later stage. Two major companies with an R&D centre next to the production plant in MSR are Visteon (Autopal) and Siemens.

Impacts of FDI on the region

The investments by Hyundai and other firms have an enormous impact on the region. First, the investments generate a large number of jobs. It is estimated that Hyundai creates 3,500 direct jobs and another 4,000 indirect jobs at supplier companies (Hyundai, 2007b). The government provides incentives to companies (like subsidies) to attract FDI. One of the requirements for the companies that want to use the incentives are job guarantee contracts, which force the companies to create a certain number of jobs. It is also important to note that some foreign companies prefer to use local employees over foreigners for cost reasons. For instance, Sungwoo Hitech aims to limit the number of Korean employees, since they cost more than local workers. The supplier trains local employees in Korea in order to give them the required skills. This is important for the upgrading of the cluster. Furthermore, it can be expected that higher skilled workers who have grown up and studied in MSR, but work in other regions, will come back to the region. Before the massive (foreign) investments in MSR, many higher skilled workers left the region because they could not find a job. For instance, when PSA and Toyota opened a plant in Kolín, it attracted many automotive engineers from MSR who did not have job opportunities in their home region. Nowadays, when Hyundai and the suppliers invest in MSR, it is expected that the employees will come back to their home region to fill the vacancies of the new companies. Second, another positive consequence of the FDI is an improvement of the area's image. The presence of Hyundai, a modern and global player, in the region can contribute to the area as an automotive region. This might help to attract other firms in the automotive industry and to improve the international network of actors in the Ostrava region. In addition, it may help to improve the general image of the region as 'the steel heart of the country'.

The FDI has also a downside on the labour market as the demand for employees, especially for higher skilled labour, is larger than the supply, and it has led to rising prices for employees (wages) and to competition between firms for the workers. In general, there is a shortage of high-skilled labour, especially for automotive engineers. The development of human resources, in order to solve the shortage of engineers, is a key issue for the cluster organisations. The government,

VŠB Technical University Ostrava and companies cooperate to train youths through technical education and promote the manufacturing sector among youths. Moreover, there is governmental support to bring highly skilled workers who left the region, back to the region. This is also possible due to the higher competitive wages offered by the foreign firms. For current students, there are more reasons to stay in the region after they finish their studies because of job opportunities and competitive wages.

6.4 Local and global networks

As previously stated, MSR is an important region for automotive production; it concentrates various production plants of OEMs and suppliers. Upgrading in the region also takes place through investments in R&D facilities and increasing cooperation between the automotive industry and universities. The actors in the automotive industry in MSR are involved in various local and global networks. In this section we investigate relevant production and knowledge networks, and the upgrading process of the region in more detail.

Production networks

The relevant region for most of the suppliers is a large 'region' consisting of (parts of) Czech Republic, Poland and Slovakia. The role of proximity differs by supplier and depends on various factors, like the weight of the product. For instance, Siemens VDO produces various sensors for the automotive industry. It produces the sensors in a plant in Frenštát pod Radhoštem in the northern part of MSR and exports the sensors to OEMs everywhere in Europe. Ten to fifteen per cent of sales are exported to countries outside Europe. For light products like sensors proximity does not matter, but for other suppliers, proximity to the customers is important. For example, Brano (a local Czech supplier) has offices near its major customers: in Wolfsburg (Germany) near the headquarters of Volkswagen and in Detroit near the headquarters of the 'Big Three', GM, Ford and Chrysler. In addition, it has an office in Bratislava (Slovakia) near a production plant of VW.

In general, OEMs and large suppliers have global supplier networks, and source from companies in the region as well as from companies in other parts of the world for cost reasons. For instance, on the one hand, Sungwoo Hitech uses local suppliers for large parts. Therefore, for logistical reasons, in MSR it searches for local tooling companies to stamp body parts, which can replace tooling companies in Korea (the use of local suppliers decreases the transport costs). In addition, Sungwoo wants to outsource some tasks to local suppliers. On the other hand, it sources smaller parts from China. Also, Siemens VDO purchases its components globally through a special purchasing company that buys parts for the entire group. Only connectors are bought in neighbouring countries like Slovakia; other parts are bought from suppliers farther away. Printed circuit boards, for example, are all sourced from Asia because of low labour costs.

Local Czech suppliers have become less dependent on the Czech OEM, Škoda. First, many suppliers used to deliver only to Škoda; now, they increasingly deliver to other brands of the VW Group (which owns Škoda) and other OEMs. Nevertheless, the VW Group is still the major customer for many Czech suppliers. For instance, the VW Group has a 36% (including 11% to Škoda) share of the sales of Massag, a supplier of various car parts located in the Ostrava region. Brano, another Czech part supplier, sells 47% to the VW Group (including 17% to Škoda). Korean suppliers that follow the Hyundai Group to Europe, are able to sell their products to other OEMs in the larger region. For example, Sungwoo Hitech, which has opened a plant in Ostrava to deliver car bodies to Kia in Žilina and Hyundai in Novisec, considers delivering to other OEMs outside the Hyundai Group. For this, it intends to open satellite plants near other OEMs like GM in Gliwice.

Through competition within the automotive sector, OEMs force suppliers to reduce costs. For instance, Brano's customers force it to reduce its costs by about 3% per year. After a certain period, OEMs evaluate the performance of their suppliers. If the suppliers' performance is too low, and the suppliers can not fulfil the requirements of the OEM (for instance, the supplier can not reduce the costs), the OEMs choose other suppliers. Furthermore, a high product quality is important for the relations between OEMs and suppliers. Suppliers ask for high quality from their lower tier suppliers and are also responsible for their products. Therefore, tier-one suppliers like Siemens VDO are using Supply Quality Management systems in which they check a certain portion of the incoming goods. Sometimes entire batches are checked. In addition, tier-one suppliers have special purchasing teams that visit lower tier suppliers to check the production facilities of their suppliers. It is important to note that tier-one suppliers also teach lower tier suppliers; for instance Siemens VDO teaches its lower tier suppliers certain production techniques.

Knowledge networks and upgrading

MSR increasingly receives investments in R&D facilities, and cooperation between companies and local universities has grown. These investments and the development of knowledge networks in the region are important to upgrade the region from a production cluster to a more advanced and knowledge-intensive one. For the knowledge networks in MSR we use two types of relations: relations between the automotive industry and universities, and relations between OEMs and suppliers for product development.

In the region, there is a long-standing cooperation between the VŠB Technical University Ostrava ('Technical University') and the industry. The Technical University, which started as a mining school, has been cooperating with the mining industry since its establishment in 1950. Currently, it cooperates also with other industries, like the automotive industry. It has three forms of cooperation with the industries:

- *Bilateral agreements*. The main topic in these agreements is the development of human resources (HR). This is an important topic given the shortage of qualified (technical) employees in the region.
- *Applied or contract research*. The technical university conducts research for industrial partners in various technical fields. In addition to the contract research, the Technical University also offers masters courses for workers in industrial firms.
- *Support of clusters*. The Faculty of Mechanical Engineering of the Technical University is closely involved in two clusters: mechanical engineering and automotive. The dean and vice-dean are on the board of the clusters and take part in the creation of the vision and planning of the cluster. The clusters receive financial support from CzechInvest. Universities can do research projects with several companies instead of with only one. In addition, the university organises workshops to show the industry what research can be done by the university.

The mechanical engineering cluster is a traditional cluster in the region. It has nearly 15 members, from old established companies like Vítkovice and Mittal Steel, but also from the automotive industry such as Visteon and Brose. The Faculty of Mechanical Engineering has various research projects (on a department level), which are divided in two flows: design (robotics, production, machines and transport equipment) and technology. It is notable that the priority in the mechanical engineering cluster is HR development and that research projects are relatively new. One of the tasks of the cluster organisations is to improve the image of the sector for youths. Within the clusters, the industry and the Technical University organise several activities, like company excursions, to introduce youths to the sector, and events like a 'day with mechanical engineering workers'.

For the automotive cluster, the Technical University cooperates with various firms in the automotive industry. It cooperates with local firms as well as with foreign companies. Via CzechInvest, foreign companies get in touch with the Technical University even before the foreign companies have invested in the region. Currently, the university has good relations with suppliers in the automotive industry like Visteon and Brano, but there is no cooperation with Hyundai yet. The cooperation with the companies includes:

- *Companies deliver research topics to the university*. Two current research projects include: (1) a project with various firms in the automotive cluster to measure noise of automotive components, and (2) crash tests for which a special lab is provided by Visteon;
- *Companies provide money for new labs*. For instance, Siemens has modernised a laboratory of programmable robots and distributed control systems which is equipped with Siemens products from the field of industrial automation systems and will serve for the practical education of students (Siemens, 2006);
- *Companies organise excursions for students*;

- *Companies support lectures, give guest lectures and guide diploma sessions.* For example, Brano supports practical lectures in which teachers of the Technical University can use Brano's machinery. In addition, managers of Brano give guest lectures at the technical university in Ostrava.

A problem for Technical University Ostrava, as with many other universities, is the lack of equipment/labs. There is often no money for laboratories. A positive development is that universities increasingly receive money from the state and the EU for investment in equipment. In addition, the Technical University tries to increase its cooperation with companies. A problem, however, according to one of our interview partners, is the lack of a good legal scheme in which companies can support universities.

Finally, it should be noted that the cooperation between firms and universities is not limited to cooperation with Technical University Ostrava. For instance, Siemens VDO has research projects with the Technical University in Ostrava for research on sensors, and with the Technical University in Žilina (Poland) it works on plastic techniques. Another example is Brano, which cooperates with a university in Lille (France) to help solve the shortage of educated people in technical fields.

Relations between OEMs and suppliers and regional upgrading

Investments in R&D facilities and participation of firms in development projects contribute to regional upgrading. In section 6.3, we have seen that Hyundai mainly invests in production facilities in MSR and concentrates its R&D and design facilities for the European market in Germany. Although Hyundai does not consider opening R&D and design facilities in MSR, many suppliers increasingly invest in R&D and design centres in the region. Often, the (foreign) suppliers start with production facilities at a first investment stage, and later they also invest in R&D and design facilities. Suppliers benefit from the tradition of the Czech Republic as an engineering country. The increasing investments in R&D and design facilities following investments in production plants are important for the upgrading of the cluster, which is changing from a pure manufacturing cluster to a more knowledge-intensive production cluster that also includes development and sometimes even research activities. R&D is seen by various actors as a 'logical step after production'. A good example of a firm that has upgraded its facilities in MSR is Siemens VDO, which started as a producer of electric motors and now has an important development centre for sensors in the region, as well as the global management for the sensor division (see Box 6.2).

Another company that plans to invest in an R&D centre is Sungwoo Hitech, a Korean company that followed the Hyundai Group to Europe. Hyundai and Kia develop special models for the European market in R&D and design centres in Germany. Kia focuses on safe vehicles (it aims to have the best scores in European crash tests), and therefore it sets high standards for its suppliers. Sungwoo needs to have an R&D centre in Europe, on its site in Ostrava, to develop the special

Box 6.2 Siemens VDO Frenštát as an example of 'upgrading'

Siemens started its operations in Frenštát pod Radhoštěm with a production hall for the production of electro motors. In the first three to five years, the products were developed in Germany and the same product was produced in Frenštát, without development of the products in the Czech location.

Later, Siemens closed a division in Dortmund and transferred 200 jobs to Frenštát. Also a development centre was closed in Dortmund and Siemens started with three employees from the Dortmund centre to open a development centre in Frenštát. This development centre uses the same high-tech machines as comparable development centres of Siemens VDO in Germany and France.

Currently, Siemens VDO has 66 R&D workers: 50 in the Sensor division, 4 in the Electronics and Drivetrain division and the other 12 are active in the Interior Control division. The major activity in the development centre is the development of sensors. In addition to development and production of sensors, Siemens VDO uses the location in Frenštát also for the global management of the sensor division.

Although, the core competence of Siemens VDO in Frenštát is still manufacturing (it has three production plants which cover 40% of the total site), it has a very complete site including production, logistics, R&D, facilities and offices for administration and management.

Source: own elaboration

parts for the European Kia models. Currently, the company has a small R&D department, but a special R&D centre will be set up in the future as soon as the production facilities are finished. The R&D centre cooperates intensively with the main R&D centre in Korea. It is important to note that Sungwoo prefers hiring local engineers above Korean engineers for cost reasons. Currently, there are about 30 Koreans in the Sungwoo site in Ostrava, but that number is likely to decrease, together with a global increase of qualified Czech engineers. To educate local workers and engineers, Sungwoo trains them at its home base in Korea. This is important for the upgrading of the region.

Relevant for the knowledge networks are the relations between OEMs and suppliers regarding product development. Often, OEMs involve suppliers in the development process of new parts. If an OEM starts with the development of a new model, it asks some suppliers to develop a prototype of a certain part. For individual suppliers, this is a risky operation, since the supplier pays all the development costs of a prototype until it is accepted by the OEM. However, also the OEM runs the risk of not finding a suitable supplier. The suppliers work with the OEMs on projects for the development of a certain part. Later if a certain product is successful, it can also be sold to other OEMs. In addition, sometimes current products can be used in new models.

6.5 Policy and cluster organisations

The national, regional and local governments have various programmes to stimulate economic development in the Czech Republic and its regions. Some of these are also relevant for the automotive industry in MSR, for instance programmes that offer additional funding to attract foreign firms to the region. Notable is the fact that in many programmes EU structural funding is being orientated towards competitiveness and economic development initiatives.

This section first discusses the role of various governments and their policy programmes, followed by the role of the automotive cluster major organisation Moravian-Silesian Automotive Cluster ('Moravskoslezský automobilový klastr', MAK).

The government and policy programmes

For the discussion of the policy programmes, we investigated the role of the government on different geographical levels: state, regional and local.

The state plays an important role in economic development of the country, in particular by attracting FDI. FDI is usually attracted by CzechInvest, a business and development agency under the Ministry of Industry and Trade. The agency supports foreign investors by providing linkages with structural funds of the EU, state aid, and by providing information to potential foreign investors. Another programme run by the Czech national government which is relevant for the manufacturing sector is the Operational Programme Industry and Enterprise 2004–2006 (OPIE),[5] which is also carried out by the Ministry of Industry and Trade. The goal of OPIE is to develop a competitive production base to strengthen economic performance and to boost structural changes in the economy. The OPIE 2007–2013 does not have special allocations for individual regions[6] or for individual sectors. Therefore, it depends on the absorption capacity of individual firms and municipalities in the regions, how much financial resources will be received in the regions. The companies in the three classical clusters (wood, automotive and engineering) in MSR will receive 45 million CZK from OPIE from 2007. In these clusters, the companies cooperate to identify common problems and lobby for funding to solve the common problems. Currently, the priority of all processing industries in MSR is to develop a technically qualified workforce to meet their needs (ARR, 2006).

The national government is responsible for attracting FDI via incentives provided through CzechInvest. Lower governments (the region and local governments) do not provide incentives, but are responsible for implementation of programmes. They prepare the land and develop infrastructure and are responsible for development of industrial zones. On the lower levels, policy is often laissez-faire and good coordination between various actors is often lacking.

One actor at the regional level in MSR is the Regional Development Agency Ostrava ('Agentura pro regionální rozvoj', ARR). The development agency acts as a consultancy firm and prepares strategic plans for municipalities. There is more than one global vision and strategy and large cities develop strategic plans by themselves. Nevertheless, this is not seen as a problem. ARR, in cooperation with the regional government and other actors, also provides information for (potential) investors outside the region who want to invest in the region or who want to do business with companies in the region. This is done via a special Internet portal (www.dynamicregion.cz) and is part of the official regional marketing strategy. Finally, ARR is involved in various regional development initiatives, like ForTech and ClusterNet (an initiative which creates a common platform for assisting cluster managers of the individual clusters, sharing information and methodology).

Local governments (cities and municipalities) can use ARR to support them in the development of local strategic planning and vision, or can do it by themselves. Often these visions and strategies are independent of the regional vision or other municipalities' strategy, with poor articulation. Even within individual municipalities/cities, there is not always a joint strategy. For instance, in Ostrava, plans to develop new technological infrastructure and business parks are not taking into account the strategy and aims of the already established Ostrava Science and Technology Park.

MAK

MAK was established in 2006. The cluster has grown from 22 members at the start to 35 in October 2007, and further growth is expected. MAK's goal is to create a competitive environment for SMEs in the automotive cluster. The cluster organisation aims to involve all actors of the 'theoretical cluster' and attempts to cover the entire value chain, including tier-one suppliers, lower tier suppliers and knowledge institutes (like the Technical University Ostrava). The cluster organisation focuses on SMEs and small suppliers, but it is also open for OEMs and larger suppliers. In addition, companies from different industries can join; companies with all activities in the automotive industry, but also companies with a lower share in the automotive industry but with potential in the automotive industry, can join. Finally, the cluster organisation is also open for companies outside MSR, like TÜV SÜD Auto which is located in Prague.

Currently, MAK's main challenge is to improve the quality of the labour pool to solve the shortage of automotive engineers. A way to do this is to offer technical education. Migration of workers from other regions to solve the shortage on the labour market has not been seen as a complete solution, only a short-term solution, because wage differences between regions decline and employees will return to their home region. For example, PSA/Toyota sources employees from MSR to solve the shortage of employees in the Kolín region. Now that Hyundai and several suppliers are opening plants in MSR, PSA/Toyota fears that the workers will go back to MSR to work in their home region. A related challenge for MAK is to improve the perception about working in automotive plants, especially among youths. In this way, youths can play a role in reducing the shortage of automotive engineers. Other contemporary challenges of MAK include: finding connections with and learning from other European cluster organisations; preventing cultural problems between European actors and actors from Korea; increasing the cooperation between industrial partners and research institutes. The largest hindrance for the cooperation between industry and research is the time frame. Time is less crucial in university research. In the industry, in contrast, R&D have short lead times because of the fierce competition between car makers. Up to now, MAK has only one research institute as a member: VŠB Technical University Ostrava.

6.6 Conclusion

The Czech Republic has a long tradition as a manufacturing and engineering country. One of the major industrial regions in the country is the Moravian-Silesian Region (MSR), the 'steel heart of the Czech Republic', for its large base of mining and heavy industries. Since the beginning of the 1990s, all major industries (coal mining, coke making, metallurgy and heavy engineering) in MSR have been declining, resulting in unemployment rates far above the national average. The government and various stakeholders from the industry and knowledge institutes have taken various measures to change the tide and to restructure the region towards a more diverse and knowledge-intensive economy.

The national government has developed two policy tools to guide the restructuring process: cluster policy and attraction of FDI (via CzechInvest). Both policy tools are implemented at the regional level, however. In its cluster policy, the region has selected three 'traditional' industries and three modern industries to spearhead the future economic base. One of these 'traditional' industries is the automotive industry, an industry with which MSR has a long tradition through the presence of Tatra and various suppliers. The region is successful in attracting FDI for the automotive industry, especially with the investment of Hyundai in a production site, but also many foreign suppliers have chosen MSR as the location for their production plants.

In this chapter, we have analysed the dynamics of the automotive industry in MSR, which has emerged as one of the Czech Republic's major automotive clusters. A specific aim of the study is to analyse the evolving functional and spatial relationships between production (manufacturing) and other activities in the value chain, in particular R&D activities.

In the remainder of this section, we summarise the main findings of our research and draw some conclusions. First, we analyse the development of the automotive industry and the link between the automotive industry and the regional foundations. Second, we look at the structure of the cluster and production and knowledge networks.

Development of the automotive industry and conclusions on foundations

The FDI has an enormous impact on economic development and upgrading of the region. First, it has an impact on the *economic base* of the region. MSR's economic base is currently relatively weak. The region has a relatively high unemployment rate and a one-sided economic structure dominated by manufacturing industries, but much progress has been made through restructuring and FDI. Many jobs are being created (especially due to FDI in the automotive industry), and the growth of the manufacturing sector also gives a boost to the service sector. Although the economic structure is still one-sided and the share of services is relatively low, there are growing possibilities to link the automotive industry with other industries like IT, steel and hydrogen.

Second, FDI has impact on the *knowledge base* of the region and contributes to upgrading. Although Hyundai mainly uses the MSR site as a production location, suppliers increasingly invest in R&D facilities and train local workers and suppliers. A best-practice is Siemens-VDO, which evolved from a 'simple' production location to a research and global management centre of the sensor division. Local firms also invest in improved products and production processes, as a result of pressure from large car assemblers and first-tier suppliers, but also with the help of these firms. Moreover, foreign and local firms increasingly cooperate with local research institutes.

The large FDI also has a downside. The large investments in the automotive industry have resulted in competition for high-skilled engineers. This increases the shortage of technical engineers, which is also caused by the negative image of manufacturing, in combination with the low job perspectives in the sector in the recent past. Furthermore, the full potential of the knowledge base is not used. Partly because the (technical) universities are still focusing on research in traditional industries; and courses are often in Czech, which keeps foreign students from studying there.

Important for the *accessibility*, is the favourable geographical location of the region bordering Poland and Slovakia, which are also part of the relevant cross-border automotive cluster. This cluster consists of seven different car assemblers and multiple suppliers located in a radius of 300km around Ostrava (see Figure 6.9). Recently, through the construction of the D47 motorway and upgrading of a main railway line, transport connections to other regions are improving, which is an important development for automotive logistics, facilitating just-in-time deliveries. In contrast, the facilities for passenger transport are still relatively low (no HST connection and regular flights from the airport to only two destinations). This makes MSR less attractive for the higher parts of the value chain, like R&D. These functions are increasingly done in international project teams, which require quick connections for knowledge workers to other regions.

We have found that, although progress is made, manufacturing negatively influences the *quality of life* in MSR. Air pollution declined steeply in the 1990s, but the region still suffers pollution from the heavy industries, in particular in the cities of Ostrava and Karviná. In addition, due to its relatively small size, Ostrava, the central city of the region, does not have large cultural attractions. This makes the region less attractive for high-skilled knowledge workers, like designers or top engineers, who are still much attracted to Prague. Nevertheless, the mountains and the hills in the region offer the possibilities to attract (foreign) workers and visitors.

For the *factor costs*, we have seen that the Czech Republic still can benefit as a location for cheap production due to relatively low labour costs. Most of the factor costs in MSR are comparable with the Czech average, also making MSR attractive as a location for labour intensive production. An additional asset for the region is the large-presence of brownfield sites, also within the city borders. These sites can be used to attract new manufacturing and service industries, if properly redeveloped.

Cluster structure and networks

The automotive industry in MSR is a mixture between old established firms like Tatra and various domestic suppliers and new (foreign) firms that are investing in the region. The cluster is growing rapidly through FDI, not only through Hyundai, but also through suppliers. Many suppliers have chosen MSR as a production location (and in later stages an R&D location), and are benefiting from the advantages of the region. First, there is a long tradition in the automotive industry, and Tatra and several Czech suppliers have created a specialised cluster. Second, there is a long-standing cooperation between the knowledge institutes and the industry. Since its establishment, the Technical University cooperates with the industry. It started with cooperation with the mining industry, but today it is increasingly cooperating with automotive and new industries. Third, there is strong support from the state for regional development. CzechInvest provides various incentives which are used by foreign companies. Fourth, the region has a favourable geographical position in the relevant automotive region; as noted before, within a radius of 300km there are seven different OEMs and there are numerous suppliers serving these OEMs. In addition, CEE is one of the growth markets for the automotive industry. Finally, transport connections have been improved; in particular, the development of the D47 highway, which connects the region with the western part of the Czech Republic and with Poland, has been seen as a major improvement. Nevertheless, it should be noted that the infrastructure in the Czech Republic is behind other countries in CEE and development takes time. Moreover, better international connections are desirable, in particular for passengers, and are a key requirement to attracting higher parts of the value chain, like R&D.

From analysing the cluster structure in our analysis framework, we have distinguished two types of networks: production networks and knowledge networks. One of the aims of this research is to investigate the role of proximity, i.e. to find out to what extent the actors in the cluster make use of regional or global networks. In *production networks*, this differs per case and per product. On the one hand firms are sourcing from countries with lower factor costs than the Czech Republic (for instance Siemens VDO sources small parts for its sensors from China), on the other hand, we see that firms want to be located near their suppliers for logistical reasons. Important for the development of the cluster is follow-sourcing; various Korean suppliers have opened production plants in the larger region to supply car parts to the Kia plant in Žilina (Poland) and Hyundai in MSR. Follow-sourcing can lead to additional economic growth when suppliers make use of local suppliers and local employees. For instance, Sungwoo Hitech, a Korean supplier of car bodies, is trying to find local tooling firms for stamping body parts, which can replace tooling companies in Korea for logistical reasons (the use of local suppliers reduces the transport costs). The company also aims to maximise the number of local employees, since they are cheaper than Korean ones. Another positive development in the production networks is that local Czech suppliers are less dependent on Škoda, although the VW Group (which owns Škoda) is still the major customer of many suppliers. This makes the local firms less vulnerable to the performance of one firm. In addition, Korean firms that followed Kia and Hyundai to Europe have the possibility to sell their products to other OEMs in the region, which makes them less dependent on the performance of the Hyundai Group.

For *knowledge networks*, which are essential for further upgrading of the region, we have investigated the relations between the (automotive) industry and knowledge institutes. There has been a strong cooperation between the local industry and knowledge institutes for a long time. The Technical University Ostrava has been cooperating with local mining firms since its establishment as a mining school in 1950. Nowadays, the Technical University still cooperates with local companies from the traditional industries through joint research projects and by solving joint problems like the shortage of technical engineers. The Technical University also cooperates with firms from the automotive industry in the form of joint research projects and education, relevant for the (future) workforce. In addition, via its membership in the cluster organisation MAK, the Technical University is involved in solving common problems like the shortage of engineers. Nevertheless, there are some barriers which could hinder the cooperation between industry and universities. One barrier is the lack of a good formal scheme in which companies can support universities. Another barrier is that education is still largely in Czech, whereas the automotive industry is very international.

Notes

1 Eurostat databases, online available at: http://epp.eurostat.ec.europa.eu/portal/page/portal/region_cities/regional_statistics/data/database (accessed 13 September 2007).
2 During the writing of this report, the first part of the motorway (between Ostrava-Rudná and Bohumín) was opened, but other parts are expected to open (much) later.
3 http://www.airport-ostrava.cz/en/page-statistics/ (accessed on 13 September 2007).
4 Sources: interviews and http://www.tatra.cz/en/spodek_en.asp (accessed 16 October 2007).
5 Sources: Ministry of Trade and Industry (MPO), 2007 and http://www.mpo.cz/en/business-support/opei/ (accessed 13 September 2007).
6 Except for the programme 'Development' which will be implemented only in the most structurally affected regions.

Bibliography

ARR (Regional Development Agency Ostrava) (n.d.) *The Moravian-Silesian Region is ready for doing business*, Ostrava: ARR.

—— (2006) '45 million CZK grant for clusters', press release, 6 December 2006.

—— (2007a) 'Car producers top the table of investors in the region', press release, 9 March 2007.

—— (2007b) 'Region sees record-breaking fall in unemployment', press release, 9 November 2007.

BBC News (2007) 'Kia investment boosts Slovakian growth', available online at: http://news.bbc.co.uk/1/hi/business/6298651.stm (accessed 6 February 2007).

Břusková, P. (2007) 'Industrial strategies – impact on the regions. The case of Morevian-Silesian Region (MSR)', presentation given during AGORA conference Brussels, 31 May–1 June 2007, available online at: www.eurada.org/doc/AG07/Pavla%20Bruskova.ppt#257.

City of Ostrava (2006) Fact sheets Ostrava.

Czechlnvest (2007) 'Regions of the Czech Republic', Factsheet No. 21, available online at: www.czechinvest.org/data/files/fs-21-regions-of-the-czech-republic-85-en.pdf (accessed 18 March 2010).

CZSO (2005) 'Regional portraits – Moravsko-Slezský region', available online at: www.czso.cz/csu/2004edicniplan.nsf/engt/8C0028FD4E/$File/136304a19.pdf (accessed 18 March 2010).

—— (2006) *Statistical Yearbook of the Moravskoslezský Region 2006*, Ostrava: Czech Statistical Office, Regional Office Ostrava.

Eckart, K., H. Kowalke and J. Mazeland (2003) *Social, economic and cultural aspects in the dynamic changing process of old industrial regions: Ruhr District (Germany), Upper Silesia (Poland), Ostrava Region (Czech Republic)*, Münster: Lit.

Eurofound (2006) 'Hyundai plans major greenfield investment', *EIROnline*, April 2006, available online at: www.eurofound.europa.eu/eiro/2006/04/articles/cz0604029i.htm (accessed 18 March 18, 2010).

Hyundai (2007a) *Annual report 2006*, Seoul: Hyundai.

—— (2007b) 'Hyundai motor manufacturing Czech starts works on new plant construction', press release, 25 April 2007.

Mittal Steel Ostrava (2007) 'Mittal Steel Ostrava meets significantly tighter emission limits', press release, 2 April 2007.

MPO (2007) *Annual implementation report for 2006*, Prague: MPO Structural Funds Department.

Nesporova, A. (1998) 'An active approach towards regional restructuring: the case of Ostrava, Czech Republic', in: Kean, M. (ed.) *Regional development and employment policy. lessons from Central and Eastern Europe*, Geneva: ILO.

Siemens (2006) 'Siemens modernizes laboratory at VSB-Technical University in Ostrava', press release, 3 April 2006.

Suchácek, J. (2005) 'Regional decline and restructuring in Ostrava agglomeration and Katowice conurbation', Paper presented at the 45th Congress of the European Regional Science Association, Amsterdam, 23–27 August.

7 Paris: automotive industry

7.1 Introduction

With a production of 3.3 million cars (2003), France is the second largest European car producer after Germany (5.1 million). In France, the Paris region (Île de France) is one of the main locations for the automotive industry. In the region, more than 1 million cars are produced per annum, and the industry employs some 100,000 people. This makes it the most important manufacturing industry in Paris.

The automotive industry is characterised by a high level of dynamism in several respects. Technological development is fast, competition is increasing (leading to immense pressures to reduce costs, improve quality and develop new models more quickly), and functional and spatial divisions of labour are under constant change. What is the position of the automotive industry in Paris in the light of these dynamics? Particularly, we address the changing role of manufacturing in the automotive value chains, as well as the linkages between production/manufacturing on the one hand, and knowledge-related activities on the other. Also, we will reflect on the 'manageability' of this regional industry cluster: To what extent can public policy contribute to the sustainable development of the regional automotive cluster?

The rest of the chapter is organised as follows. Section 7.2 analyses the regional-economic situation of the Paris region. We describe a number of 'foundations' on which the region's automotive industry rests: the economic base, the knowledge base, quality of life, accessibility and factor conditions. Section 7.3 describes the activities of the most important 'players' in the automotive value chain of the Paris region: OEMs (mainly PSA and Renault), suppliers, and research and educational institutions. Also, it shows pictures of the geography of the cluster. Section 7.4 turns to the analysis of networks: we highlight the dynamics of the various local, national and global networks in which the Paris region is embedded, in light of current technological and market trends. Section 7.5, finally, summarises our main findings.

7.2 Foundations of Paris

Île-de-France (Paris and its wider region) is the most important geographical concentration of the automotive industry in France. The region is home to 11.3

million inhabitants, which is 19% of the total French population. Over 2 million inhabitants live in the core county (the so-called 'département')[1] of Paris (CRCI, 2006). The region is the undisputed economic, political and cultural centre of France.

Economic base

With a gross regional product (GRP) of €42,712 per inhabitant, the Paris region is rich compared with the French and EU average (CRCI, 2006). The unemployment rate is 9.8% (2005), slightly below the French average (10.1%). Compared to other metropolitan regions in Europe, the labour productivity in Paris is very high (measured as output per worker). Also, labour productivity growth has been strong through the last decade (see Figure 7.1).

The economy of the Paris region is highly diversified and service orientated. The vast majority (80%) of the economically active population works in service industries. The region is specialised in knowledge-intensive services (financial services, other business services) and headquarters functions. In terms of value added, the largest economic sector is business-to-business services (28.4%). Other large sectors are creative industries, tourism and R&D (CRCI, 2006).

The manufacturing sector is small compared to the service sector, but still employs more than 500,000 people. Manufacturing has declined rapidly in the last years, however. In the period 1990–1999, manufacturing employment decreased by 15% (Bernard *et al.*, 2002), and by 2002, represents 20% of the economy. Still, the Paris region is the main manufacturing region in France in terms of number of workers: 16% of all manufacturing workers of France are in the Paris region. Key manufacturing industries are aeronautics/space/defense, 'eco-industries', ICT, life sciences (such as biotechnology, pharmaceuticals and medical equipment),[2] and, last but certainly not least, automotive. The diversity of the region's economic base is an asset for the regional automotive cluster, as automobile production is closely linked up to other branches and disciplines such as plastics, electronics, mechanics, new materials, etc., and also to the service sector.

The Paris region is hardly known for its automobile industry, but the automotive industry is a significant branch in the region. It produced a value added of €4.3 billion, which is 1% of the total value added created in the region (CRCI, 2006). The region ranks among the largest automotive clusters in Europe (with Bavaria, Stuttgart, Lower Saxony and the Turin region as the other big ones).

Knowledge base

Paris has an excellent knowledge base. The region's economy is highly knowledge intensive, in various respects. First, the region has a relatively high share of knowledge-intensive sectors. More than 7% of the total employment consists of high-tech services, making Paris one of the few regions in Europe which such a high percentage (CRCI, 2006).

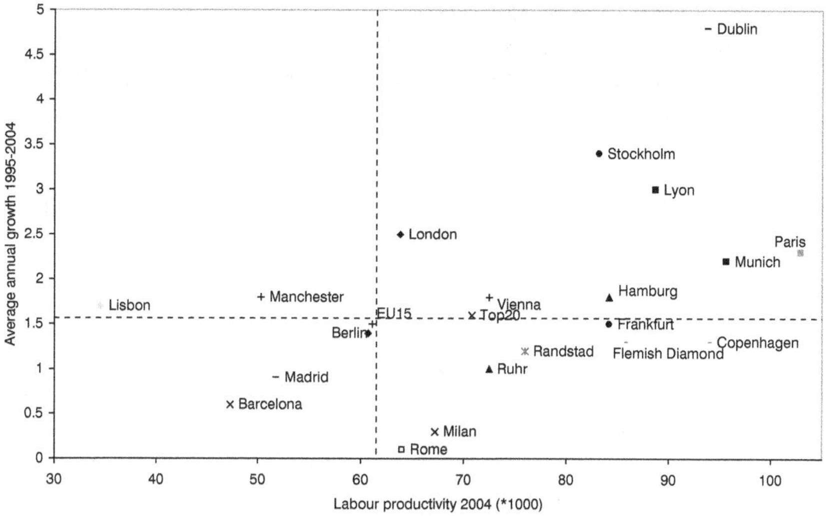

Figure 7.1 Labour productivity in the Paris region, 2004, and average annual productivity growth, 1995–2004 (source: Manshanden *et al.*, 2005)

Second, R&D expenditures are relatively high in the region; 3.3% of the GRP is spent on R&D.[3] The Paris region has about 2,500 public and private research laboratories that represent 8% of the European expenses in industry (PRDA, 2004). The region stands out as a research location: 46% of all researchers in France work in Paris, but the region counts only 25% of total employment (IAURIF, 2004). The automotive sector is a leading contributor to the R&D profile of the region: 24.1% of all corporate R&D workers in the Paris region were involved in automotive research (19,559 jobs) (PRDA, 2004). Of all French R&D workers in the automotive industry (researchers and engineers), 67% work in the Paris region. Since 1992, the dominance of the region in this respect has increased. In 2000, automotive companies in the Paris region spent over €2 billion on R&D. This is 63% of the French total (IAURIF, 2004).

Another relevant indicator for the knowledge base of the region is the educational level of the population. Here, Paris scores very high as well. In a recent ranking, Paris scored second (after London) in terms of availability of qualified staff (Cushman & Wakefield, 2006). Around 50% of all French engineers live in the region (interview). For automotive companies, this makes the region attractive.

Accessibility

The Paris region is internationally very well connected. It has several airports for national, regional and intercontinental flights, with Orly and Roissy-Charles-de-Gaulle as the main ones, handling over 75 million passengers in 2004 (PRDA, 2004). International connections are increasingly relevant for automotive companies in Paris, given the ongoing internationalisation of the sector. Furthermore, high speed train services link Paris with the larger urban centres in France, the UK, Belgium, the Netherlands and Germany. Rail connections are also important for the transport of vehicles and parts produced by the large car production plants of the region and by suppliers. The region scores second, after London, in a European city ranking of best cities in terms of access to markets, external transport links, internal transport and quality of telecommunications (Cushman & Wakefield, 2006).

Within the region, there is a large and well-developed public transport network, with multiple bus lines, regular rail links, metro lines and trams for internal as well as a good external accessibility (CRCI, 2006). Furthermore, the Paris region is the heart of the French motorway network.

Congestion is a large problem, however, in particular on the internal roads and on the ring roads and during rush hours, and the problems are getting worse. Several of our interviewees suggested that congestion is one of the threats for the car manufacturing plants in the region, in two respects. First, it undermines the reliability of the delivery of supplies. This is problematic, as JIT (just-in-time) production methods require very strict delivery schedules. Second, the workers of the plants are stuck in traffic and often arrive late. The bad traffic situation not only affects production plants, but also the many automotive-related import, export and trade functions that are very significant in the region.

Quality of life

Paris is a real global metropolis, very rich in culture, amenities, architecture, parks and leisure facilities, and it has a reputation of being globally trend-setting in many domains. The city is a magnet for tourists (36 million per year, 60% from abroad (Île-de-France, Regional Council, 2006). In the Cushman & Wakefield survey of 30 cities (2006), Paris scored 4th in terms of quality of life for employees (behind Barcelona, Zurich and Madrid). In terms of pollution, the city did not score well: 29th (out of 30). It should be noted, however, that air pollution decreased by 32% in the period 2002–2006, mainly thanks to reduced emissions by cars (Le Monde, 2006).

The picture is not completely rosy. Whereas most of downtown Paris looks nice and wealthy, in some of its suburbs ('banlieus'), the situation is grim. There are large deprived neighbourhoods with an accumulation of interrelated socio-cultural problems, including high levels of (youth) unemployment, high crime rates, worrying school drop-out figures, poor housing conditions, low education levels, etc. Thus, Paris – like any other large metropolis – is a dual city, with large social, economic and cultural divides. The potential destabilising and undermining forces of this duality came to the surface with the outbreak of large-scale riots in many Parisian suburbs in the autumn of 2005.

Factor conditions

Conglomeration brings costs and benefits. Paris is one of Europe's most expensive locations for living and for doing business. Living in Paris (especially downtown Paris, or the better-situated suburbs with good transport connections) is very expensive, from a French and international perspective. But buying or renting property in Île de France means buying access to a large job market and a rich variety of amenities, and many individuals and families make this choice. Also for business, the Paris region is expensive. In a Cushman & Wakefield survey of 30 cities (2006), Paris scored 20th in terms of costs of staff, and 17th in terms of value for money for office space. For companies, the high costs are largely compensated, however, by the high labour productivity of workers in Paris (see Figure 7.1). Also, the availability of specialised staff is a very strong point of the region.

The question is whether this trade-off is still positive for the automotive industry. For high-level functions such as management, R&D and design, the answer is probably positive. For manufacturing, the situation may be different. First, manufacturing is space-intensive, and space is extremely valuable in the Île-de France region. Second, manufacturing mainly requires low- to medium-level skills. For people with these skills, the Paris region may be too expensive. In manufacturing, it is questionable whether workers in the Paris region have a higher labour productivity than elsewhere.

7.3 The automotive cluster in Île-de-France

The Paris region is home to one of the leading automotive clusters in Europe, besides Munich, Stuttgart and Wolfsburg in Germany, and Piedmont (Turin region) in Italy. Some key figures illustrate the importance of the automotive cluster in the region (see Box 7.1).

In the period 1990–2002, the value added of the sector increased by 70%, with particularly strong growth in the period 1996–2002. In terms of value added, the sector outperforms other industrial sectors in the region (+26%). The share of the automotive industry in the total manufacturing industry increased by 5% (in terms of value added) and reached 13.8% in 2002 (IAURIF, 2004).

At the same time, the number of workers in the automotive industry has declined. In the period 1993–2002, the number of employees dropped by 10%. This drop is caused mainly by a reduction of employment at production facilities and in the supplier industry. Furthermore, there have been major qualitative changes in the labour market within the automotive sector. In general, the share of highly qualified jobs (engineers, managers) has grown, whereas the share of lower level jobs has declined. This holds in particular for unskilled labour: due to automation technology, the amount of unskilled labour had dropped significantly. Production work is still important, however: 43% of the employees in the sector work in production (IAURIF, 2004).

Box 7.1 Key figures of the automotive industry in the Paris region

- Production: 1.1m vehicles in 2002, which is 6.5% of the total European production.
- Employment: 100,000 (in 2003), in almost 400 companies (OEMs, suppliers and automotive research centres). This is 13% of the total manufacturing industry.
- R&D staff: about 18,400.
- Exports: € 59b (18% of total French car exports, 2002).

Source: IAURIF, 2004

Professionally skilled workers are more in demand. Experts in maintenance, equipment handling, metal work, mechanics, quality control, etc. are in short supply, although the share of this group in the total employment structure is declining slightly. Our interviewees suggested that there are too few young people who want to take these jobs. This is an image problem: working in a factory is not popular, although working conditions and salaries have improved strongly over the decades.

The higher echelons in the automotive industry (management, engineers, etc.) represent some 30% of total employment in the sector, and this share has been rising over the last decade, showing the increasing knowledge intensity of the Paris automotive industry.

In the remainder of this section, we introduce the various types of actors that play a role in the automotive clusters: OEMs, suppliers and knowledge institutes. Also, we describe the geography of the cluster.

OEMs

The automotive industry in the region is dominated by two large OEMs: PSA (well known from the brands Peugeot and Citroen) and Renault. Both are active mainly in the 'volume segment' of the automotive industry. Taken together, they employ 65% of the total number of workers in the regional automotive industry. Their activities in the region span the entire value chain: headquarter functions (13,000 employees), R&D and design, production, logistics, finance and sales functions. For PSA and Renault, the region is still a significant production hub. Taken together, the two companies employ 20,000 people in their production facilities, which is 31% of their combined total staff in the region (estimation IAURIF, 2004).

In the last years, there has been a tendency to produce more vehicles abroad, and fewer in France. Figure 7.2 shows clearly that, since 2004, fewer cars are produced in France (after a period of growth, during the late 1990s). In terms of the number of workers, the picture is more dramatic. Figure 7.3 shows the development of the number of workers in the French automotive industry in the period between 1980 and 2005. In that period, the number of workers dropped from 321,000 to 184,000. Note that the decline occurred mainly in the 1980s.

In the Paris region, the two OEMs are the spiders in the web of the regional automotive industry. Many suppliers depend on them, and many knowledge institutes are linked to them as well. The fate of these two firms thus deeply affects the regional automotive cluster at large.

PSA is the second largest carmaker in Europe, with a market share of 13.8% in Europe (2006). Worldwide, it employs 211,700 employees (2006). In 2006, the company sold 3.37 million cars, mostly in the small and medium segments, and 70% of the cars are sold in Western Europe. PSA also owns several other sizeable automotive companies. The main daughter company is Faurecia, a leading first-tier supplier that produces seats, cockpits, doors and other parts. This firm has 60,000 employees worldwide, 160 production sites, sales of €11 billion (2005), and 28 research centres. Another daughter company is Gefco, one of Europe's largest transport and logistics firms, with revenues of €3 billion. This company plays a major role in PSA's logistics (PSA, 2007a). PSA's headquarters are in Paris, as are its main R&D activities.

Globally, the PSA group has 16 assembly plants and 15 mechanical component plants. Most production takes place in France and Western Europe, but there are several indications that the company aims to reduce its capacity in these regions. In 2006, PSA closed a large plant in Ryton (UK), and 7,000 workers lost their jobs. Furthermore, PSA's CEO announced a reduction of 10,000 jobs in France and Western Europe (mainly in production). One of the factories at risk is in Aulnay, in the Paris region (Capital, 2007).

Meanwhile, in the last years the company has invested heavily in Central and Eastern Europe. In 2005, PSA opened a plant in Kolín (Czech Republic), where it produces the Citroen C1 and the Peugeot 107, in partnership with Toyota. In 2006,

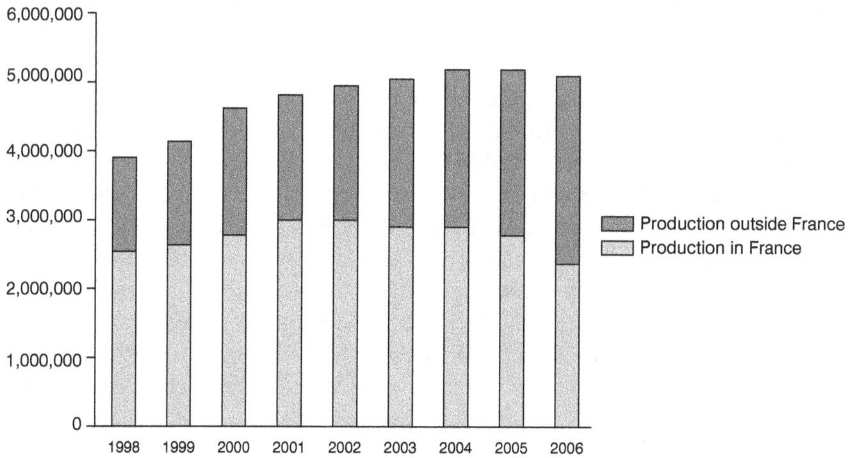

Figure 7.2 Production by French OEMs, in France and abroad (source: www.CCFA.fr)

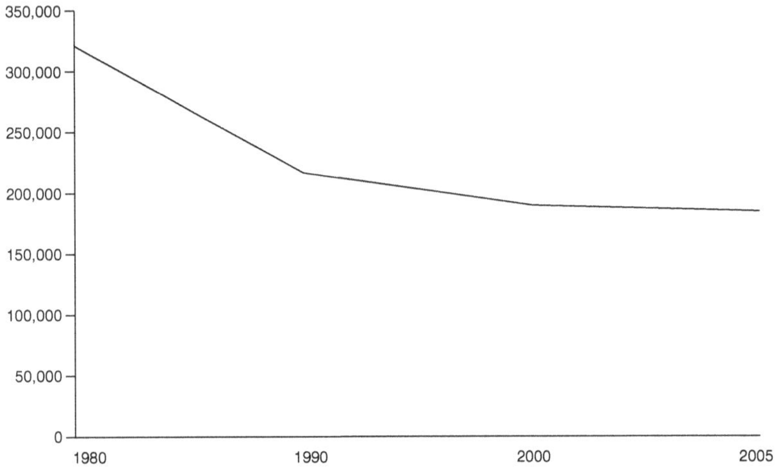

Figure 7.3 Number of workers in the French automotive sector, 1980–2005 (source: www.CCFA.fr)

it opened a plant in Trnava (Slovakia) to produce the Peugeot 207, with an annual capacity of 300,000 vehicles. The company plans to further expand the capacity of this plant.

In relative terms, France is becoming less important as a production location. In 1998, PSA produced 70% of its cars in France. In 2006, this percentage had dropped to a little over 50%. In absolute numbers, the figure is less dramatic. PSA's production in France rose sharply during the late 1990s (from 1.3 million in 1998 to 1.9 million in 2002), but since then dropped to 1.5 million in 2006 (www.CCFA.fr).

The Paris region still plays a significant role in production. PSA has two large plants in the region: Poissy (329,200 units) and Aulnay (283,100 units). Both plants produce cars in the smaller segment. Also, there are some smaller production facilities (PSA Saint-Ouen, PSA Asnières).

PSA spends some 5% of its turnover on R&D (€2.25 billion per annum, 2003; PRDA, 2004). It has identified two research priorities: to limit greenhouse gas emissions and to improve road safety. Much of the group's research efforts are channelled in these directions. To reduce emissions, PSA focuses on the development of hybrid-diesel technology and fuel cell technology. To improve road safety, the company conducts research in primary safety (to prevent accidents), secondary safety (protection in the case of collision) and tertiary safety (post-accident). PSA has recently developed an Emergency Call System, offered on every Peugeot or Citroen model. In section 7.4, we show that PSA does not conduct all its research in-house: it is involved in a multitude of cooperation with others.

The Paris region is by far the most important location for PSA's R&D activities. The company has four major research centres in the region, employing over

10,000 people. The largest ones are the Technical Centre in La Garenne-Colombes (3,360 employees) and the technical centre in Vélizy (5,360 employees). The two others are somewhat smaller: the technical centre in Carrières-sous-Poissy (close to the large plant) has 500 employees, and the design centre in Vélizy employs 1,000 staff. As said above, developing cleaner technology is a research priority. Recently, PSA inaugurated a new unit at its site in Carrières-sous-Poissy for the development of fuel cell technology, their integration into vehicles and their related technologies. In this development work, PSA closely cooperates with CEA, a research institute specializing in hydrogen-related technologies (PSA, 2006b).

More than ever, the quality of car design is essential in the commercial success of cars, and OEMs have developed a number of methods to improve the design process. In 2004, PSA opened an Automotive Design Network (ADN), in which the design activities of Peugeot and Citroen were concentrated. Over 1,000 people of 21 different nationalities work in this design centre, in Vélizy, close to downtown Paris. For the development of a new model, ADN has to compete with outside design bureaus. In 2006, a new element of competition was introduced. To stimulate creativity for each brand (Citroen and Peugeot), two competing teams of stylists vie to develop a new car model. Teams are asked to come up with strong, innovative and highly distinctive styling designs (PSA, 2006c).

PSA has increased its R&D spending in the last decade, and the Paris region has benefited. Recently, however, PSA announced it would cut costs in its headquarters and its engineering departments: 4,800 jobs would be lost in those two domains (Le Figaro, 2007).

PSA has a policy to be involved in local communities where the company is active. PSA plants support local projects and, in some places, it has close ties with local educational institutes. The company considers this as part of its corporate social responsibility (PSA, 2006a).

Renault ranks among the world's largest car producers, with a production of 2.5 million cars in 2005. Worldwide, it employs 126,000 people. It has a sales network in 118 countries, counting over 12,000 dealers. Furthermore, Renault has a cross-share holding with Nissan Motors (Japan) and the two OEMs cooperate ever more closely. In 2005, the global sales of the alliance amounted to 5.9 million vehicles (9% of the world market). The Renault-Nissan Alliance ranks among the world's leading four automakers. It includes five brands: Nissan and Infiniti for the Nissan group, and Renault, Dacia and Samsung for the Renault group.

Globally, Renault has 27 production sites; France is still the main production location, but its importance is declining in both relative and absolute terms. In 1998, the firms produced 1.2 million cars in France, 60% of its total production. Eight years later, in 2006, this number had dropped to 890,000 cars, which was only 42% of its total production in that year (CCFA website).

In Paris, Renault has a big plant in Flins (5,000 workers), and some smaller ones. Manufacturing employment in its Flins site is still significant but it has dropped dramatically in the last decades, from 20,000 people in 1980 to 5,000 in 2006 (interview). The number of cars produced at the site has grown, reflecting the large gains in productivity realised in that period, and the shift of outsourcing tasks to suppliers.

The company's explicit strategy is to install production sites close to final markets (Renault, 2006), mainly to reduce logistics costs. Therefore, Renault is expanding its production capacity mainly in Eastern Europe and Asia, due to fast growth in these markets. In 2005, a new factory was opened in Moscow for the production of the Logan (a low-cost car). In that same year, Renault announced a joint venture agreement with Mahindra & Mahindra Ltd (an Indian company) for the production and sales of the Logan, in 2007. In Morocco, Renault has set up a plant with an annual capacity of 30,000 cars (also Logan), and in Pitesti (Romania), the company invested €100 million in its Dacia plant (Logan) to increase the capacity from 235,000 units to 350,000 units (Le Figaro, 2006).

Renault spends 4.9% of its turnover on R&D. This amounts to €1,737 billion per annum (2003). In 2005, the company filed 895 patents. For Renault, as for PSA, the Paris region is by far the most important region for research and development activities. Its main R&D site is the Renault Technocentre in Guyancourt. Here, 12,000 specialists work on R&D projects, including 2,500 from suppliers. The aim of the Technocentre is to improve the efficiency and effectiveness of R&D: the development time for a car should be reduced from 5 years to 2.5 years. This target was reached for the Renault Modus: several hundred engineers developed this new model in 29 months and the total development costs amounted to €535 million (Die Welt, 2004). The centre works on the development of engines, chassis, interiors and exteriors. Also, it has prototyping facilities, a laboratory for materials research, as well as manufacturing and logistics facilities. The centre has the capacity to develop five cars simultaneously.

Renault also has a design centre in Central Paris (Bastille), where a team of 15 top designers are creating the cars of the future. Renault has four focus areas for research: safety; environment and energy; travelling comfort; and mobility and services.

In 2007, Renault opened a styling centre in São Paulo, near its plant there. This centre employs 10 engineers, whose role is to monitor local tastes in Latin America and to develop/implement new styling ideas for this market. The centre is closely linked to the Technocentre in Paris.

Suppliers

The supplier industry is very well developed in the Paris region. There are around 250 establishments in the region, and they employ some 30,000 people, which is 30% of the total employment of the regional automotive industry (estimation IAURIF, 2004). The 'great names' are all there. To name a few: Valeo (9 sites, among which are headquarters and R&D; 4,500 employees in the region), Faurecia (headquarters, production and R&D sites; 2,000 employees), Siemens (2 sites; 1,500 employees), Bosch (production and R&D; 1,300 employees), Eberspacher (German), GKN (UK), Autoliv (specialises in airbags; production and R&D sites; 1,100 employees), Johnson Controls (production, R&D and services; 2,000 employees), Delphi (European headquarters and R&D; 1,600 employees), Visteon, and Lear (technical centre; 900 employees).

The foreign groups represent between 60% and 70% of the supplier sector in the region (in terms of employees, IAURIF 2004), and in recent years there have been substantial investments by foreign suppliers. In the period 2002–2003, they created around 600 new jobs.

Recently, Delphi opened its European headquarters in Tremblay-en-France, north of Paris. The establishment contains its testing and development facilities that used to be spread all over the Paris region. The new site employs 900 people from 33 nationalities (Le Figaro, 2005a).

Suppliers are increasingly involved in research and development. In the Paris region, their share in total automotive R&D spending increased from 1.8% in 1992 to 6.2% in 2000. In absolute terms, the increase is more dramatic: from €23 million to €127 million (IAURIF, 2004).

Our interviewees suggested that for smaller suppliers, it is increasingly difficult to meet the ever-growing demands of the OEMs in terms of quality, logistics, organisation and ICT. To help smaller suppliers, the RAVY network has been created (more on this in section 7.4).

Education, research institutes and other actors

The Paris region hosts an impressive number of educational facilities with a substantial 'automotive' focus. Table 7.1 lists the educational institutes for higher education (bachelors and masters). The Paris Regional Development Agency identified another 215 courses and training programmes that are dedicated to the automobile industry, on the level of vocational training and technical high school (Region Île de France website).

Also, there are several training centres operated by private companies. PSA runs a programme for vocational education, in cooperation with a private technical school (27 apprentices per year); Renault takes 90 apprentices each year in its Alternance-programme. Renault also developed a masters course titled 'Transport & Sustainable Development', in cooperation with three institutes of higher education. In a 17-month course, 30 students learn how to reconcile economic, social and environmental issues in their decisions. The course is funded by the Renault Foundation (Tribune, 2004).

Further, there is the Institut Supérieur de la Vente Automobile (dedicated to automobile sales sources), run by BMW, Peugeot, Citroen and Fiat, with 500 apprentices per year.

Research relevant to the automotive industry takes place in universities and dedicated research institutes. Many of the universities mentioned in the previous subsection also conduct automotive-related research.

Many national research institutes are located in Paris; they are funded mainly by the state, but are increasingly encouraged to participate in joint research projects with industrial partners. Some highly relevant institutes are:

Table 7.1 Higher education relevant for the automotive sector

	Number of graduates per annum
Specialized schools	
ESTACA (Ecole Supérieure des Techniques Aéronautiques et de Construction Automobile); private engineering school	180
ENSPM (Ecole Nationale Supérieure de Pétrole et des Moteurs)	290
Other engineering schools relevant for the automotive industry	
ECP (Ecole Centrale Paris); multidisciplinary engineering	450
EPF, Ecole d' Ingenieurs (five-year engineering school)	200
ENSAM (Ecole Nationale Supérieure des Arts et Métiers); mechanical and industrial engineering	1,000
CESTI (Centre d'Etudes Supérieure des Techniques Industrielles); Master of Science in Engineering	130
ENSMP (Ecole Nationale Supérieure de Mines de Paris); materials, robotics, logistics, ICT	120
ESME–SUDRIA (Ecole Supérieure de Mécanique et d'Electricité); mechanical and electrical engineering	280
Management schools	
Fondation d'Entreprise Renault; international MBA for 25 Japanese students per year	25
ISTM (Institut Supérieur de Téchnologie et Management Marne-la-Vallée); advanced materials and integrated engineering	20
Universities	
FIUPSO (Formation d'Ingenieurs de l'Université Paris–Sud Orsay); electronic systems, processing, materials (masters)	204
Institut Galilée (part of Université Paris XIII); materials and applied mathematics (masters)	125
Université Marne-la-Vallée; maintenance of systems and production management (graduate)	48
Université Pierre et Marie Curie; mechanics/electronics	8

Source: www.paris–region.com

- INRIA (Institut National de Recherche en Informatique et Automatique). This institute Specialises in ICT research. It cooperates closely with Renault to develop control systems for cars.
- INRETS (Institut National de Recherche sur les Transports et leur Sécurité). It is active in three applied scientific domains: 1) people's safety in transport; 2) optimising the use of transport networks and reducing the dependency on oil; and 3) increasing the reliability and sustainability of transport systems,

optimising energy consumption and reducing the impact on the environment. In many projects, INRETS cooperates with industrial partners, including the leading car manufacturers and suppliers in the Paris region.

- CEA (Commissariat a l'Energie Atomique), specialises in research on new types of energy.
- CNAM (Conservatoire National des Arts et Métiers). This institute participated in a project of Renault and Peugeot to develop a state-of-the-art wind tunnel.
- CNRS (Centre National de de Recherche Scientifique).

The French government supports R&D activities in many ways. In total, it spent €120 million on automotive related R&D in 2006 alone. Also, the government has designed a favourable fiscal treatment of R&D expenditure in the automotive sector (Plan d'Action, 2006). Some of our interviewees indicated that a major problem in the public research infrastructure is the lack of interest in economic matters and commercialisation. This is still considered by many as 'dirty' or unworthy; cultural differences between public and private research are enormous.

Other relevant actors in the cluster include service providers, such as consultants, engineering bureaus, etc. In total, they employ some 2,000 people (2% of the total employment). Furthermore, many foreign OEMs have operations in Paris. Examples are Daimler-Chrysler (headquarters France, and a training facility), Nissan (European headquarters) and BMW (headquarters France, and a logistics centre).

The geography of the cluster

The automotive industry is scattered over the region, but there are some major concentration areas and axes. One 'subcluster' is situated near Paris, on the west side. Here, we find a concentration of head office activities. Second, there is an axis from the outskirts of Paris towards the southwest. Here, we find major concentrations of R&D activities. A third concentration area is found a bit more to the northwest of downtown Paris, in the lower Seine basin. This area is dominated by production activity, see Figure 7.4. The picture also shows that Paris is the heart of a larger automotive region: the north-western part of France. Several large production facilities are located in Normandy and eastern Brittany.

7.4 Local and global networks

The actors in the region's automotive cluster are heavily involved in a number of networks. In this section, we will focus our attention on these networks. Many types and configurations of networks can be identified: some of them are regional, others stretch far beyond it; some networks are aimed at innovation, others at production and logistics. Furthermore, these networks are not stable but subject to constant change. Our aim is to identify tendencies and trends in the dynamics

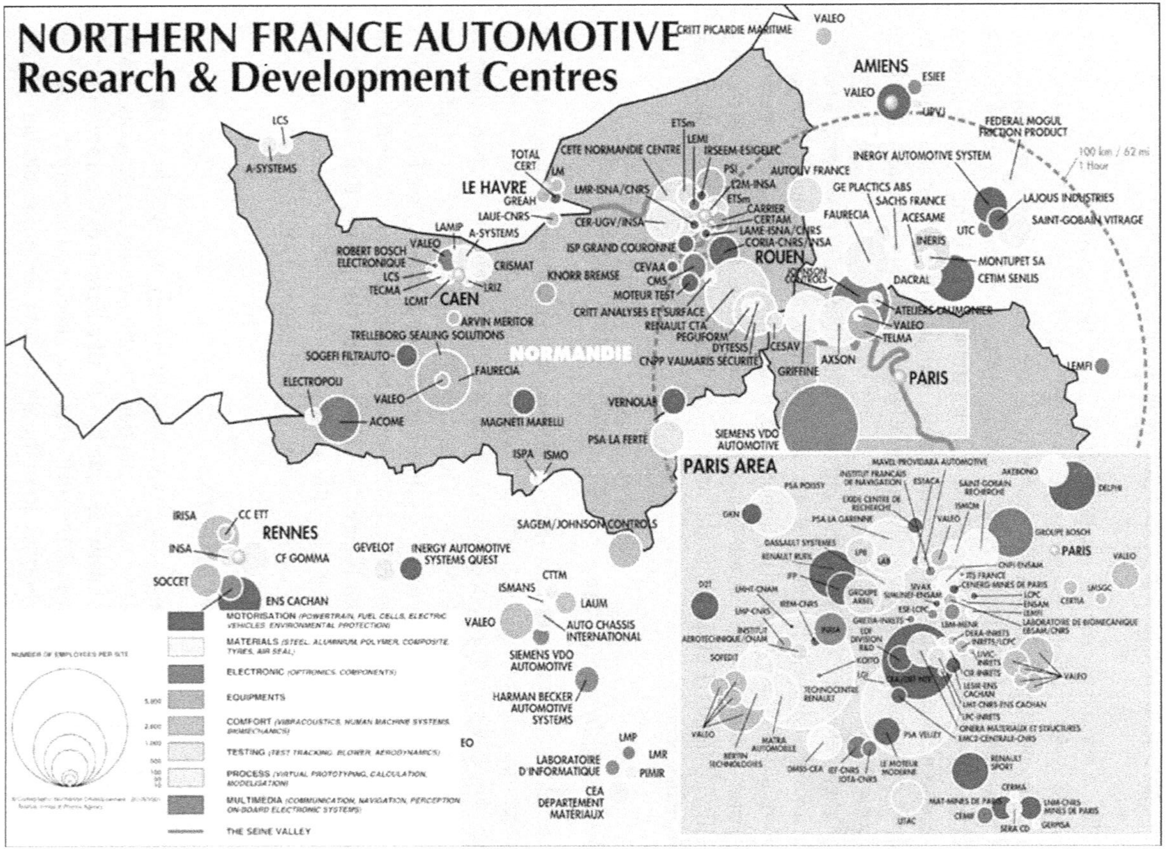

Figure 7.4 Geography of the cluster (source: IARIF, 2004)

of these networks. We will discuss networks among OEMs, with special attention for the Renault-Nissan alliance, networks between OEMs and suppliers, and networks aimed at R&D cooperation. Finally, we discuss the management of the cluster, and the role of the public sector therein.

Networks between OEMs

To share risks and pool resources, PSA and Renault increasingly work together with other OEMs for specific purposes. Renault's main partner is Nissan, with which it has a strategic partnership. PSA works with several OEMS.

PSA's strategy is to develop 'platform agreements' with other OEMs. In these cooperations, PSA jointly develops and manufactures platforms or chassis/power train components. Currently PSA produces 70% of its cars on the basis of only three platforms. This way, PSA shares the capital costs and risks with other OEMs, and benefits from complementary technological competences. Joint production and development also yield economies of scale and production.

PSA has a close relationship with Fiat, too. For instance, the two companies jointly produce MPVs (multi-purpose vehicles) at a plant in northern France (Sevelnord). In Valenciennes, large MPVs from Fiat, Lancia and Peugeot are manufactured. Together, the groups are developing a new minivan that will be manufactured in Turkey (Bursa). With Renault, PSA produces automatic gearboxes and V6 engines. With Ford, PSA jointly manufactures diesel engines (in Tremery, eastern France).

With BMW, PSA developed a new 1.4 litre gasoline engine in 2006. The joint development of the engines did not take place in Paris but in Munich, at BMW's R&D centre, with teams from both groups. The R&D costs were shared by both partners. Manufacturing is planned to take place at two locations: in Douvrin (northern France) for the PSA cars, and in Hams Hall (UK) for the Mini brand of BMW (PSA, 2007b). This example illustrates that in partnerships between OEMs, there is often a geographical separation between R&D and manufacturing. Geographically, the networks of PSA are focused on European partners.

PSA and Renault are competitors, but they also work together in pre-competitive domains. An important project in Paris is the joint development of an aero-acoustic wind tunnel. This tunnel (investment: €36 million) was developed together with the highly prestigious research institute CNAM (Conservatoire National des Arts et Métiers).

The Renault-Nissan alliance deserves attention in this respect. Unlike PSA, Renault focuses on one strategic partner: Nissan. The two firms have been allied since 1999, when Renault bought a 36.8% share in the then troubled Japanese company. Since then, the firms have worked more closely together. The alliance aims to be recognised among the best three car makers in the world in terms of quality, technology and operating profit. The cooperation takes shape in several domains: in engineering, R&D and design activities, the two firms are moving towards each other. They are engaged in the following common activities:

- *Development and use of common platforms.* The 'B' platform, already in use by Nissan, is now also used by Renault for the Modus (since 2004) and the Clio III (September 2005). Another platform (the 'C' platform) is used by Renault for its Megane II model and by Nissan for the Lafesta.
- *Interchangeable components policy.* This policy aims to ensure that parts can be exchanged among models from both partners. This requires a coordinated development of standardised components.
- *Joint development of power trains/engines.* The partners have jointly developed a range of new engines, used in cars from both partners.
- *Research and advance engineering.* Renault and Nissan seek to optimise the allocation of R&D resources of both groups. They have identified common areas of research: fuel cells, hybrid vehicles, materials, electronic, power trains and active safety. Also, solutions developed by one partner can be applied to cars of the other. These cooperations can drastically reduce the development time of cars.
- *Electric and electronic systems.* In 2004, Nissan and Renault have jointly developed a navigation and communication system that is applied to new models of both partners.

Renault and Nissan also achieve significant cost reductions with joint purchasing and a common supplier base. The Renault-Nissan Purchasing Organization (RNPO) was established in April 2001 as the first alliance joint venture. The scope of joint purchasing activities and the geographic responsibility of RNPO were regularly enlarged. In 2006, the share of purchasing under RNPO responsibility represented 75% of the alliance's worldwide purchases. In manufacturing, the alliance offers Renault the opportunity to produce cars in Nissan plants, and Nissan in Renault plants. Renault returned to the Mexican market with Scénic, which was produced at the Nissan Cuernavaca plant from December 2000. This was the first cross-manufacturing operation under the Renault-Nissan Alliance. It was followed by Clio and Clio Sedan (Nissan Platina), built at the Nissan Aguascalientes plant. Since October 2002, Nissan has produced some versions of the Renault Trafic/Opel Vivaro compact van as the Nissan Primastar at its Barcelona plant for Renault, Opel and Nissan. Still in Europe, Renault Kangoo was adapted and sold by Nissan under the Nissan Kubistar name as of October 2003 (www.renault.nl).

Furthermore, Renault and Nissan are harmonising information systems and R&D information technology, to ensure efficient technical cooperation between the two groups. In July 2002, Renault-Nissan Information Services (RNIS) was set up to deliver cost-effective systems and optimised infrastructure for the two groups' information and IT departments. The alliance has created a three-year information systems master plan, as well as the 'Alliance Worldwide Backbone' (AWB) broadband network. This is a high-level private network infrastructure for the alliance, connecting Renault and Nissan's four main telecommunications centres in Japan and France.

Networks between OEMs and suppliers

Suppliers produce the largest part of the value added of a car. In the relations between OEMs (PSA and Renault) and suppliers, we identify two main trends: 1) an ongoing pressure on suppliers to reduce costs and improve quality, and 2) an increasing involvement of suppliers in development.

For Renault's relations with suppliers, the alliance with Nissan has a global impact. As described, Renault and Nissan have set up a joint purchasing organisation to obtain economies of scale and improve quality standards. The purchasing department plays a large role in the selection of suppliers, and increasingly in their development and upgrading. The department has a presence in all of Renault's factories outside Europe, with the explicit aim to increase the use of local suppliers: this will lower production and logistics costs. Local purchasing teams visit, select and monitor local suppliers to ensure that they comply with the required quality. Renault has introduced standardised work processes so that teams all over the world share the same knowledge. Renault also encourages its suppliers to behave in a socially responsible way (i.e. prohibiting the use of forced labour and child labour; the requirement to limit industrial hazards, and the pursuit of good employee relations).

Japanese suppliers increasingly supply Renault's plants, also in Europe. For instance, in 2005, Kyowa Metal Works opened a branch in Paris to supply parts of the transmission system to Renault (Nihon Keizai Shimbun, 2005).

Renault also wants to intensify the involvement of suppliers in innovation. It has set up three committees that have identified eight priority areas for co-development with suppliers. Also, it has clarified rules regarding intellectual property rights to facilitate cooperation projects (Renault, 2006).

Furthermore, to cut costs, Renault is constantly increasing the share of supplies from low-cost countries, not only for its factories in those counties but also for its factories in Europe. This implies that the supplier industry in Europe (and particularly in France and the Paris region) will lose business to rivals in Central and Eastern Europe, and Asia.

PSA is France's biggest purchaser, with a total procurement budget of €29 billion. A small number of suppliers dominate: 50% of its series purchases are sourced from just 19 main suppliers. To highlight the importance of supplier performance, each year PSA awards Best Supplier Trophies in four categories (quality, innovation, parts and services, and cost performance). The group also involves suppliers in its development activities. Recently, PSA signed a partnership agreement with Plastic Omnium, a supplier of plastic components. In the partnership, engineers from the two companies will work together 'upstream' in the development process, clarifying and optimising the relationships between the partners. The partnership will focus on the development of opening modules, roof modules, fenders, front ends and fascia. All these parts are deeply integrated in the design of a car.

In the innovation process, cooperation between OEMs and suppliers is people-orientated, highly interactive and unpredictable; it is the 'logic of services', and proximity is very important. This is different from the world of production: there, what matters is logistics, efficiency, quality, predictability. The fabrication of components is globalised: the same components are produced at every plant of an OEM wherever it is located. However, the creation of new complex systems (cars) is not globalised. Here, interaction in a region is still high.

Networks for R&D and design

Many of the larger firms in the automotive industry are involved in research programmes and projects that are sponsored by the EU (framework programmes), the French government and/or other funding organisations. Furthermore, they subcontract tasks to engineering bureaus and other technical experts.

Renault takes part in 61 research projects in total, on a wide range of topics, including fuel cells, multimedia, acoustics and accidentology. The firm identifies four advantages in engaging in joint research: 1) it saves costs; 2) it leads to a broader understanding and an expanded skills base; 3) it yield insights into new approaches; and 4) it provides opportunities to learn from and benchmark with other leading industrial groups. The company has identified four cooperation areas on which it wants to focus: environment and energy, safety, vehicle features (acoustics, ergonomics, etc.), and engineering services (for optimising vehicle design and development). These areas are consistent with its own in-house R&D activities (see section 7.3).

Cooperations of Renault Technocentre

Renaults' Technocentre is the main location where new models are developed. It cooperates with a number of partner organisations and subcontractors:

- With knowledge institutes, including laboratories of the University of Orsay and of the polytechnic school INRETS (Research Institute for Transport); it also has links with secondary education institutes in the region to offer courses adapted to the needs of the Technocentre.
- With all kinds of subcontractors: it is estimated that 2,500 to 2,900 jobs are 'subcontracted' to external partners. This concerns support functions (restaurants, cleaning, etc.), engineering services (1,200 jobs) and information technology. Most of the engineering bureaus are located close to the Technocentre, as this facilitates frequent interaction.
- With Idestyle, a large technical group (700 employees) close to Guyancourt where Renault has its main offices. They also work with PSA and other main suppliers in the region.

For the development of fuel cell technology, PSA works closely with CEA, a French government-funded technological research organisation specialising in nuclear energy research. Together, they developed a fuel cell stack to be used in PSA cars.

Managing the cluster?

The preceding sections have made clear that the automotive industry in the Paris region is highly complex, diverse and volatile, and linked up to many types of networks. Also, the case has demonstrated that the automotive cluster is highly significant for the regional economy, as a source of innovation and as a major employer. Nevertheless, it is clear that no single actor or organisation can ever claim to 'manage' the development of this regional conglomerate of actors and networks. The governance of the cluster occurs in the interaction of a multitude of actors (firms, knowledge institutes, etc.) that make decisions for their own benefit. The systemic nature of the cluster means that even in the absence of market relations or partnerships, the fate or actions of one actor affects the other. But there are good reasons to assume that the individual decisions of the cluster members do not always result in an outcome that is optimal for the region or cluster as a whole.

The private companies in the cluster are primarily driven by profit. As long as it makes business sense, actors 'use' or 'consume' the assets of the Paris region. The two large OEMs (PSA and Renault) are particularly important for the fate of the regional automotive industry. They play a structuring role by steering large supplier networks and engaging in regional, national and international networks of knowledge partners. The behaviour of not-for-profit actors (national research institutes, universities, etc.) in the cluster is influenced by all kinds of criteria used to measure their performance.

One organisation that plays a role in the promotion of the automotive industry is the Paris Regional Development Agency (ARD). This organisation is in charge of helping French and international companies that seek to start, expand or relocate business activities to the Paris Region. ARD works on five key sectors, of which the automotive industry is one. On its website, it offers information that may be of use for automotive firms that are considering opening an establishment in the area. It has web pages with general information about the regional automotive industry, but also specific ones, for instance about suppliers in the region, or relevant knowledge institutes and addresses of industry associations.

In the Paris region, there are two main initiatives to improve the alignment and interaction of different actors in the cluster, with the ambition to promote the development of the regional automotive cluster: RAVY and Mov'eo.

Ravy (Réseau Automobile Val d'Oise Yvelines) is a network organisation that aims to promote the automotive cluster at a subregional level. The RAVY network is geographically confined to the department of Val d'Oise and Yveline (to the west of Paris). These departments employ half of the automotive workers in the region, and have the two largest productions plants (PSA in Poissy and Renault in Flint). Ravy is active in four fields that were selected, not by RAVY's management but by its participating firms. First, it helps smaller suppliers to obtain assignments from OEMs and to meet their high demands. For instance, RAVY organises sessions in which purchasing managers from large OEMs explain their policies and their way of selecting and dealing with suppliers. Also, RAVY provides advice. For example, when a supplier works for an OEM, the OEM can demand that the supplier follows the OEM when it opens a plant in another country. For small companies, this is a big step, and RAVY helps them out. Second, RAVY aims to improve the alignment of the educational system to the needs of the automotive companies. For instance, it initiated a study on the (mis)match between the regional skills demand and supply. Third, Ravy is active in communication and marketing. One priority is to improve the image of the industry. Few young people see the automotive industry as an attractive place to work: many of them have seen their parents or other relatives struggle in the factories, working long shifts for low salaries, and this image is slow to disappear. Finally, Ravy seeks to improve the transport infrastructure in the region, for the benefit of the automotive firms. As noted in section 7.2, the deteriorating accessibility is increasingly considered as a problem, and undermines the viability of production functions.

Mov'eo was created in response to the 'Pôles de competitivité' initiative of the French state. It is a cluster organisation active in the Paris region and the Normandy region (where a significant car industry can be found). It is a merger (created in 2006) of two regional poles: Vesta-Polis (Paris) and Normandy Automotive. Its aim is to enhance the innovation of the automotive industry in both regions. Mov'eo has two objectives: 1) to generate R&D projects in which firms, public knowledge institutes and educational institutes cooperate, and 2) to 'anchor' R&D activities to the region. Mov'eo is part of a national scheme to create a strong regional 'centres of excellence' in France (see Box 7.2). In this scheme, each region could come up with strong clusters to be eligible for state support. Mov'eo was selected by the state as one of the 'world class' clusters. It has the flexible structure of an association in which the main actors in the regional automotive sector take part.

Mov'eo facilitates research and development projects that are conducted by public actors (laboratories, universities, etc.) and private partners (OEMs, suppliers). Typically, 30% of the funding comes from public sources, 70% from private firms. Mov'eo focuses on four lines of R&D: 1) energy and environment. This comprises projects to make cars less polluting: electrical cars, new types of engines, recycling of cars, etc.; 2) road safety. This contains a variety of research projects on active safety (obstacle recognition, prevention), safety of pedestrians, interaction driver/car/infrastructure, signalling systems, etc.; 3) mobility. This is about traffic management, accessibility, 'intelligent crossroads', toll systems, but also accessibility for handicapped people, and ICT systems to guide tourists, etc.; 4) mechatronics. This is about the integration of electronics and mechanics. This is relevant in the car industry, where the two domains interact a great deal.

Projects can be initiated, submitted and performed by many actors; a group of actors can make a proposal, and a committee will judge if it gets funding from Mov'eo. The organisation also wants to promote the quality of automotive orientated education, to improve the coordination between the many institutes, and to strengthen the link between industry and education. One of the challenges for Mov'eo is how to make public and private research cooperate, and how to valorise the knowledge in the public research infrastructure.

Box 7.2 Mov'eo: part of France's new industrial policy

In the field of research and innovation policy, France has a strong centralist tradition but many changes have occurred in recent years. Public funding of research has decreased as a percentage of total research expenditures, and, more importantly for this paper, the regions have entered as participants in research and innovation policy (Heraud, 2004). For instance, the regions are now involved in the financing of university buildings and in the support of innovative SMEs. In 2004, a new step was taken: the government announced an ambitious policy initiative to promote regional knowledge-based clusters, as a means to stop the exodus of companies and jobs, and to improve the innovative performance of the country. The government expects that the formation of regional clusters will contribute to national competitiveness (Loos, 2005). A national selection committee CIACT (Comité Interministériel à l'Aménagement et à la Compétitivité des Territoires) was set up, consisting of representatives of several ministries and experts. Regions were invited to submit 'cluster projects' in which firms, universities and public bodies would create collaborative networks, preferably based on existing strengths and competences (Bommelear and Chavane, 2005).

The following criteria were designed to evaluate the proposals:

- The strategy of the cluster should be founded on a dynamic and internationally oriented local economic structure. There should be a strong link between the cluster strategy and the more general economic strategy of the area.
- The cluster should have international visibility and credibility and should thus be state-of-the art in its field, from a global perspective.
- The proposed partnerships and governance structure is a crucial aspect. In particular, the quality and effectiveness of links between R&D, education and firms is a major criterion for 'recognition' of the cluster.
- The projects should lead to substantial synergies and value added in terms of R&D output and innovation. This should contribute to the final aim of the scheme: to improve the competitiveness of France and to create high-level employment.

In the end, 105 proposals were submitted (a lot more than expected), from virtually every French region. In 2005, the national committee approved 67 projects, a lot more than the 10–20 that the government initially planned to support. The projects are divided in three groups with a different 'status'. The first group consists of six leading knowledge clusters with world-class competences. These clusters will receive considerable financial support to further improve their position as international centres of excellence. The other groups are less 'high profile'. The French government invests heavily in this new regionalized innovation programme: the total public expenses to support these poles amount to a substantial €1.5bn for a 3-year period. The amount is to be spent on tax exemptions (€ 300m), loans (€ 400m) and subsidies for R&D institutes and agencies (800m).

Source: *Le Figaro*, 13-7-2005

In Mov'eo, there are two types of cooperation: pre-competitive/open research, and confidential/closed research. In the latter case, beforehand, partners make a contract on how to deal with the results, what may be published and what should remain secret. This happens, for instance, in a project on the development of a new engine. In principle, foreign companies can also participate (Siemens for instance), however, in practice, most Mov'eo projects are with French partners only. The aim of Mov'eo is also to include smaller firms but this is hard to realise in practice. They are less involved in research, in general, due to a lack of budgets and dedicated research staff.

Some of our interviewees doubt whether the actions of Mov'eo are sufficient. One unresolved issue is the very low 'job mobility' between the two sectors. Currently, engineers either have a career in industry or in academia, and this is not good for mutual understanding. One problem is the wage difference between industry (high) and academia (lower). Some improvements are already being made. One example is joint PhD projects, in which the PhD student is supervised by both INRETS and companies. Other interviewees warned that the speed of action of Mov'eo may not be sufficient: bureaucratic procedures to launch a project take too long, and this is a serious problem in the automotive sector where time pressure to introduce new technologies and concepts is high and increasing.

From 2005–2011, Mov'eo has a budget of €150 million. In 2006, it funded 50 projects, at a value of €40 million; 80 enterprises are involved, as well as 50 public research/education institutes.

7.5 Conclusions

The automotive industry is the most important manufacturing industry in the Paris region in terms of employment and value added, and it is one of the very few remaining industries with large-scale factories in the Paris region. To some extent, it has escaped the massive de-industrialisation that characterised the Paris economy in the last decades. The automotive industry in Paris is 'full-fledged', including all parts of the value chain, and it has a high level of knowledge intensity.

The Paris automotive industry is developing but not growing: it is a cluster with 'jobless growth'. The number of jobs in the regional automotive industry has decreased by 10% in a decade (1993–2002), and more job cuts are expected to be announced by important companies. Production (in terms of number of cars produced) has increased, however, as well as the added value (+70%, in the period 1990–2002), reflecting enormous productivity gains. Our case study confirms that the automotive industry in Paris is going through significant qualitative changes. It is becoming more knowledge intensive and technologically sophisticated, the role of manufacturing has decreased in relative terms, skill levels have gone up significantly, and cooperation with far-away knowledge partners has increased, not in the least because of Renault's strategic cooperation with Nissan. Local knowledge networks are becoming more firmly embedded in international ones. In this turbulence, the industry has proved to be resilient. The main assets of the regional cluster are the pool of skilled workers, the well-developed knowledge base and the excellent international accessibility of the region.

The fortune of the Paris cluster strongly depends on the success of its two giants: PSA and Renault. In a sense, this implies a vulnerability, but we should note that few car-producing regions in the world can boast two main OEMs inside their borders; many would be happy with one. The Paris cluster has strong indigenous research capacities, an excellent skilled labour pool, a leading role in knowledge networks, and headquarters functions of OEMs and suppliers.

Although services are by far the dominant sector in Paris, the upstream parts of manufacturing industries (R&D, management, marketing and finance) are strongly represented in the Paris region as well. This is an asset for the automotive industry, as this industry is strongly and increasingly linked to other branches (such as electronics, materials, design). The city has an enormous 'pool' of skilled engineers (50% of all French engineers live in the region), and a multitude of high-level (public) research institutes relevant to the automotive industry. This makes the region a hotspot for automotive-related R&D, and also attracts foreign direct investment.

The worsening congestion in the region adversely affects the operations of production sites. Longer and more unpredictable travel times hamper an efficient delivery and staffing of the factories. On the other hand, the region's international accessibility (airports, high-speed trains) is excellent, and this is a highly relevant asset given the ongoing internationalisation of the industry. In terms of quality of life, Paris has a lot to offer with high-level amenities and jobs, which makes the region a magnet for skilled people from France and abroad. However, this comes at a price: costs of living and factor costs for companies (labour, land and office space) are the highest in France and are also high from a European perspective. This seems to be no problem for the upstream parts of the automotive cluster (managerial, engineering and R&D functions), as high prices are compensated by a high productivity of the staff. For production activities, however, the trade-off is less positive.

In the last decade, the automotive industry in the Paris region has been in a continuous process of upgrading. This is reflected in the composition of the workforce: the share of engineers and managers has increased significantly in the last decade, at the expense of the share of manual workers. The number of employees at production sites has dropped dramatically in the last decades, due to outsourcing of activities to suppliers, automation and the opening of new plants abroad. This process is likely to continue, but the nature of outsourcing is changing: the Paris factories are increasingly supplied from low-cost countries. This may imply a gradual loss of production jobs amongst suppliers in France, but it also means that an eventual future downsizing of production capacity in the Paris region has less indirect effect on suppliers in the region. The process of upgrading also takes place *within* the Paris plants: the need for unskilled workers has dropped to almost zero, and the demand for skilled people at the professional level (in logistics, quality control, maintenance, etc.) is rising. The factories have difficulties hiring these people, partly due to the negative perception of many young people about working in a production plant (they associate it with long shifts, dirty work, low pay and job insecurity).

The ongoing consolidation and globalisation of the automotive industry has a profound impact on the composition of the Paris cluster. The OEMs (Renault and PSA) increasingly locate their plants in emerging consumer markets, and actively promote the development of a 'local' supplier base to save costs and improve the logistic process. Some development work (adapting models to local tastes and regulations) is also transferred to these countries. The management of operations is increasingly decentralised, and the geographical link between manufacturing and design/R&D becomes weaker. Paris is developing as the 'brain' of global networks, specialising in management functions (with an increasing emphasis on designing and implementing quality control systems, purchasing standards and managing global relationships), complex research and development (new platforms, models, engines, powers trains and complex systems), and finance.

An important tendency in the industry is the emergence of international cooperation in research and development. Teaming up with partners is a popular strategy to address the challenges of increasing technological complexity, time pressures, and associated risks and capital costs. The global race to develop and use alternative energy sources has added a new dimension to this trend. The process of joint innovation leads to the development of hyper-competence networks, consisting of very specialised knowledge-intensive firms and organisations. These networks have their centre in specific localities with ample availability of skilled staff and renowned knowledge institutes. Paris figures as prominent amongst these localities, with specific strengths in road safety and clean diesel engines. This explains why so many foreign companies (OEMs and suppliers) have substantial R&D activities in Paris. New, specialised poles are arising, especially in emerging economies where skill levels and competences rise rapidly. For instance, it may well be that some regions in India will develop as competence centres in ICT-related automotive research.

To maintain a leading position in global R&D networks, the quality of research and development in Paris should be absolutely world-class. An excellent region is needed, with prolific cooperation between research and business. Here, there are still deficiencies: our interviewees confirm that academia and the business sector are still worlds apart, and incentives to work with the industry are insufficient. The creation of Mov'eo, a centre of excellence to promote research in the field of automotive technologies and to link public research to private companies, is an important instrument to resolve these issues. Its projects help to bring business and academia closer together, and the organisation channels substantial new financial resources to innovative research. However, given the tight time pressure in the automotive industry to deliver, Mov'eo should seek to shorten the time between project proposal and implementation, and simplify its procedures.

A central question in this study concerns the nature of the linkages between manufacturing activities (car production in plants, or the production of car parts by suppliers), and upstream and downstream activities. Several observations can be made.

On the one hand, the link between R&D and manufacturing is weak: 'complex' research and development takes place in cooperation projects from leading technical and design centres across Europe or worldwide, and there is no need to be 'near' a manufacturing site. Even if all factories close down in the region, the sophisticated parts of R&D will still be carried out within the region, the main reason being that this type of activity depends strongly on the regional knowledge base. The 'spatial link' between top-level R&D and manufacturing has become much weaker in the last decade (with many factories opening in emerging markets), without significant reductions in R&D activity in the Paris region.

For some other types of development, the link between development and manufacturing is stronger. This is true for 'adaptive' development, i.e. adapting cars to local tastes or regulations. Slight adaptations typically have immediate implications for the manufacturing process and they depend on the local market, so it makes sense to carry out the development work near to the factory. This type of development is increasingly transferred to emerging economies, where the lion's share of new plants is opened.

On the other hand, the functional link between research and manufacturing is always there. The design of a new car should be such that the car can be produced very efficiently on a large scale. The quality of the manufacturing process (including logistics) is extremely important for the reputation of any car brand. Thus, knowledge of manufacturing processes is needed in the teams that design new cars or parts. Also, when OEMs set up factories in emerging markets, quality is of prime concern. The design and operations of factories will remain a core activity of OEMs. The knowledge and experience in these domains in the Paris region is very well developed, not least because several 'state-of-the-art' factories are still there.

The manufacturing plants in Paris rely heavily on deliveries from suppliers, but many of them are not located in the region, and increasingly, they are in low-cost countries. This negatively affects the production activities of suppliers in the region, and it implies that the regional economic spin-off of the Paris car factories, though still significant, will decrease. Many suppliers in Paris, however, are not merely 'producers' of parts but are also involved in R&D, often in cooperation with PSA and/or Renault, and many of them are foreign. In the Paris region, they find skilled staff, an excellent research infrastructure and many cooperation partners. This will not change when factories are downsized.

If we relate these findings to the spatial development of 'networks of exploitation' (production networks) and networks of exploration (knowledge/innovation networks), we can make a number of observations. First, for the networks centred around basic, pre-competitive research (new energy, for instance), Europe and France are the 'relevant regions', mainly because here, firms can benefit from national subsidies or European framework programmes. The resources provided by Mov'eo will probably bring some regionalisation to these networks. For the commercial development of complex systems, platforms and engines, networks are developing on a global/European scale on the one hand (PSA with BMW and Fiat, Renault with Nissan), and on a regional scale on the other (cooperation between OEMs and first-tier suppliers in the development of components often take place in the Paris region). Finally, 'simple', adaptive development work is partly transferred to suppliers, and is located in the vicinity of the plants. Thus, these knowledge networks are becoming more localised, and more strongly related to production networks. For these production networks ('networks of exploitation'), the picture is also mixed. On the one hand, production networks are becoming more localised in emerging markets (due to efforts of OEMs to develop a local supplier base); on the other, the networks around the plants in France and Western Europe are becoming more global, as the share of sourcing from low-costs countries is increased.

Notes

1 Other counties in the region are Seine-et-Marne, Yvelines, Essone, Hauts-de-Seine, Seine-Saint-Denis, Val-de-Marne and Val-d'Oise.

2 For more general info about the industrial structure see: www.ile-de-france.drire.gouv.fr/di/dev_eco/indidf050415.htm.

3 http://www.paris-region.com/ard_uk/upload/document/D101.pdf.

Bibliography

Bernard, C., F. Bauchard, C. Schaub, L. Steinmann and C. Vincent (2002) *Vers une France sans usines?* Enjeux, November 2002.

Bommelaer, C. and L. Chavane (2005) 'Villepin dessine le nouveau visage de la France industrielle', *Le Figaro*, 13 July 2005.

Capital (2007) *10,000 emplois supprimes chez PSA et ses sous-traitants?,* 20 April 2007.

Cushman & Wakefield (2006) *European cities monitor 2006*, London: Cushman & Wakefield.

Die Welt (2004) 'Verkaufsfertig in 29 Monaten', 24 September 2004.

Heraud, J.A. (2004) *Science, technology and innovation policy: new governance schemes in Europe, in France and at regional level.* Paper written on the basis of a contribution to the symposium 'Developments and policies on advanced technologies', Centre of Advanced Policy Science (CAPS), Tokyo University, Tokyo, 24 September.

IAURIF (2004) *La Filière Productive Automobile en Île-de-France: Etat des lieux et enjeux*, Institut d'Amenagement et d'Urbanisme de la Region Île-de-France.

IAURIF/INSEE/CRCI (2006) *Chiffres-Cles de la region Île-de-France*, 2007.

Île-de-France, Regional Council (2006), Île de France, a users guide, Regional Council Communications Department, 2006.

Le Figaro (2005a) 'Delphi a inaugure son siege Europeen de Tremblay-en-France', 8 April.

—— (2005b) 'Le plan de Villepin pour réveiller la France industrielle', 13 July.

—— (2006) 'Renault investit 100 millions dans l'usine Dacia de Pitesti', 12 December.

—— (2007) 'PSA supprime 4,800 emplois', 27 April.

Le Monde (2006) 'La pollution de l'air a baisse de 32% a Paris depuis 2002', 20 December.

Loos, F. (2005) *L'ambition de la France gagnante*, Industries, September, p. 9.

Manshanden, W.J.J., W. Jonkhoff, P.J.M. de Bruijn, S.P. Newrly and R. en Koops (2005) '*De top-20 van Europese stedelijke regio's 1995–2004; Randstad in internationaal perspectief*', TNO, Delft.

Nihon Keizai Shimbun (2005) 'Les fournisseurs de Nissan accroissent leurs ventes a Renault', 11 July.

Plan d'Action (2006) *Plan d'Action pour la filiere automobile annonce par le Premier Minister,* 29 November.

PRDA (2004) *Automobile industry Paris region: land of innovation*, Île-de-France: Paris Regional Development Agency.

PSA (2006a) PSA Peugeot Citroen: the group in 2006.

—— (2006b) Press release, 9 January.

—— (2006c) Press release, 20 September.

—— (2007a) PSA Peugeot Citroen: the group in 2007.

—— (2007b) Press release, 19 January.

Renault (2006) *Annual report 2005.*

Tribune (2004) 'Lan fondation Renault finance un Mastere sur les transports', 30 November.

8 Porto: metal-electro sector

8.1 Introduction

Portugal has traditionally been an importer of equipment and machinery. In the last decades, those imports supported industrial development in other manufacturing sectors like metal, textiles or footwear. However, recently Portugal has become a relevant exporter of those items, in niche areas. The large urban region of Porto has the largest concentration of actors in the sector. It represents a large number of direct and qualified jobs and its output value has been growing.

The broader metal-electro sector has been facing a transformation worldwide and competition is fierce. In Europe, countries like Germany or the UK are well-known producers, but other countries, such as Finland, have recently positioned themselves as very strong players in specific niches. Emerging economies simultaneously constitute exploitation opportunities for European manufacturers, and serious threats, pushing up raw material prices and smashing margins for customised metal products and tools. What is the position of the metal-electro sector in Porto in light of these dynamics? In this case study we explore the composition of the sector in its large urban region and depict the main networks in which the sector and its players are embedded, as well as its geography. Moreover, we analyse the role of public policies.

This chapter is organised as follows. Section 8.2 analyses the regional-economic situation of the Porto region and the 'foundations' on which the region's metal-electro sector rests. Section 8.3 describes the composition and geography of the sector as well as its most important 'players'. Section 8.4 turns to the analysis of networks: we highlight the dynamics of the various regional and global networks in which Porto's urban region is embedded in the light of current technological and market trends. We propose our conclusions in section 8.5.

8.2 Foundations of Porto's urban region

Porto is the capital of the Portuguese north-western urban region. Its metropolitan area has 1.5 million inhabitants, and more than 3 million if we consider the greater urban region (30% of the total Portuguese population). Increased accessibility makes it a highly integrated urban region, which we take into consideration for this study.

During the 1970s, the city of Porto started a de-industrialisation process, and the tertiary sector grew, namely through services for the industrial firms' needs in the broader urban region. Today, services make up the largest share of employment and added value in Porto; however, the industrial tradition is still noticeable. The broader region around the metropolitan area has large industrial concentrations, namely in traditional sectors that are facing sharp international competition (textiles, shoes, metal, furniture and wood). Some of the metal-electro industries have a strong presence in the region (like specific machinery and electrical equipment) and are one of the few growing manufacturing sectors present in the region.

Economic base

During the 1970s and 1980s Porto was an important growth engine. However, its economic condition has been deteriorating recently. Its GDP per head, even if measured at purchasing power parities, has had difficulties catching up with European averages and even with the capital Lisbon (see Figure 8.1). The highly industrial north-western region, a strong exporter, has been showing poor economic growth, and since late 1990s its manufacturing sectors have been facing competition from Eastern Europe and Asia (CCDRN, 2007). In this context, the unemployment rates in Porto and in its wider region have been growing (INE, n.d.).

Porto's metropolitan area has a diversified economic structure though manufacturing still makes up a sizeable share of its economic structure (see Figure 8.2). Nevertheless, and if we consider the broader urban region, the share of manufacturing is much higher – more than 33% in 2004 (INE, n.d.). Public

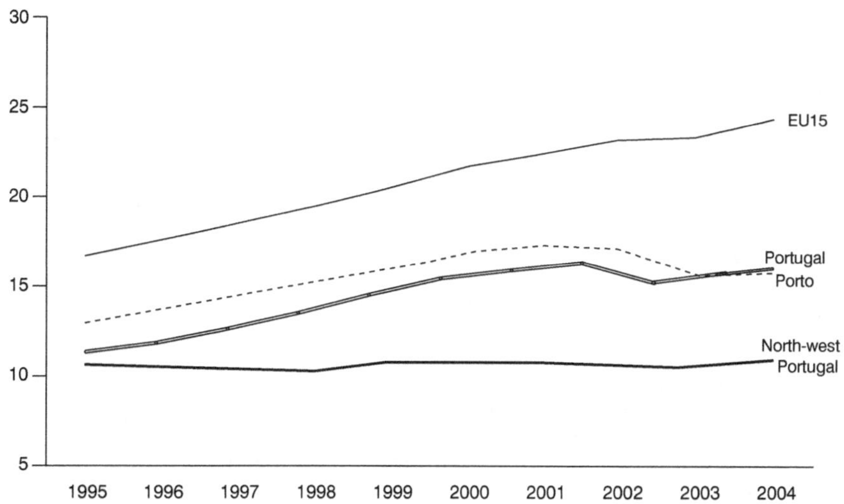

Figure 8.1 Gross domestic product per capita, €1,000 purchasing power standard, 1995–2004 (source: INE, Portuguese Statistics Bureau, n.d.). Note: Porto and Lisbon data correspond to its metropolitan areas; north-western Portugal corresponds to the aggregation of the NUTS III Grande Porto, Ave, Cavado, Tamega and Entre Douro e Vouga.

services (namely education and R&D) and other private services (including advanced services) make up the largest share of employment. Despite the growth in the private service sector, there is evidence that some high value added services (finance, telecommunications, software) are relocating towards Lisbon.

The economic base of the metropolitan area is mainly composed of small and medium enterprises (SMEs). Nevertheless, it still houses the headquarters of some of the largest and most international Portuguese groups, like Sonae (in retail, telecommunications, construction, real estate and wood), EFACEC (electrical machinery and equipments) or Unicer (beverages).

The economic base of Porto is influenced by the economy of the surrounding region. There, some of the most representative sectors are textiles, footwear and furniture, organised in traditional industrial districts. Despite some highly competitive textile niches (technical, automotive, habitat), the evolution and image of the sector has been unfavourable. Simultaneously, new sectors like pharmaceuticals, biotechnology and ICT still lack critical mass to be independently considered new growth clusters.

However, as Figure 8.3 illustrates, the industry of electrical equipment and machinery is relevant and has been highlighted as an example of a potential growth cluster. It has at least one important key player (EFACEC group), several SMEs, new high-tech start-ups and excellent R&D from regional knowledge institutions. Different industries within the sector are representative and have been growing; together they overtook the share of the textiles added value in the region (see Figure 8.4).

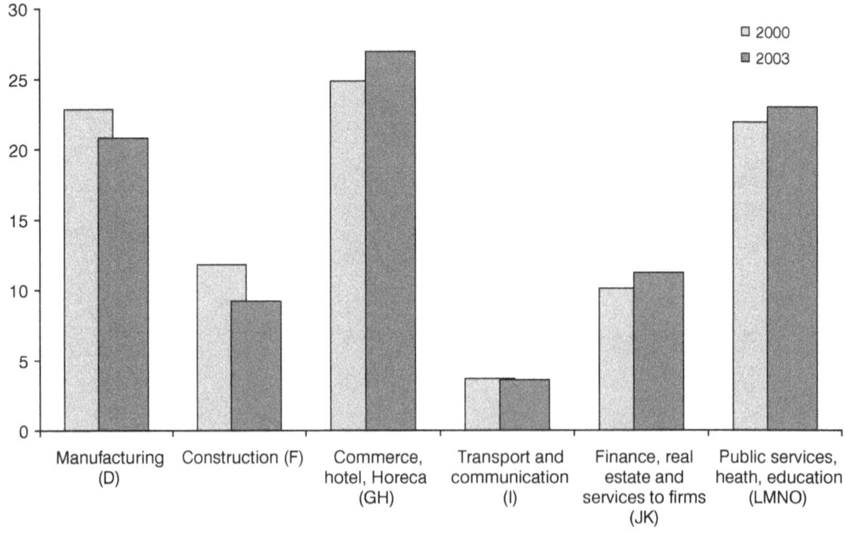

Figure 8.2 Employment share in Porto (metropolitan area), % (source: INE, Portuguese Statistics Bureau, n.d.)

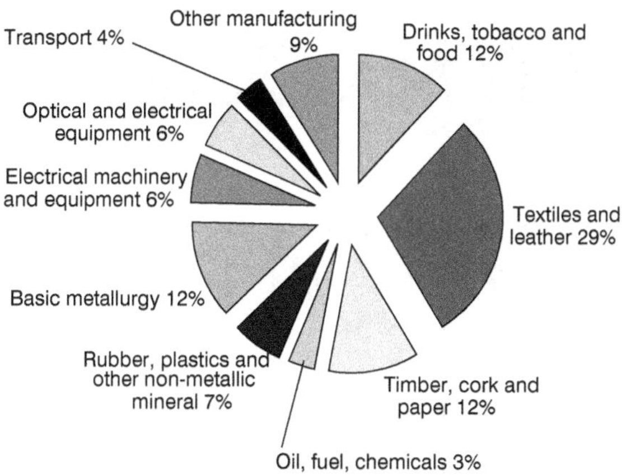

Figure 8.3 Employment share in manufacturing sectors, Porto (metropolitan area), 2003 (source: INE, Portuguese Statistics Bureau, n.d.)

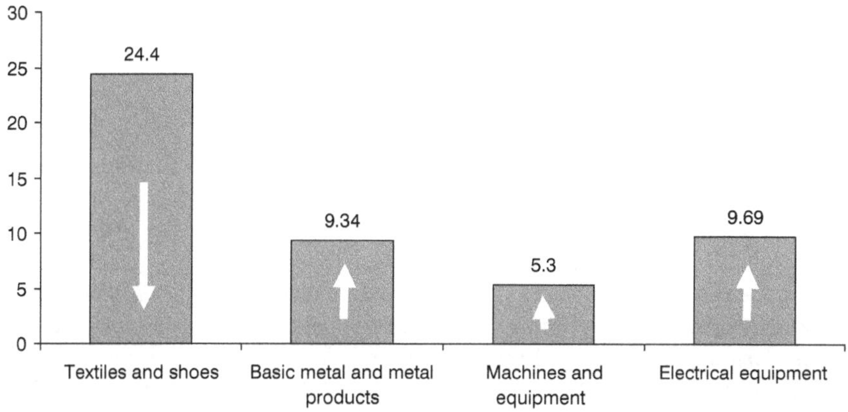

Figure 8.4 Share of turnover, north-western Portugal, 2004 (source: own elaboration, data from INE, Portuguese Statistics Bureau, n.d.). Note: arrows represent the direction of evolution in the last 5 years.

Knowledge base

Porto's education levels are higher than the Portuguese average and catching up fast to the EU averages. Around 14% of the population in the metropolitan area has a higher education degree (Figure 8.5). Convergence has also been significant in terms of graduates in science and technology in the population aged 20–29: 11% of Portuguese graduates are in these fields, against 12.6% in the EU-25 (EUROSTAT, n.d.).

The number of publications in international referenced knowledge databases has been growing fast in Portugal. Starting from a rather low position, Portuguese universities published almost 6,000 scientific articles in 2005 (OCES, 2006). The

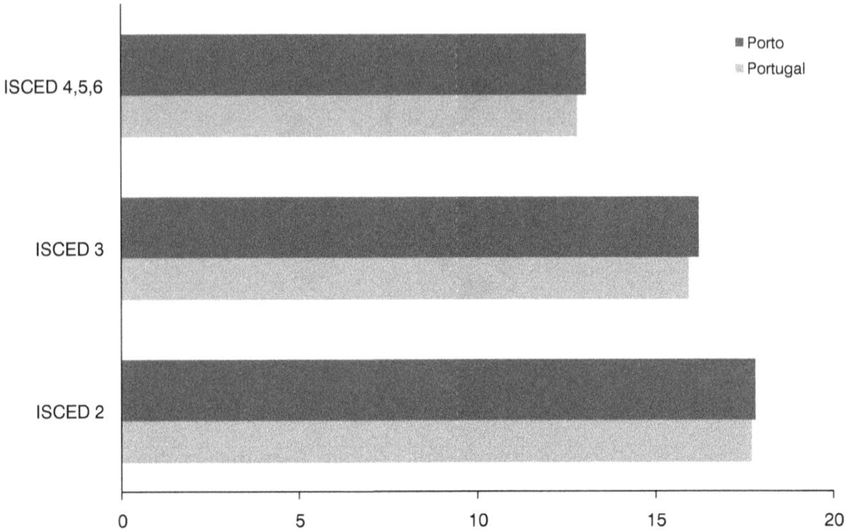

Figure 8.5 Educational attainment, 2004, % (source: INE, Portuguese Statistics Bureau, n.d). Note: ISCED: International standard qualification for education, ISCED 4, 5, 6: high education

University of Porto hosts top international research teams in medical sciences and engineering.

In Portugal a large share of R&D has been carried out by the institutional sector (government, higher education institutes and non-profit organisations), leaving the private sector with a relatively modest share. Overall, this suggests a weak link between R&D efforts and the industrial and productive base. However, in the *Norte* region, firms and private non-profit institutions are responsible for the largest share of R&D execution (see Figure 8.6), suggesting better results than the Portuguese average in terms of transforming knowledge in innovation. Moreover, in this region, R&D efforts have been systematically growing and more than doubled in the period 1997–2003 (OCES, 2005).

Porto's urban region is endowed with research infrastructure and facilities. Besides the University of Porto, other universities in the broader region are growing increasingly important. The universities of Aveiro and Minho (less than 50km away from Porto) supply a large number of graduates in science and technology fields and perform applied industrial research. These three important universities are presently developing science parks and incubation facilities in the vicinity to accommodate the growing demands of start-ups and spin-offs that recently started to emerge, like UPtec (University of Porto) or AvePark (University of Minho) hosting the European Institute for Tissue Engineering and Regenerative Medicine (Carvalho, 2009). The region has many qualified engineers working in manufacturing, and excellent educational institutes in various engineering and technical fields.

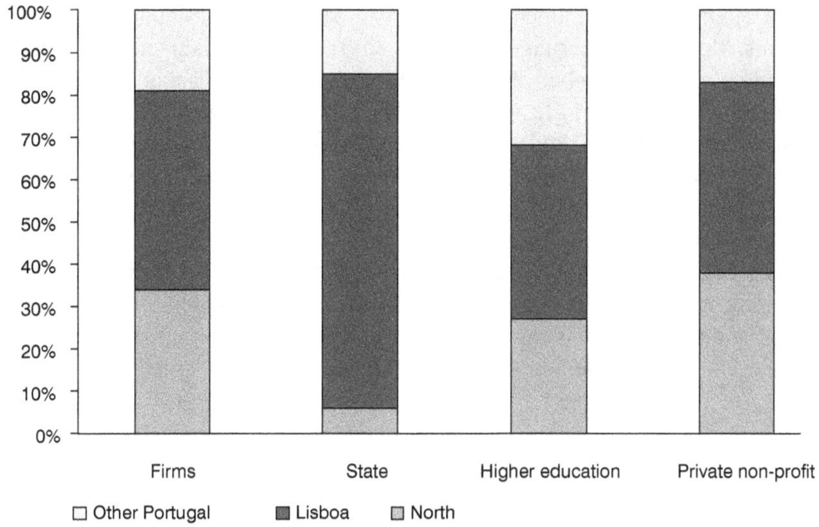

Figure 8.6 Share of R&D expenses, by execution sector, 2003 (source: OCES (Portuguese Ministry of Science and Technology), 2005)

Accessibility

During the last years, accessibility has improved in Porto's urban region. The urban region is well linked through many highways and fast roads. Porto's international airport has been recently enlarged; it has cargo facilities and it connects with several international destinations for all the major European hubs and other European cities. Ryanair recently opened one of its hubs in Porto's airport. Regional road accessibility ensures cross-border influence towards Galicia/Spain, and will improve even more with the HST link Porto-Vigo. Porto has the second largest cargo port in the country, despite some bottlenecks in the development of a logistic platform and inter-modal infrastructure.

Porto is connected to Lisbon in less than 2.5 hours by rail and road, and with all major cities in Portugal and the north-west of Spain. Internally, together with city rings and several bus connections, the metro light rail is bringing more integration to the metropolitan area, reducing congestion and improving residents' quality of life. However, for cargo, some problems remain. Some important industrial areas in the urban region put pressure on the main roads, bringing congestion and nuisance, while many industrial plants are scattered in the region leading to logistical difficulties, inefficiencies, pollution and congestion.

Quality of life

Porto, as a relatively large metropolitan area, is endowed with relevant cultural facilities, like diverse top-level cultural venues, museums, international schools, as well as environmental amenities like urban parks, and coastal and waterfront locations. Moreover, its heritage (UNESCO site) adds to a certain 'cultural' image, attracting a growing number of tourists. Nevertheless, some major problems are endangering its quality of life and image. The city's downtown is facing a serious housing decay problem linked with social exclusion and (lack of) perceived safety. Nowadays, some suburban areas are fully fledged cities, with their own amenities and labour markets.

In the surrounding urban region, the duality is even more evident. To counter the high unemployment rates and a 'crisis' image, public investments led to improvements in all types of amenities in surrounding cities like Guimarães or Vila Nova de Famalicão. Those (growing) cities receive international music and theatre events and Guimarães will be the European Cultural Capital in 2012.

Factor costs

After entering the European Union in 1986, low labour costs were a major magnet in attracting strong foreign direct investment (FDI) and orders in sectors like textiles, automotive and metal-electro. Porto's urban region benefited from this situation. Presently, although labour costs are still low by European standards, new members of the EU and Asian countries tend to be much more competitive in mature industries and non-customised manufacturing, even considering productivity gaps. This context is unfavourable, for instance for the metal and equipment industries, since other important players like the Czech Republic present still much lower labour costs.

Nevertheless, Porto and its urban region have been quite successful in attracting and retaining FDI and research centres of corporations like Siemens, for relatively low-cost but high-quality research. Industrial space is not scarce in the region and all the municipalities present industrial-devoted areas, with infrastructure and moderated land prices. Porto also has moderately priced living and office costs, especially when compared to Lisbon or Madrid.

8.3 The metal-electro sector in Porto

Portugal does not have a strong tradition in equipment and machinery production; during the last decades, it has been a net importer (see Figure 8.7). These imports were crucial to support industrial development in different manufacturing fields, like textiles, footwear or metal industries, but also, for example, in construction.

The sector in Porto goes back to the nineteenth century, starting with metal and welding artisans and with the production of tools and basic equipment for the wood industry and small-scale shipbuilding repairs (like welding and helixes). Companies were almost exclusively family-owned SMEs, developed on the basis of individual competences (like a talented crafter or a qualified engineer). Some of the surviving firms enter the twenty-first century with a diversified portfolio in specialised and high-tech niches. Most of them are still medium-sized firms, internationalised and linked with R&D institutes. Although lacking 'visibility',

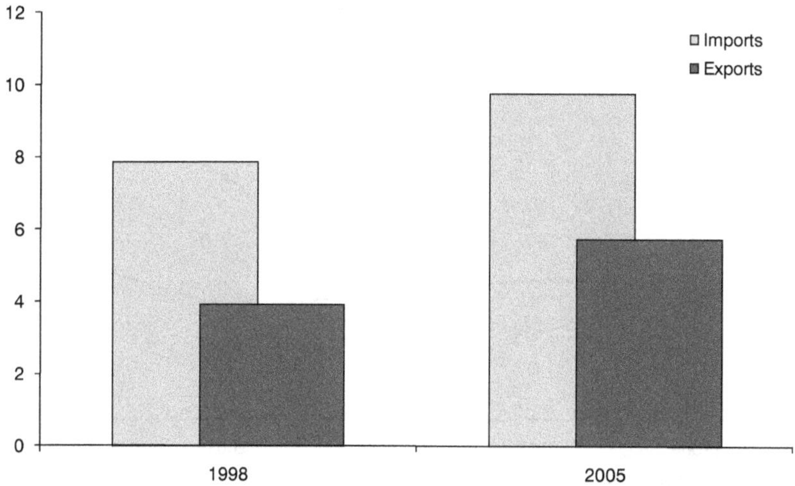

Figure 8.7 International trade of machinery equipment and apparatus (section XVI), Portugal, 10 billion euros (source: INE, Portuguese Statistics Bureau, n.d.)

some firms are regular partners of important world players in different sectors (like automotive or aeronautics); some others provide electrical solutions and enabling technologies for big infrastructures worldwide (like airports, rail transport, power stations and energy) or growing sectors in the regional economy (like medical equipment). Despite that their markets and network connections are increasingly global, the management decisions are still regionally based, sometimes rooted in personal and familiar relations and long-term partnerships.

Recent trends and evolutions

The metal-electro sector in Portugal has a relevant share in the industrial structure.[1] In 2004 (according to the national statistics bureau), it represented 24% of the total manufacturing firms, 18% of total manufacturing employment and 14% of total manufacturing turnover. Figure 8.8 depicts the sector's evolution in the last years, in Portugal as a whole. Although employment generated by the sector has been declining slightly, the number of firms in the sector has been relatively stable, while turnover shows positive growth. It is interesting to note that the relevant growth in turnover happened in a period of stagnation (and even recession in 2002) in the Portuguese economy, indicating that the growth may be strongly related with exports of equipment and machinery.

Within Portugal, Porto's urban region houses the most relevant firms in the sector and the largest share of employment. The concentration of the sector in the region is 1.8 times higher than the average for the whole country (see location coefficients in Figure 8.9). That concentration level is not dissimilar for the various metal-electro industries. The highest concentration in 2004 was in the fabrication of metal products, tools and heating apparatus, and the lowest relative concentration

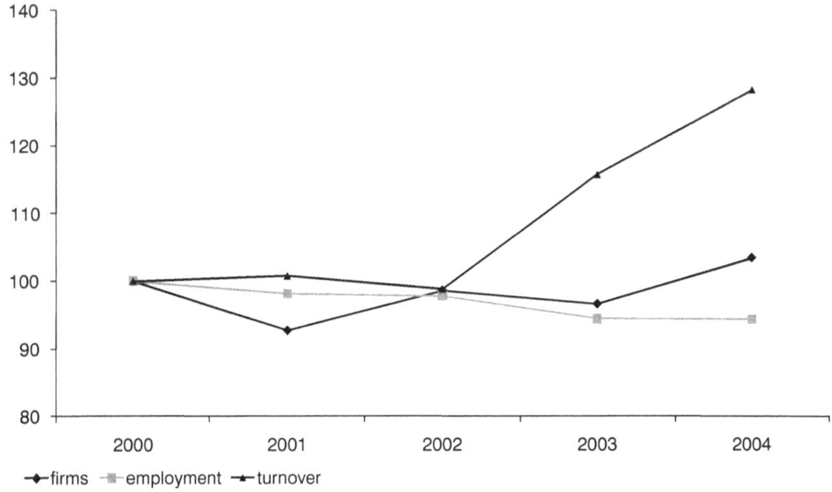

Figure 8.8 Firms, employment and turnover in the Portuguese metal-electro sector, 2000–2004 (year 2000 = index 100) (source: INE, Portuguese Statistics Bureau, n.d.)

(although still 1.7 times higher than in the rest of the country) was in the machinery and equipment industry. Despite the relative decline in the concentration of machines and electrical equipment in the region (measured by employment), all other industries became spatially more concentrated in the larger urban region of Porto. In absolute terms and for the larger urban region of Porto, the sector is representative, accounting for more than 5,000 firms in 2004, and employing almost 87,000 workers and generating roughly €6 billion in turnover (see Table 8.1). The industry of metal products, tools and heating devices (International Standard Industrial Classification of All Economic Activities (ISIC) 28) is the most representative (in firms, employment or turnover), followed by machines and equipment (ISIC 29) and machines and electrical equipment (ISIC 31).

In 2004, the Norte and Centro Regions revealed a noteworthy shift in the sector trade pattern. From net importers of machinery, they become net exporters, reflecting new competitive advantages in specific niches (see Figure 8.10). The foundations for this aggregated change can be found in micro-level stories of innovation and new R&D partnerships (Carvalho, 2008) (Box 8.1).

Sector composition and main players

What and who are the main elements and players of the metal-electro sector in Porto's urban region? Figure 8.11 aims to give a simplified and general image of the sector, its players and most important linkages in the metal-electro 'system' under analysis.

In the remainder of this section we explore the various types of actors that play a role in the sector: firms in the considered industries, knowledge institutes

Table 8.1 Key figures on the metal-electro sector in broader Porto (north-west of Portugal), 2004

	ISIC 28 *Metal products*	*ISIC 29* *Specific machines and equipment*	*ISIC31* *Machines and electrical equipment*	*Total*
Firms	3,616	1,406	361	5,383
Employment	45,118	25,659	16,177	86,954
Turnover (€ million)	2,537	2,055	1,345	5,936

Source: INE – Portuguese Statistics Bureau, n.d.

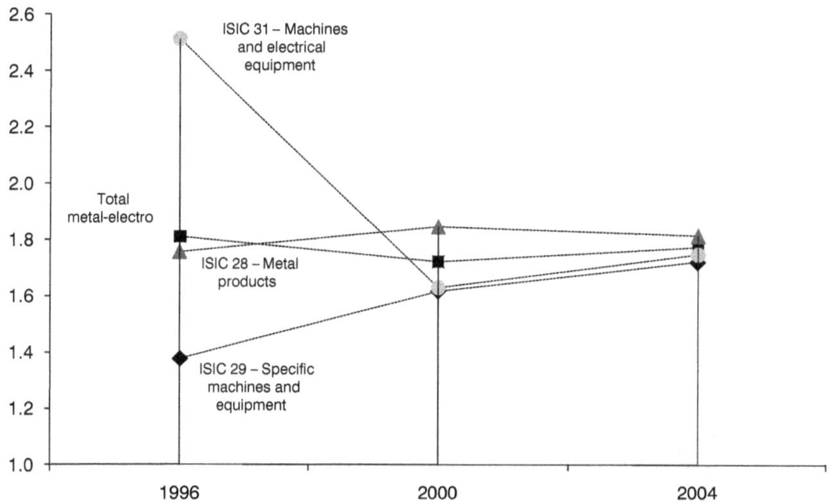

Figure 8.9 Location coefficients of the metal-electro sector and its different industries, north-west of Portugal, employment, 1996–2004 (source: own elaboration with data from INE (Portuguese Statistics Bureau), n.d.)

Notes: (1) Location coefficient is an indicator of sector concentration. It compares the share of the metal-electro sector in the north-western region with the share of that sector in the whole country. (2) North-western Portugal corresponds to the aggregation of the NUTS III Grande Porto, Ave, Cavado, Tamega and Entre Douro e Vouga.

and educational facilities, services and suppliers, and public agencies and sector associations. Finally, we elaborate on the geography of the sector in the urban region.

Leading firms and a vast fabric of SMEs

The metal-electro sector in Porto's urban region encompasses very different types of firms, mainly nationally owned (despite some important FDI), and mainly

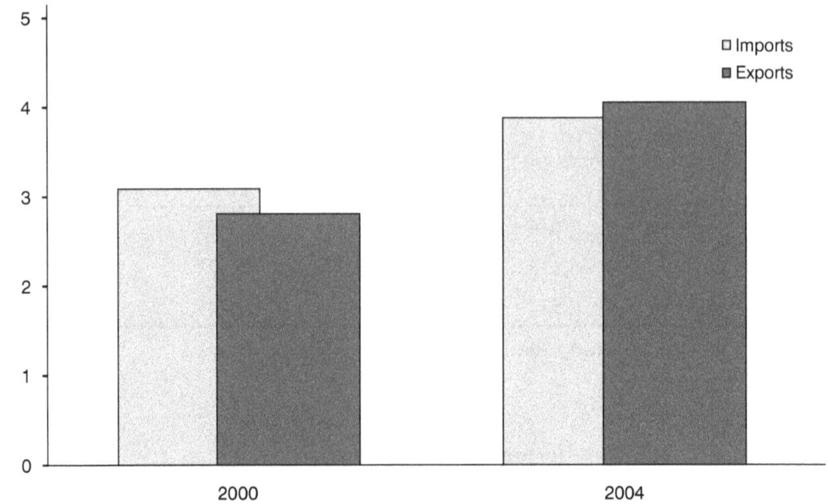

Figure 8.10 International trade of machinery equipment and apparatus (section XVI), Norte and Centro regions, €10 billion euro (source: INE – Portuguese Statistics Bureau, n.d.)

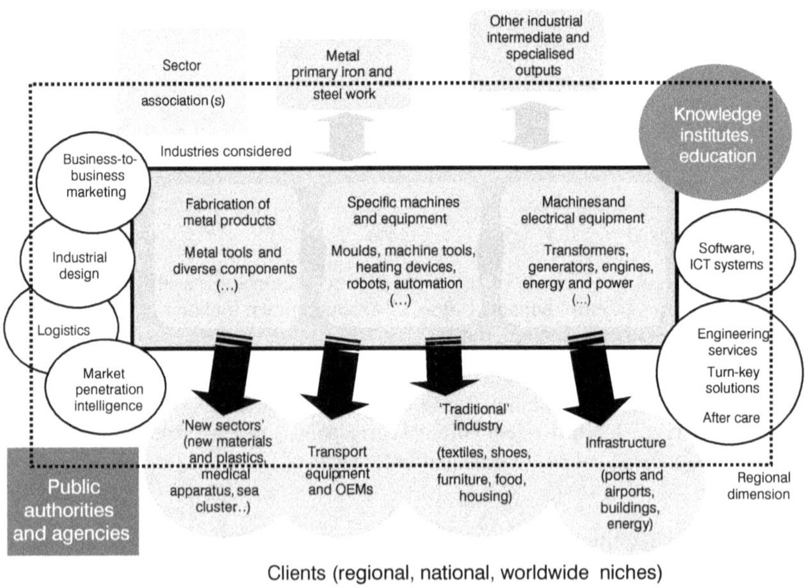

Figure 8.11 General overview of the sector (source: own elaboration)

Box 8.1 New machinery and equipment technology in niche sectors – the footwear case in north-western Portugal

The footwear industry in Portugal is clustered around three major poles, all in the vicinity of the Porto's urban region – Felgueiras and Guimarães and the industrial agglomeration of Entre Douro e Vouga. The sector used to import machinery for its production processes, typically for large and non customized series in labour intensive firms, frequently outsourced by global corporations like the Italian Ecco.

During the 90s Asian and Eastern European countries increasingly took the lion's share of this production. By that time the industry realized that a strategic move was needed, namely a shift towards the manufacturing of small series for higher value market niches. However, the sector lacked the technology to do so. As a response, in 1996, a first R&D consortium led by CTC – National Footwear Technological Centre – and by the footwear sector association, gathered several partners towards the development of new technologies for the footwear sector – footwear and machinery firms, R&D institutes. This initiative showed to the several partners the advantages of joint work and risk sharing for solving common problems, crossing different competences towards the development of industrial solutions aimed for international markets. During 2002–2005, a new and larger project (FATEC – Footwear Active Technology) followed the same methodology, counting 5 R&D institutes, 5 machinery and equipment producers and 9 footwear firms. The project budget was 13 million euros, joint funded by the European Union and by the different partners involved and gave rise to a series of new high-tech machinery for prototyping, laser cutting, logistics, automatic painting, robots, weaving technology, ICT solutions, etc.

Nowadays, these technologies are exported to several countries, not only for footwear production but also for automotive and food production. For the success of the project, interaction between the client and the supplier of the machinery was crucial. Geographical proximity allowed for permanent interaction, but also for cultural proximity and trust, while the presence of the sector association as an intermediary and trustworthy institution also played a central role.

Source: own elaboration

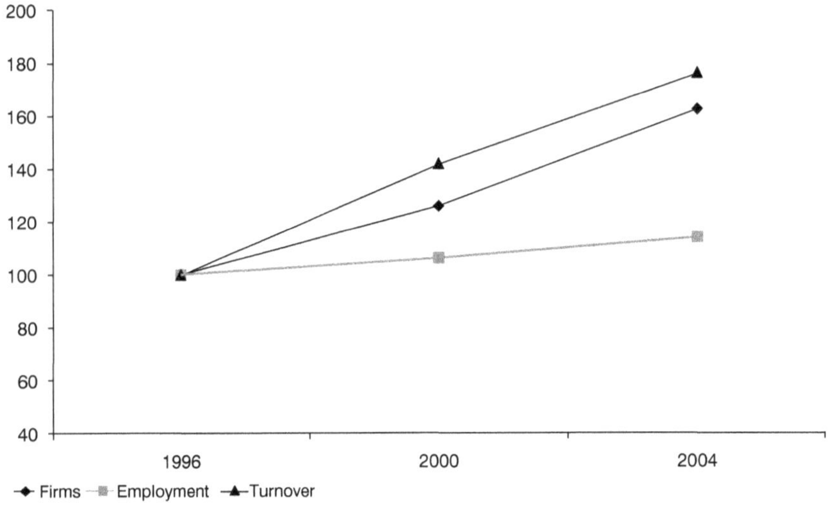

Figure 8.12 Firms, employment and turnover in the north-western region, ISIC 28 – Metal Products, 1996–2004 (year 1996 = index 100) (source: INE, Portuguese Statistics Bureau, n.d.)

SMEs, but with the presence of some large players. Although some subsectors are close to being 'worlds apart' (like metal tools versus energy apparatus), there are important linkages between them, which are also visible at the regional level (for example electrical equipment and components used in the production of specialised machinery).

The fabrication of metal products sector (mainly in ISIC 28) includes the manufacturing of products and structures for habitat and furniture (like taps, metal components and containers), metal tools and hard components for industries (like chemicals, textiles, packing, automotive and wood). The evolution of this subsector has been quite favourable in the last years (see Figure 8.12): not only have turnover and number of firms increased, but also employment.

Old family-owned SMEs co-exist with university spin-offs. One example is Alto, a spin-off from the R&D institute INEGI (see next section), working in the habitat segment with composite materials, as well as other innovative SMEs in the same sector, like Cifial. Amtrol-Alfa is an example of a firm with German capital that developed an innovative light gas cylinder; it was awarded several prizes for innovation and has the support of local R&D institutes.

To cope with global pressures and innovation requests, some former metal-based firms became engineering firms. That is the case of Frezite, a firm located in the urban region of Porto, with 200 employees and a turnover of €11 million (*Exponor News*, 2006). With selling points in Spain, Brazil and the United Kingdom, the firm produces high-tech and state-of-the-art precision and cutting tools, tailor made for specific client's requests (for example in automotive, aeronautics, moulds and wood).

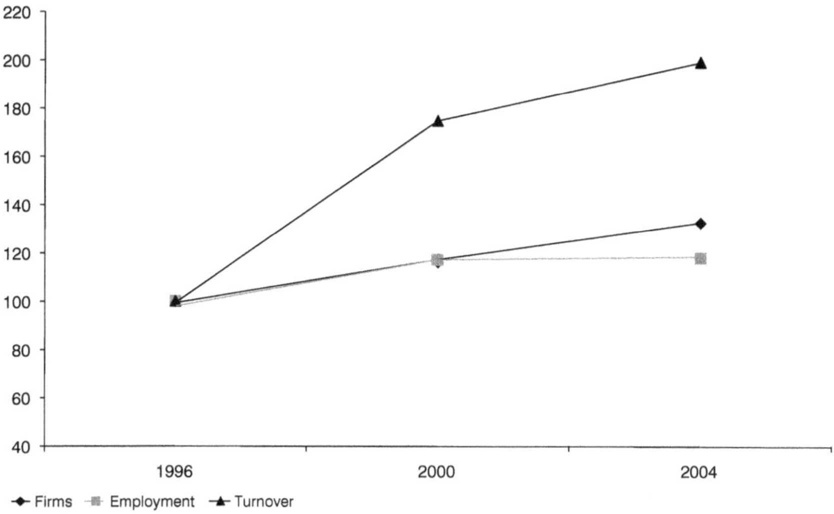

Figure 8.13 Firms, employment and turnover in the north-western region, ISIC 29 – Specific Machines and Equipment, 1996–2004 (year 1996 = index 100)(source: INE, Portuguese Statistics Bureau, n.d.)

The growing subsector of machines and equipment (mainly within ISIC 29) is also rather diverse (see Figure 8.13). It encompasses manufactured products like heating and refrigeration devices, machine tools for cutting, handling and precision, moulds (namely for plastic injection), robots and automation.

SMEs are the main players; many have family origins, although large foreign firms exist. One example is Vulcano, a Portuguese firm producing high-tech heating devices, with more than 1,000 employees, now part of the Bosch group. Vulcano takes part in many R&D consortiums in the region; it is the world's third largest producer of heating devices and the European leader. Moreover, robots and machine tools for precision tasks are some of the most relevant 'enabling technologies' of the sector, and important firms have been growing during the last decade. One of them is Adira, a 50-year-old firm that is still directed by family members. Adira is the Iberian leader in its niche: producing machinery with laser technology to cut and fold metal. It exports more than 75% of its production, which is manufactured in Porto. The firm distinguishes itself not only by its superior products but by having a vast range of services, like strategic niche studies, applied research, marketing and design, as well as superior ICT tools for internal and external organisation and supplier management.

Design skills, marketing and management of external relationships are increasingly important. Although the majority of the final outputs are intermediate goods, it has been pointed that industrial design increasingly makes the difference, as does business-to-business marketing. However, there is evidence that a vast majority of the firms in the sector do not take yet marketing and design efforts very seriously. In this subsector, the image of the product (namely in terms of

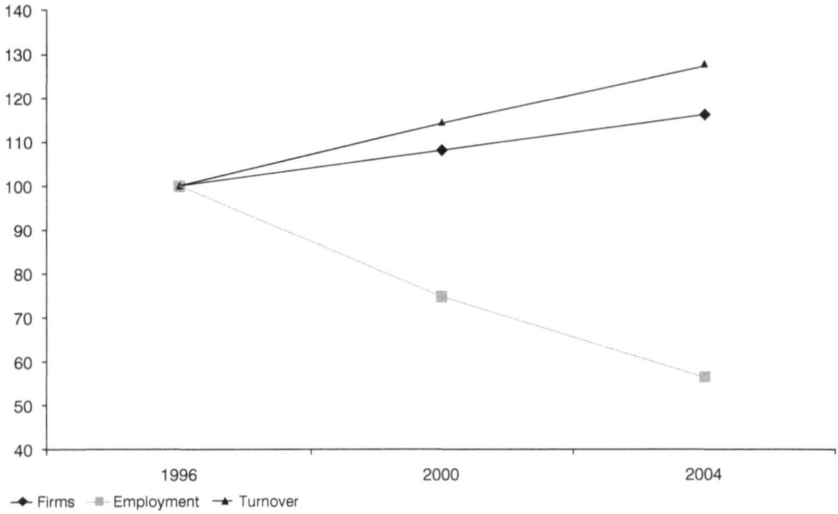

Figure 8.14 Firms, employment and turnover in the north-western region, ISIC 31 – Machines and Electrical Equipment, 1996–2004 (year 1996 = index 100) (source: INE, Portuguese Statistics Bureau, n.d.)

quality) is strongly associated with the tradition of the country as a producer. Germany, Northern Italy and the UK have clear advantages as quality engineering and machinery countries. Some Portuguese firms indeed export top machines produced in Porto through distributors in other countries, in order to avoid the 'image disadvantage'.

Another relevant exporting niche is metal moulds (such as for automotive plastics), clustered in the south of Porto's metropolitan area. To supply this sector, firms need to cope with high standards and top logistics. Simoldes is the largest and most international, but others, like Olesa, are growing fast.

Although growing in turnover, the number of jobs in the region has been declining for the subsector of machines and electrical equipment sector (mainly within ISIC 31) (see Figure 8.14). This is associated with the exit of FDI in simple electrical components for the automotive industry, moving eastwards. Besides those products, the subsector produces transformers, power generators, engines, energy devices and machinery for energy plants. In these segments, Porto has a very strong global player, a large group with family origins, headquarters and R&D units based in the metropolitan area: the Efacec group. Turnover is growing in Portugal and abroad (see Figure 8.15), and the company has production and sales sites throughout the world. One of the most relevant growing areas is related to 'green innovation' and diverse energy solutions. Besides in-house R&D, Efacec also has many R&D protocols in the region and a highly qualified labour force: 44% of the staff have secondary school qualifications and 38% have higher education degrees (Efacec, 2006).

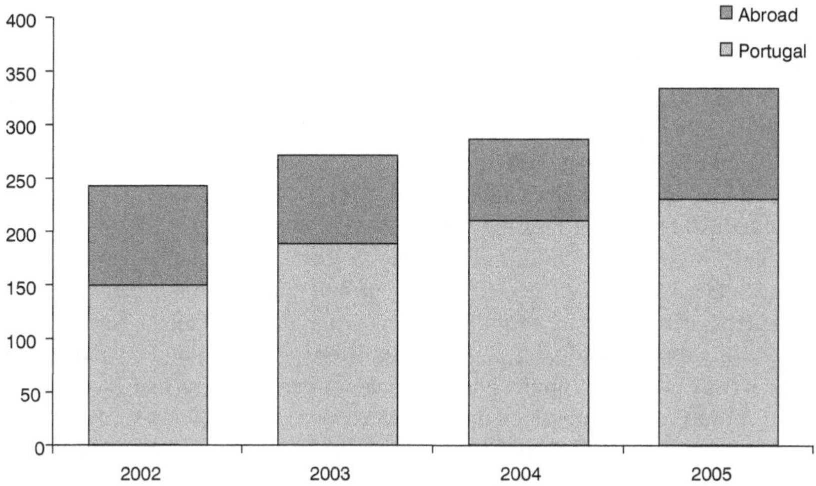

Figure 8.15 EFACEC – turnover evolution (source: EFACEC, 2005)

Presently, Efacec has six large speciality areas, using cross-competences from its different fields to produce tailor-made solutions:

- Energy, related to power generation, transmission and distribution, automation and telecontrol, as well as power supply systems;
- Transport, capitalising on the competences of other firms within the group, providing solutions in metro light rail systems (it recently won a bid for the metro in Jerusalem), railways and heavy metros, road transport and airports (it recently won a bid for new solutions provision in Charles de Gaulle Airport);
- Industry and large buildings, such as electric installations, handling, telecontrol, surveillance, solid waste treatment, power supply and safety;
- Logistics, for heavy industries and processes, like warehouse automated storage systems and baggage handling;
- Telecommunications, mainly integrated engineering systems and engineering solutions;
- Environment, covering two main areas: air (air conditioning, heating) and water (treatment stations).

Almost all products are tailor-made in cooperation with the client, including design and prototyping, installation and after-care services (also part of the company's core competences). Efacec recently sold its automation division (elevators) to the Schindler Group, and the managers believe other parts of Efacec might get an offer from large multinational corporations due to Efacec's strong competences.

Education and research institutes

The availability of skills and the presence of top research institutes are two of Porto's main strengths. An important part of new technology competences comes from regionally located research interface institutes between universities and firms.

INESC and INEGI (see descriptions below) are among the best engineering interface institutes from the University of Porto. Founded during the 1980s by Portuguese PhD-holders returning from the USA and the UK, both institutes struggled (within the academic bureaucratic structures of the University of Porto) to constitute true interfaces between university and industry. Today, both institutes are cornerstones of the innovation of regional (and national) metal-electro firms; they are highly international, active in transferring technology to the private sector. INESC is located on the university campus at the Faculty of Engineering. INEGI moved to the same location, and now more than 400 researchers in the metal, mechanic and electronic fields are there. Their actions are national and international, though they keep a very strong regional dimension.

INESC Porto is the Institute for systems, engineering and computers; it is considered an excellence 'national laboratory' with permanent research funding (telecommunications, energy engineering and production, ICT, optical and electronic technologies). ICT and electronic solutions are increasingly incorporated in machinery and equipments. INESC Porto employs more than 200 researchers, from different nationalities, and has been responsible for a diverse number of spin-off firms in Porto, especially in the fields of sensor systems and ICT. The funding sources of INESC Porto are diverse. In 2006, 19% came from European research projects, 35% from national research projects, 20% from direct research contracts with firms/consultancy and the rest come from diverse sources, like training (INESC, n.d.). INESC leads the largest European consortium for developing solutions for electric batteries to be used in the automotive sector.

INEGI (Institute of Mechanical Engineering and Industrial Management) (comprising instrumentation, automation, robotics, rapid tooling, sheet metal forming technologies, coatings, vibrations, energy, thermal engineering and composite materials) has roughly 160 researchers (see Figure 8.16) and a scientific output of about 300 publications per year, plus technology transfer activities. The larger component of its revenues comes from applied consultancy, followed by direct R&D contracts. The strategy of the institute is to get the largest share of its revenues from direct contracts with firms (many of them regional). Clients belong to almost all manufacturing sectors. Faced with the relative decline of the basic metal industry in the region, INEGI has been diversifying its activity towards the need of sectors like aeronautics (like developing R&D with the European Space Agency), transport and energy.

There is evidence of researchers from INESC and INEGI being absorbed by regional manufacturing firms, showing the increasing innovation awareness in local firms. In the case of INEGI, a very interesting trend is the increasing number of shareholders in its capital structure, namely regionally based firms, now with a relevant share and stake in the vision and mission of the institute. Both institutes

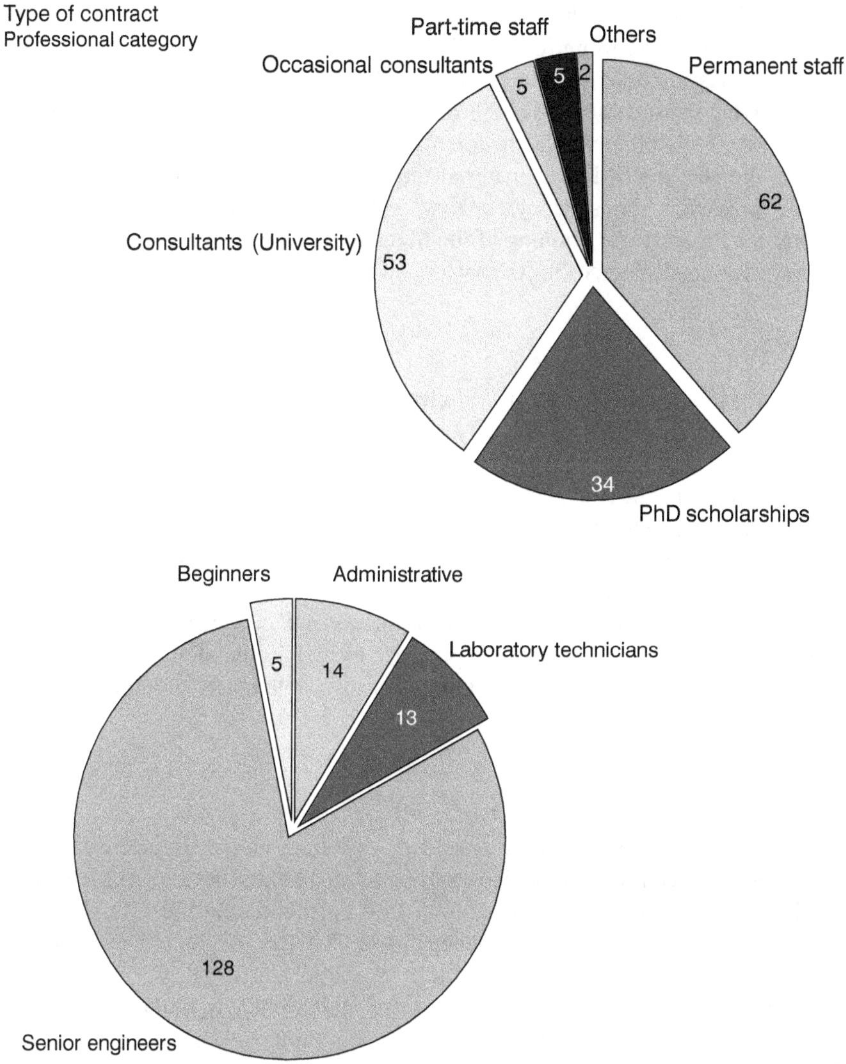

Figure 8.16 INEGI's staff (source: INEGI, n.d.)

are active in R&D platforms regionally, nationally and abroad. In the region, they face multiple calls to upgrade traditional sectors and steer the consolidation of new growth clusters (like ICT, medical apparatus, marine/maritime, 'green' innovation and energy). However, many researchers face the tension between working closely with regional firms and tapping into international networks of excellence and scientific research, increasingly supported by the National Science and Technology policy (Carvalho, 2009). This is a threat for the development of a sound regional innovation system.

Other departments in the Faculty of Engineering, like the robotics institute, work with the sector. Moreover, PIEP (new materials and polymers, University of Minho) has important industrial links and has been cooperating with INEGI in joint R&D consortia. Besides the R&D interfaces, the sector has the support of CATIM (Technological Centre for Metal Industries), located in Porto. They support the sector's SMEs in material testing, calibration, training, as well as service provision (consultancy, quality, environment and safety). CENFIM (Centre for Professional Training of the Metal and Mechanics Industry) provides several tailor-made courses for the sector.

Suppliers

As previously mentioned, the metal-electro sector is characterised by strong inter-industrial relations. Thus, many of the sectors analysed are suppliers and customers of each other, and regional links are very important (like the moulds industry supplying the energy solutions sector, or metal tools supplying equipment goods machinery). Other suppliers of the sector are global, namely for basic metal (globally sourced) and very specialised components, and are not found in the region (e.g. laser technologies).

Simultaneously, the need to cope with clients and fast technological change requires increased speciality services provided by and for metal-electro firms (like marketing, design, internationalisation and sales intelligence, procurement and after-care services).

Other relevant actors

Other actors in the region play a role in fostering the sector. Here we can distinguish between sector associations (like AIMMAP, Association of Metal and Mechanical Entrepreneurs of Portugal and AEP, Portuguese Business Association) and national and regional public agencies (like CCDRN, Regional Development Coordination Bureau and ADI, Portuguese Innovation Agency).

AIMMAP is the sector association for metal and mechanics, providing services and technical support for the sector. It is located in Porto and has a strong regional dimension. Located in the same campus as CENFIM and CATIM, it provides training and technological support for the firms in the sector. Besides services like law and consultancy support, AIMMAP is active in fostering the internationalisation of firms. AEP, with a national span but also located in the region, provides similar services but is not focused on any specific sector. Nevertheless, it recently held international seminars and EMAF (International Exposition of Machine Tools and Machinery) training courses for SME management, and provides innovation support. Other sector associations, for shoes, footwear, cork or wood have been acting as 'innovation pulls' for the sector, promoting research protocols with machinery and equipment producers.

Other associations exist for specific niches within the metal-electro sector, with similar aims (internationalisation, training, market intelligence, service support), also working at the national level. One of them is CEFAMOL, for the moulds industry, founded in 1969 by seven firms. Other is ANIMEE, covering the niche of the electrical and electronics system, energy and telecommunications (located in Porto).

ADI, the Portuguese Innovation Agency, is a joint agency from the Ministry of Economy and Ministry of Science and Higher Education, willing to strengthen the link between 'knowledge-technology-innovation'. Although transversal to all sectors, the regional metal-electro sector has been benefiting from different governmental and European incentives, especially for the development of R&D centres within the firms and research protocols for innovation.

CCDRN, the Regional Development Bureau of the Norte Region, has an important role in managing EU structural funds in the region. It manages incentives (for micro and small firms) towards innovation. In its recent strategic document for 2007–2013, metal-electro was highlighted as a transversal and key sector.

The geography of the sector

Although the specific boundary of the metal-electro sector has not been mapped before, our interviewees have little doubt about the geographical span of the sector around the large urban region of the north-west of Portugal (used in this case study).

The metropolitan area of Porto houses the the major support institutions. Besides important firms (like Efacec or Adira), this concentration hosts the core of the educational, R&D and knowledge facilities (like the University of Porto, CATIM, INESC and INEGI) as well as the sector association (AIMMAP), the national business association (AEP) and regional development bureaus (like CCDRN). The concentration stretches north, where relevant firms can also be found, namely in the municipalities of Trofa, Guimarães, Braga and Esposende. FDI has some relevance in these areas and concentrates the second biggest knowledge infrastructure of the region, namely the Faculty of Engineering of Minho and its research centres, like PIEP. This subregion will house two major R&D centres in the near future, in the fields of nanotechnology and new materials. Towards the south, the heart of the moulds industry can be found in the municipalities of Oliveira de Azemeis and Santa Maria da Feira; firms like Vulcano and others in the 'habitat' niche are found around the city of Aveiro, in municipalities like Águeda or Ovar.

We collected data on employment per sector and industries, at NUTS 3 level, in order to measure the concentration of the sector and its industries in the region. Then we measured the concentration (through location coefficients, Figure 8.17) of the sector and its industries compared to Portugal over time, on the basis of two different spatial units: the metropolitan area of Porto and the broader urban region of Porto. The results show that the relative concentration of the sector is much higher when the broader urban region is considered instead of only the metropolitan area (location coefficients are higher for the sector as a whole and for its different industries). The results also point towards the loss of relevance

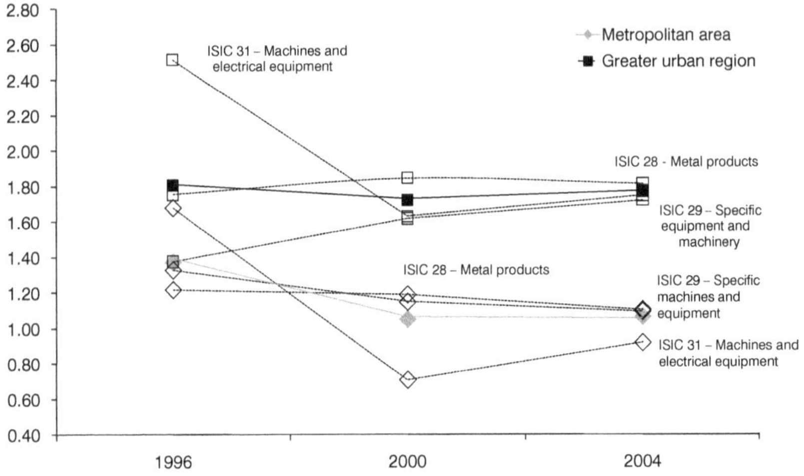

Figure 8.17 Location coefficients for the metal-electro sector, Porto – metropolitan area and broader urban region (source: own elaboration, based on data from INE – Portuguese Statistics Bureau(n.d)). Note: metropolitan area encompasses the 9 municipalities of the former administrative delimitation; broader urban region encompasses the 9 municipalities and the surrounding NUTS III of Ave, Cavado, Tamega, Entre Douro e Vouga and Baixo Vouga.

of the metropolitan area of Porto as industrial core, at least for industrial firms' location. The location coefficients decrease over time in the metropolitan area and increase (slightly) when the broader urban region is considered. Our interviews confirm the density of networks between the sector players within this spatial unit. This is particularly true for networks between firms and R&D institutes in order to benefit from the support and services provided by public agencies and associations. Despite this major national concentration, other locations also accommodate important firms and facilities, namely towards the south of the country, in Marinha Grande (moulds) or Palmela (automotive and machinery suppliers), with whom some of the region's players also tend to interact.

8.4 Local and global networks

Overall, Porto's urban region tends to show weak levels of collective efficiency and institutional fragmentation (Figueiredo and Silva, 2004; CCDRN, 2007). The metal and electro sector is not a total exception, despite interesting particularities. In reality, strong inter-industrial relations exist and there are relevant networks between firms and other players. We devote the next section to analysing the character and major trends in those networks, both for knowledge exploration and exploitation of present capabilities, as well as their spatial scope and the role of policy.

Networks between firms

Some important networks evolve around joint projects between competitors. This type of network is not very common within firms with similar profiles, producing the same type of product. The risk of spill-over of sensitive information is high, hampering cooperation. Moreover, the older generations of entrepreneurs tend to be very competitive and culturally refuse project involvement with 'rivals'. However, there are exceptions, like in the moulds industry. Led by young managers, some of the strongest firms share common resources and information on international markets in order to survive in a very competitive sector, namely supplying OEMs and first-tier suppliers. Other firms buy metal together and support each other to meet demand in case of stock failures.

When cognitive distance between firms increases, like belonging to the same industry but not directly competing, networks for joint projects emerge. There have been different examples in the last years, namely for the development of logistics systems and equipment solutions to deal with small series in the footwear sector (see Box 8.1), or to work towards a single product for which diverse competences are needed (like car seats or a complex machine). These networks tend to emerge at the regional level, and proximity is relevant. Partners can easily meet and know each other well. In the case of collaboration for industrial upgrading of other industries, geographical proximity is crucial.

Other firms participate in international projects, usually with high-profile research institutes. The EU framework programme has been funding diverse networks (for example for the development of techniques for flexible manufacturing). Some of those networks have derived new projects in which national firms access other markets and networks (aeronautics). In these cases, the networks have a European scale. Moreover, larger firms and groups are increasingly active in resource and competence sharing. Efacec combines competences of its different firms and competence centres towards integrated and tailor-made solutions.

Other networks mix firms, suppliers and clients. A decade ago, Efacec used to internalise a majority of the functions needed to provide complete solutions, from the manufacturing of the big pieces to 'last mile' services. Today, although a large share of manufactured outputs and services are in-house, non-core services are outsourced (usually in the region), while the company maintains the design, prototyping, manufacturing, complex assembling and core services. Acquisition of metal, steel and simple components are outsourced globally. Nevertheless, there is some criticism within the group on whether some services (like basic assembling) are compromised in quality, something which could unduly affect the reputation of the group. Ideally, those tasks should be reserved for specialised staff, at least with secondary school education. Efacec has also started to outsource the metal moulds needed for their pieces; before, they were manufactured in-house. The regional moulds industry gets some of the work, but relations are not stable.

Other firms in the sector also work with some of their suppliers for non-standard solutions. Efacec works with 'Fibersensing', a spin-off of INESC to develop monitoring systems and state-of-the-art sensors (for energy, transport infrastructure and logistics solutions). Other machinery and equipment firms in the region look for specialised suppliers to develop imaging systems; almost all machines need images and screens. When technology is available, companies contract with the suppliers and work together to meet the specifications. When it is not, R&D protocols are arranged with regional institutes and other firms. Here proximity is once again relevant to assure trust.

However, some other firms in the region look for solutions abroad. Adira has important suppliers abroad, specifically for hydraulic systems, software systems or photovoltaic cells, in Australia, Switzerland and Germany. However, joint development and testing is usually done in the firm plant on a temporary basis.

The character of the sector implies frequent interactions between the producers and its clients to meet specifications and produce customised solutions. In Porto's metal-electro firms, networks with clients are used for exploring new capabilities and to exploit present competences; both happen at different spatial scales. When the client is not located in the region, some innovation networks tend to happen anyway, linked with previous contacts and proven capabilities (for example with aerospace or automotive OEMs). Some firms establish networks to meet the clients to fine tune specifications. Some firms have engineers 'temporarily' close to the clients' manufacturing facilities. Networks in this case are mainly directed towards developed economies or the location of the OEM or first-tier producers. This may bring dependence for the firms in the region (and for the region itself). They are supplying for highly competitive markets and need to be constantly updated, and that capability may depend on partners outside the region.

Networks for exploiting a firm's present capabilities are increasingly relevant and several firms have foreign distributors or partners in other countries; others have production facilities. Efacec is a manager of external networks: they subcontract and are subcontracted, and are usually part of different multinational consortia. For the strategy of the Efacec group, global and local interaction for exploration and exploitation are increasingly intertwined. Its sources of competitiveness and competence are not only constructed within the region but globally, through interaction with clients and by exploring new production capabilities (like transport and energy systems). This increases its competitiveness and competences both internally and globally. This bi-directional interaction, gathering and applying knowledge locally and globally, makes the firm stronger. Despite its international approach, the group is strongly rooted/anchored in Porto (decision making and capital, knowledge, competence centres) and does not intend to relocate.

Networks between firms and knowledge institutes

Knowledge institutes have several research protocols with firms and with other R&D institutes, and are responsible for some of the most visible and evident internationalisation endeavours of the region. They are responsible for different types of networks and different spatial scopes.

Networks with firms are getting stronger and more stable. Several firms in the sector have regular research protocols with R&D institutes for knowledge exploration. Research institutes are particularly useful for SMEs lacking scale and internal R&D facilities. Our interviewees acknowledge an increasing shift in firms' mentality concerning innovation efforts. In the recent years, firms are increasingly taking the (regular) initiative to approach knowledge institutes for the provision of technological solutions. As a consequence, even more firms, large and small, are directly funding some of the R&D institutes, not only by direct contracts, but also by participating in the associative capital. In the last five years, several firms of the metal-electro sector joined INEGI as associated entities, ranging from large firms to local and regional SMEs. Roughly 50 firms and institutions are presently members of INEGI, and thus have power to discuss its mission and strategy. This situation heralds good news for the link with R&D and the local and regional economic base.

However, R&D institutes also have their own networks for exploration of new capabilities. In this sense, previous generations of EU support have been responsible for the establishment of important research networks. Moreover, knowledge gathered from outside has already proved to be relevant for some firms, namely in very specific networks in growing sectors, like aeronautics or aerospace. In order to achieve critical mass to participate in global networks, R&D institutes sometimes develop partnerships internally, crossing research competences (for example to develop new solutions for the automotive industry). The Portuguese government, together with different research institutes, has recently signed partnership agreements with the Massachusetts Institute of Technology (MIT) and with the Fraunhofer Institute, in different fields, for technology transfer projects and joint education. One example is the recently signed 'Alliance for science-based innovation in the manufacturing industry'. It results from a partnership of regionally based research institutes (INESC, INEGI and PIEP) and the Portuguese Association of Technological Centres with the Fraunhofer production alliance towards industry-focused R&D. The main objective of the alliance is to cooperate towards industry-orientated research activities.

Networks between sector associations

The relevant sector associations act at the national level, on a general basis (like AEP), on a broad basis (like AIMMAP) or are focused on specific niches of the sector (like CEFAMOL for the moulds industry). All are involved in supporting networks and cooperation between different firms in the sector, specifically through meetings and seminars, international visits or the development of joint projects. The role of the sector associations is understood by the firms as relevant to steer more cooperation between the sector's players. Firms use the events organised by the associations to exchange information on, for example, market evolutions and technologies.

Sector associations are also active in building international networks. Despite some criticism on whether these cooperation projects really benefit the associates, it has reinforced external networks worldwide. Generally, these associations also participate in European platforms, like ORGALIME, to discuss and lobby for the sector at a European level. On a national level, there has been interaction between AIMMAP and other sector associations, like the association for wood and furniture, in organising business fairs. This may yield good results in facilitating the interaction between different industries in the region and the metal-electro 'enabler' providers. There is presently some criticism on whether the institutions could better align their activities. Although activities are diversified and tend to specialise in different niches, some redundant support and activities might occur and overlap.

Public policy and thematic platforms

To what extent are public policies being directed towards the sector's development? In this last point we shed some light on past actions developed by public authorities and on forthcoming policies, regional or local, which may have an impact on the sector. In this point we analyse it through three levels of policy making: national, regional and local/supra-local.

In Portugal, the national government has traditionally had resources and competences to develop industrial and innovation policies. In the last two decades of incentives, those have been associated with three generations of EU co-funded programmes. Overall, those incentives were general (not targeting a specific sector) and, until recently, mainly focused on equipment acquisition. National policy was crucial to steer the development of technological centres and research facilities. Recent evaluation studies (AM&A, 2005) find evidence that the incentives provided by the central government towards economic development between 2000–2005 had moderate effects in productivity gains at the firm and regional levels. However, 'niche' policies have been quite successful, like R&D consortia and public–private partnerships. Several projects funded on this basis benefited the regional metal-electro sector, its associations and knowledge institutes, and thus are likely to be supported in the future. Taking the footwear experience as a benchmark, the same model is being replicated for the textiles and other traditional sectors, with predictable positive effects on the equipment suppliers of Porto.

Some years ago, the Ministry of Economy and the Ministry of Science joined forces to steer innovation and economic development, creating ADI, following the model of the Finnish TEKES to manage incentives for firms and R&D institutes (e.g. joint R&D projects, entrepreneurship and implementation of private R&D labs). In this context, project support has been bottom-up: central government does not specifically define priority sectors, but lets the projects emerge and funds them on the basis of quality criteria, not choosing any sector beforehand.

A new generation of innovation policies (following the political agenda of 'Competitiveness Factors') and a number of incentives for entrepreneurs target the competitiveness of firms in a systematic fashion, like encompassing incentives in R&D, branding, design, logistics and eco-efficiency and setting up competitiveness poles. These investments favour the metal-electro sector in the Porto region. The region has the strongest players that are prepared to shift towards more demanding policy incentives. Moreover, there will be new approaches at the regional level. Some sectors and clusters will be positively discriminated for public incentives.

The Regional Development Bureau of the Norte Region (CCDRN), an 'arm's length' institution of the national government, has very limited competences to influence industrial and innovation policy at the regional level; however this is slowly changing. With the support of the national government, CCDRN is for the first time managing direct incentives to firms, R&D institutes and sector associations, as well as promoting incentives for network building and international

Box 8.2 NORTE 2015 and Operational Programme, 2007–2013

During 2006 the Norte region prepared a participated development strategy – NORTE 2015 – to guide the 2007–2013 period of EU Cohesion Policy, for the first time with a very strong component of economic development, innovation and competitiveness. This strategy gave rise to the Operational Programme for the Norte Region. It devotes a specific investment axis for 'Competitiveness, Innovation and Knowledge' (see CCDRN, 2007), including the set up of the regional innovation system, consolidation of emergent clusters, upgrading of traditional industries, infrastructural facilities and science parks, etc. Within those initiatives, it discriminates positively some economic sectors for future funding: sectors in which the region has 'tradition' – textiles, footwear, wood and furniture – and 'emergent sectors' – electronic, machinery and electrical equipment, health, medical apparatus and pharmaceuticals and food. The regional metal-electro sector is likely to strongly benefit: firstly, it is considered one of the positively discriminated sectors, i.e. for cluster consolidation, spin off and start up promotion; secondly, it is a provider of 'enabling technologies', namely for the upgrading of traditional sectors; thirdly, together with R&D institutes, it may support the consolidation of emergent clusters (e.g. health, medical apparatus, ICT).

Other spearhead clusters are also supported, like energy and sea/water-related innovations. Also here the metal-electro sector will be called to play a role, not only by EFACEC, with strong competences in the energy fields, but also other machinery and equipment producers, namely robotics and automation. The research facilities of the region are already deeply involved in research for those areas and have protocols with firms (for example, to develop steel state of art cables for maritime platforms).

cooperation. Metal-electro players are considered crucial agents for the success of the strategy (see Box 8.2).

Despite its administrative power, the Portuguese local government has no tradition or competences in managing economic development issues. So far, its involvement has been confined to legislation and land use planning, as well as provision of infrastructure and industrial parks. There is evidence of some constraints caused by land use planning. Some industrial zones have poor accessibility and/or are strongly intertwined with the urban fabric, making it difficult for logistical operations, causing congestion and pollution, as well as becoming a nuisance for the local population.

Municipalities usually compete fiercely to attract (every kind) of investment on a local basis, and lack integrated efforts. The logic of competition has led to suboptimal results for the entire region since it failed in the past to develop a strong technological park, which could have strongly benefited metal-electro firms, spin-offs and the R&D facilities of the region. Although several efforts are under way towards the provision of science parks and top infrastructure, the region lost important opportunities in between. So far, it is reasonable to conclude that local actions, despite the provision of undifferentiated cheap land for industrial locations, have done little if anything to support or make the most out of the sector. Also the fact of it being a geographically spread sector hampers the action of concrete local entities, since they cannot reap the benefits of it within its borders and thus do not see much interest in spending money on it; moreover, it does not belong to the municipality's direct competences.

8.5 Conclusions and perspectives

In this case study we have analysed the dynamics of the metal-electro sector in Porto and in its broader urban region. Metal-electro is presently the only sector that simultaneously fits three important conditions: 1) installed critical mass (high number of firms in diverse industries, with superior routines, rich industrial tradition in the region and responsible for a vast number of direct jobs); 2) recent growth (increasing turnover and exports); 3) endogenous knowledge creation potential (rooted in firms and also in excellence R&D engineering institutes in Porto).

The sector under consideration is very diverse. It encompasses firms of different sizes, ownership models and innovation potential. Firms are 'technology suppliers' of different manufacturing sectors ranging from wood industries to aeronautics and space. This makes the sector a key technology provider and an important feature for the broader region of Porto, presently facing a sharp structural change towards more knowledge-intensive strategies. The metal-electro sector is already one of the main Portuguese and regional exporters.

Concerning its performance, the sector is performing quite well overall. We analysed performance through three different lenses: growth, development and dependence. In terms of growth, quantitatively, the sector did well in the last years. Despite slight variations across industries, the turnover volume has been growing as a whole (a substantial part due to exports), while the number of firms in the region also slightly increased. At the same time, however, the number of direct jobs has been declining. The sector has also been upgrading. Firms are increasingly involved with R&D projects to meet international demand, generally in niche sectors. While some parts of the production and routine tasks are increasingly outsourced, firms concentrate on their core competences. This led many of them to provide several associated services requiring strong engineering and management capabilities across the value chain. Increased openness and interaction with other firms, clients, suppliers and R&D institutes is contributing to this upgrade. It is also clear that the sector does not depend on single firms or even on single clients. Most firms are still family owned and their decision centres are in the region. The sector's development has been rather endogenous and with regional knowledge production capabilities. However, it presents sources of external dependence: some suppliers are not in the region (namely metal and other high-tech components) and although firms do not depend on other single firms, sometimes they may have some dependency from other sectors as clients (like automotive for the moulds industry) or innovation sources (client interactions for new product development, for instance in aeronautics).

What are the sources of this performance? They seem to rest on some critical regional features (*foundations)* as well as on the *structure* of the sector and key players. *Price factors* and *accessibility* do not seem to determine the performance of the sector; they do not assure outward investments in the sector's development stage *per se*, and, simultaneously, the region is still quite peripheral to the main production markets. Logistic conditions have still much room to improve. *Quality of life* is considered by some players as a relevant asset in retaining qualifications in the region. This links the asset to a critical foundation: *knowledge base*. Indeed, a supply of good engineers, top R&D institutes and the increasingly qualified workers give the sector the ability to produce innovative and globally competitive products and solutions. The *economic base* of the region, namely its industrial background, has also been central in the sector's development. Proximity to other industries gave rise to dynamics and interactions with equipment providers that could hardly be possible in other conditions. However, there is one important threat: in the context of regional industrial decline, the sector might lose one of its sources of advantage – the proximity to potential clients.

In terms of structure, the sector counts diverse players from different industries, but also top R&D institutes and sector associations, all of them located in the region. This gives the sector a good ground to develop (for example, through the emergence of high-tech spin-offs and start-ups). The sector, including firms and R&D institutes, is increasingly embedded in diverse networks, in different spatial scales, and in exploration and exploitation. Indeed, the combination of both types of networks has been important for the sector's performance, both in terms of efficiency and continuous upgrading.

National policy has been supporting the sector on a project basis, linking industrial and innovation policy. However, despite land use and the provision of undifferentiated industrial parks, local tiers have not been active in contributing to the sector's development. Regional policy in the near future is willing to take a role, but the approach and results are yet to be seen. Alignment of the action of sectoral platforms, international image enhancement and availability of the best researchers to cooperate with the sector are important challenges for the future.

Note

1 For data collection, we considered the statistical industrial delimitations of ISIC 28 + 29 + 31.

Bibliography

AM&A (2005) *Estudo de Revisao da Avaliacao Intercalar do Programa Operacional da Economia, [Mid term evaluation review of the Operacional Programme of Economy],* Lisbon: DGDR.

Carvalho, L. (2008) 'Footwear advanced technologies in the north of Portugal', in OECD (ed.) *A review of local economic and employment development policy approaches in OECD countries – WAG project.* Paris: OECD.

—— (2009) 'Four challenges of a new science and technology park – AvePark in Guimarães, Portugal', *Urban Research and Practice*, 2:1, 103–8.

CCDRN (2007) *Estrategia Regional Norte 2015. Versao de Trabalho [Norte's Regional Strategy 2015, working version],* Porto: CCDRN.

CTC (2006) *FATEC project report*, Felgueiras: Footwear Technological Centre.

EFACEC (2005) 'Relatorio e Contas' [Accounts report], available online: www.efacec.pt (accessed 3 June 2007).

—— (2006) 'Relatorio e Contas' [Accounts report], available online: www.efacec.pt (accessed 3 June 2007).

EUROSTAT (n.d.) Online database, available online: http://ec.europa.eu/eurostat (accessed 23 April 2007).

Exponor News (2006) 'Supplement 12 November 2006', available online: www.emaf.exponor.pt (accessed 7 February 2007).

Figueiredo, A. and M.R. Silva (2004) 'Factores Territoriais e Institucionais da Internacionalização da Cidade Região' [Territorial and institutional factors of the city region's internationalisation], paper presented on the Conference Porto City Region, Porto, 11–13 February.

INE (n.d.) Online database Portuguese Statistics Bureau, available online: www.ine.pt (accessed 23 April 2007).

—— (2001) 'Portuguese Statistics Bureau Census 2001', available online: www.ine.pt (accessed 23 April 2007).

INEGI (n.d.) Corporate website, available online: http://inegi.inegi.up.pt/inicial.asp (accessed 20 May 2007).

—— (2005) *Relatorio de Contas 2005,* Porto: INEGI.

INESC (n.d.), Corporate website, available online: www2.inescporto.pt (accessed 20 May 2007).

ISIC (n.d.) International Standard Industrial Classification – D28. Available online at: http://unstats.un.org/unsd/cr/registry/regcs.asp?Cl=17&Lg=1&Co=28 (accessed 25 March 25, 2010.

OCES (2004) 'Alunos Inscritos (1990–2004) – Inquérito estatístico anual aos establecimentos de ensino superior' [HEI Enroled Students, 1990–2004 – annual statistical survey to HEI], available online: www.oces.mctes.pt/?id_categoria = 21&id_item = 81958 (accessed 20 December 2004).

—— (2005) *Inquérito ao potencial Científico e tecnológico nacional, [Survey of the national scientific and technological potential*], Lisbon: OCES/MCTES.

—— (2006) *Portuguese Scientific Production 1981–2005 – Bibliometric Indicators*, Lisbon: OCES.

9 Rotterdam: food industry

9.1 Introduction

Manufacturing is still a dominant part of the Rotterdam economy, and many manufacturing activities are strongly linked to the city's port function. The transhipment function of the port makes it an attractive location for several types of industries. Rotterdam is specialised in capital- and space-intensive process industries, but the manufacturing of end-products is less developed. In the last decades, the share of manufacturing in the Rotterdam economy has steadily declined, especially in terms of employment. At the same time, manufacturing has become more knowledge intensive; there has been an increasing need for technical and managerial skills, and less demand for unskilled labour. From this perspective, the manufacturing industry has become part of the 'knowledge-based economy'.

In this chapter on Rotterdam, we focus on the development of the agro-food industry. The region has considerable capacity in the production of food and food-related products. Examples are the manufacturing of butter and dairy products, but also the horticultural 'production' of vegetables in the greenhouse cluster (in this chapter, we consider this activity also as 'manufacturing'). Rotterdam is a major transhipment hub of food products: many port-related activities are dedicated to transport, storage (including cooling) and distribution of food and related products. This includes major installations for storage and cooling of perishables and a large fruit and juice transhipment cluster. The region also hosts important food distribution hubs, and the region has substantial food-related R&D; the best known is Unilever's R&D activities in Vlaardingen. Unilever, one of the world's leading food companies, has its headquarters in Rotterdam.

In the remainder of this chapter, we sketch the industry context, the regional context, and the tendencies in the agro-food industry in general and in the Netherlands. Section 9.2 sketches the regional context, using the categories that were identified in our framework of analysis: the economic base, the knowledge base, accessibility, quality of life and factor conditions. Next, in section 9.3, we describe the development of Rotterdam's agro-food cluster, focusing on three subclusters: primary production (or 'horticulture'), food logistics, and the food and beverages sector. In section 9.4, we describe evolving networks relations and governance in the agro-food industry. We do this to get insights into the relations between manufacturing and knowledge-based activities, and the geographical dimension of these relations. We present our conclusions in section 9.5.

9.2 Regional-economic context

Economic base

Rotterdam is best known – in economic terms – for its port, the largest logistical hub in Europe, and until 2004, the largest in the world. The port is a gateway to the European market, and by 2005 it reached a throughput of 370 million tonnes. Despite steadily growing cargo transhipments, the employment in the port (in sea shipping, transport and logistics companies) is steadily declining, from 65,000 in 1995 to 57,000 in 2006, due to automation, containerisation and process innovations.

The port has attracted various manufacturing industries, and the transport and production industries each have a share of about 10% in the total employment of the region (see Figure 9.1). So, despite the reduction in employment, the port and manufacturing industries still account for one-fifth of the total employment. One of the major clusters is the petrochemical cluster which involves 16,000 employees, making it one of the largest petrochemical clusters in the world. This capital-intensive cluster generates substantial added value and indirect employment. Furthermore, the metal and electro-technical industries are still sizeable. These industries employ 35,000 people in this region, and between 60,000 and 65,000 if the installation branch, and the technical wholesale and the engineering bureaus are included (Kuijpers, 2006).

In the last decades, Rotterdam is facing a process of de-industrialisation: transport and manufacturing are Rotterdam's biggest losers in terms of jobs. In transport and distribution, over 7,000 jobs were lost in the period between 1996–2005. In manufacturing, another 6,000 jobs disappeared. The share of the transport sector in the local economy (GRP) dropped from 15% in 1992 to 12% in 2002. The industrial sector is weak in high-tech. In Rotterdam's industrial sector, only 6.3% of the jobs are 'high-tech', against 12.3% for the Netherlands. The productivity of workers in Rotterdam grows relatively slowly: in the period 1996–2002, labour productivity grew 0.5% on average, against 0.9% for the Netherlands as a whole (OBR, 2005).

Service industries have grown strongly, both in terms of value added and employment, and the share of business services in the economy now exceeds the national average. Knowledge-intensive services (software development, R&D, public relations, management and communication consulting) are growing rapidly in the city; in the 1996–2005 period, employment in this sector almost doubled, to 8,400 in 2005 (OBR, 2005).

The share of the port and its related industries in the urban economy is declining, and the city has experienced a shift towards a more diversified economic structure. For more than a decade, the city has pursued active policies to support and speed up economic diversification, mainly by stimulation of the leisure industry,

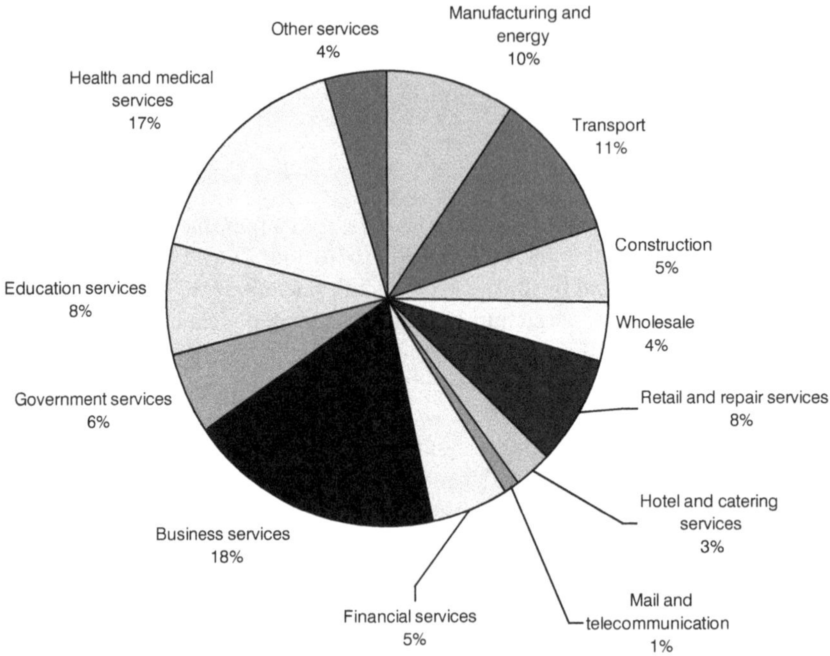

Figure 9.1 Economic structure of Rotterdam (01/01/2007) (source: own elaboration, data from Gemeente Rotterdam, 2009)

creative industries and the medical sector. The leisure industry is a growth sector, explicitly targeted by the municipality. Between 1997 and 2004, employment in this industry rose by 28% to a level of 18,000 jobs, or 6% of the total employment. In Rotterdam, the leisure sector is dominated by the business segment which accounts for more than half of the turnover (Bulterman and Van der Vet, 2006). The creative industries are another important target cluster for the municipality. This cluster comprises art, media and entertainment, and creative business services, e.g. design, architecture, etc. With almost 9,500 employees, Rotterdam is close to its 12,000-employee target for the creative sector (Bulterman and Van der Vet, 2006). Finally, the medical and healthcare industries are relatively well developed, and figure prominently as a promising growth sector in the city's overall development strategy. More than 16% of the active population is employed in health and care activities. Erasmus University's medical school has been rated the best on the national level. The Erasmus MC (Medical Centre) alone spends €20.8 million on research (Erasmus University accounting).

In 2002, the city council set the promotion of the 'knowledge economy' as one of its main goals. The Development Corporation Rotterdam (OBR) elaborated this theme in a 2003 report entitled 'Programma Kenniseconomie' (Program Knowledge Economy; PKE), stressing three main elements to stimulate long-term balanced economic growth: 1) enhancing the quality of life in the city (to attract and retain knowledge workers); 2) diversifying the economy via cluster policy – a continuation of the efforts described above; and 3) attracting and tying knowledge-intensive research and education institutes and companies to the city (OBR, 2003). According to a recent study (van den Berg *et al.*, 2005), Rotterdam has difficulties attracting and retaining knowledge workers, mainly due to quality of life and image deficits.

Knowledge base

The urban region has a strong knowledge infrastructure. Rotterdam houses several polytechnics (over 36,000 students enrolled) and one university, the Erasmus University Rotterdam (EUR), with over 20,000 students. In the neighbouring city of Delft, some 13,500 students are enrolled in the outstanding Technical University (TU) Delft. Its scientific staff numbers 2,400 people.[1] The EUR and TU Delft are both in the top 20 European universities with the biggest impact in scientific publications.

The region performs quite well, too, in terms of R&D expenditures. It accounts for 7% of total R&D expenditures in the Netherlands (fourth in all Dutch regions). The province of South-Holland ranks second behind the province of Noord-Brabant in R&D spending. The large firms contribute slightly more than universities (see Table 9.1). Most private R&D expenditures are done by large players in the food industry (e.g. Unilever, Sara Lee) and chemical companies (such as Shell and DSM). South-Holland ranks second in the number of patents applied for: 509 patents, which is 13.1% of the Dutch total number (OBR, 2003).

In 2005, the EUR and TU Delft, the municipalities of Rotterdam and Delft, and the province of South-Holland agreed to cooperate in developing the 'A-13 Kennisboulevard' (Knowledge Boulevard). This includes a science and business park and an airport-related business park. The city aims to become one of the premier European innovation centres.

A key prerequisite for developing a knowledge-based economy is a highly qualified labour force. However, the educational level of Rotterdam's working population is relatively low: see Figure 9.2 for a comparison of the four largest

Table 9.1 R&D expenditure in South-Holland province, 2002–2003

	€ million		*% development 2002–2003*		*% distribution 2003*	
	2002	*2003*	*South-Holland*	*Netherlands*	*South-Holland*	*Netherlands*
Business	577	660	14%	6%	41%	57%
University	597	563	–6%	2%	35%	28%
Research institutes	319	372	17%	4%	23%	15%
Total	1,493	1,595	7%	4%	100%	100%

Source: CBS, 2005, Edited Rabobank Nederland

cities in the country. However, this situation has improved: between 2003 and 2005, the city's low-education active population declined from 85,000 to 73,000; its mid-level and high-educated segments grew by 3% and 2.5% respectively. In the Rotterdam region ('Groot-Rijnmond'), 32% of the labour force have a tertiary education; 28% have a low-education level (mostly vocational education). The same picture applies to the municipality.

The larger Rotterdam region is endowed with many knowledge institutes in the food domain. Prominent among them is the Unilever Research Centre in Vlaardingen, which employs about 875 researchers. This centre is one of Unilever's six centres of excellence. It focuses on R&D in the domains of nutrition, health and well-being. The Unilever Food and Health Research Institute (UFHRI) conducts most food-related research activities. It aims to produce innovative new products and support innovation projects. UFHRI cooperates with external partners such as universities. Unilever Foodsolutions generates innovation for Unilever's worldwide food-service activities. This centre is geared to the business-to-business market. In the Spreads and Cooking Products division, 200 people work in the field of margarine and spreads. In 2005, the Consumer Centre was established in the Rotterdam region. It tests consumer behaviour and experiences.

Both regional universities – EUR and TU Delft – provide knowledge that is directly or indirectly relevant for innovation in the food sector: logistics, technology, management, medicine and law. The Erasmus Medical Centre has a good reputation for its medical research in the field of virology, cellular biology and immunology. The Medical Centre's main focus is curative, while the food industry is in need of preventive medical innovation. Law studies may also be relevant to the agro-food industry, in particular due to the increasing weight of international food regulations and the growing threat of claim suits. However, there are few if any links between Erasmus Law School and the food industry. The EUR's 'Erasmus Food Management Institute' is a business school for the Dutch food industry. Its training and research focus on management issues such

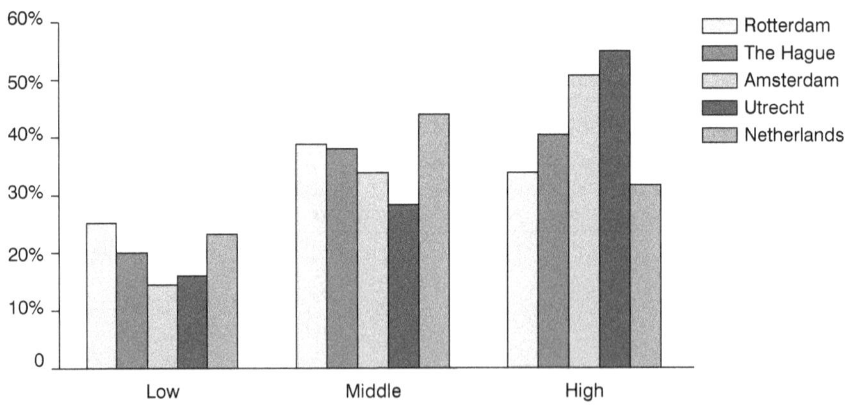

Figure 9.2 Education level, active population in the four big cities and in the Netherlands, 2005 (source: ETIN Consultants, 2006)

as 'shopping behaviour and food trends', 'brand and formula management' and 'category management and trade marketing'.

In Delft, DSM Food Specialities is a strong innovator. DSM develops, produces and markets food ingredients and it processes aids for the dairy, savoury, beverage, bakery and nutritional markets. The DFS Research and Development organisation is a strong engine of innovation (www.agriholland.nl).

Finally, several highly innovative 'seed houses' are located in the Rotterdam region. These firms run state-of-the-art laboratories in which they create new seed varieties for horticultural production for the world market. There is strong cooperation between the seed houses and producers under the direction of The Greenery BV.

Despite substantial R&D activity related to food in the Rotterdam region, Wageningen remains the leading region in the Netherlands in this domain, and is even acknowledged as one of the greatest clusters for food innovation in the world. The Wageningen University and Research Centre comprises a multitude of highly innovative knowledge institutions. The cluster as a whole brings together large innovation assets; between 4,000 and 4,500 people are working in the state-of-the-art facilities of the cluster (interview). In Wageningen we also find the Top Institute Food and Nutrition (TIFN, formerly known as the Wageningen Centre for Food Sciences), a national pre-competitive research platform which involves the major food companies such as Unilever and DSM (see Box 9.1).

Many food firms and labs in the Rotterdam region maintain very good working relationships with actors in Wageningen, and several of our interviewees did not see an obstacle in the 100km that separates Wageningen from Rotterdam (see section 9.4).

Accessibility[2]

Rotterdam is Europe's largest logistical hub. The port is the gateway to a European market of 450 million consumers. More than 500 scheduled services link Rotterdam with over 1,000 ports worldwide. Throughput in 2005 amounted to 370 million tonnes. Due to the size of the operations, the port offers significant advantages of scale. The European market is accessible from Rotterdam via five competing modalities: road, rail, inland shipping, coastal shipping and pipelines. The latest addition to its transport network is the Betuwe Line, a 160km railroad that links the port of Rotterdam to the Ruhr area. Goods that arrive in Rotterdam in a morning can be in, for example, Germany, Belgium, France or the UK the same afternoon. One of the main advantages of Rotterdam is its location on the estuary of the rivers Rhine and Maas. As a result, efficient and economical transport inland into the heart of Europe is possible.

Rotterdam's strategic location makes it also an important hub for agro-food logistics. The Merwehaven/Vierhaven area in the port of Rotterdam, also known as Rotterdam Fruitport, is completely geared to the handling of vegetables, fruits and fruit juices. Fruitport is one of the largest clusters in the world for 'perishables'. Within Europe, it dominates the field of fruit. The terminal operators integrate

Box 9.1 Top Institute Food and Nutrition (TIFN)

TIFN aims to promote and conduct top-level research with potential to be applied and commercialized. The research should result in 'the development of innovative products and technologies that respond to consumer demands for safe, tasty and healthy foods' (www.tifn.nl). TIFN has a research budget of €40m. Fifty per cent of the budget is generated by the Ministry of economic Affairs, 25% by public research institutes, and 25% by industrial partners. The research priorities are defined by the Program Council, consisting of leading university professors and R&D directors of the participating firms, and the actual research is conducted in knowledge institutes across the Netherlands. Once a certain project is finished it is then taken over by the private laboratories of the participating firms. The industrial partners are the 'big names' of the Dutch food and beverages sector, such as DSM, Unilever, Friesland Foods, etc. TIFN is a 'virtual' organisation: its 200 employees are located at the various participating knowledge institutes. Only the project leaders are located at the TIFN office in Wageningen. The institute's research focuses on three (interdisciplinary) research areas: Nutrition and Health, Sensory and Structure, and Bioingredients and Functionality. The three programme directors draw up research projects that are subsequently put forth for approval by the programme board. The projects are researched by the TIFN with resources that it rents from its members. TIFN plays a major role in providing its industry partners with leads for the development of new, healthy foods with regard to major health concerns such as obesity and metabolic syndrome.

Source: own elaboration

their activities; successive links in the chain are brought together within one company. Rotterdam Fruitport also increasingly accommodates activities geared to distribution, re-packaging and trade. A growing number of fruit-related companies operate in or near Rotterdam Fruitport. Another part of Rotterdam Fruitport, the Vierhavens area, is fully geared to the handling, cold storage and distribution of concentrated or fresh fruit juices. Products enter and leave the port as bulk cargo and in containers, tanks, drums and bins. Companies in this cluster offer sophisticated laboratory services for quality control. They also have a high degree of expertise in terms of production processes such as blending, (de)drumming and the crushing of frozen juice. The approach has been fully integrated. In addition, the port is a well-known market for fruit juices and concentrates.

Fruit, vegetables and fruit juices are increasingly being transported in special containers: so-called cool containers, or reefers. Rotterdam has a European distribution function when it comes to reefer cargo. Large quantities of fresh and frozen fish, meat and ready-to-eat meals are handled by the specialised cold stores that are located in the Waal/Eemhaven area and at the Maasvlakte. At the moment, there are some 2.5 million cubic meters of cold store capacity in and around Rotterdam and this capacity is growing. Rotterdam has roughly 3,500 so-called reefer points, special connection points for reefers, which is more than 40% of the total capacity of the ports in Northern Europe. The companies handle all aspects related to the distribution process for their clients, such as paperwork and veterinary inspections. The customs offices, which are integrated in the companies, guarantee a quick handling.

Road transport is by far the dominant mode for transporting food products, especially fresh products: it beats the other modes (barge, train) in terms of speed and flexibility. However, road transport is under pressure because of congestion and increasingly stringent regulations by several European countries to reduce its negative environmental impact.

While the goods transportation network is evidently important for the performance of the agro-food cluster in the region, passenger accessibility is equally important to attract high-skilled workers and to draw human resources from adjacent labour pools. Schiphol Airport is less than one hour drive, or 45 minutes by train from Rotterdam Central Station. The connection of Rotterdam on the HST is expected to decrease travel time between Rotterdam and Amsterdam/Schiphol, and will also substantially decrease the travel time between Rotterdam and Brussels. The regional airport (Rotterdam Airport) has surpassed 1 million passengers a year.

Rotterdam is equally well connected through several good quality highways. The accessibility by car to firms in Rotterdam is somewhat better than the other G4 cities;[3] also the possibility for parking cars is better in Rotterdam. On the regional level, the Randstadrail has been launched; this light-rail network connects the south-wing municipalities in Randstad.

Quality of life

Quality of life is becoming a significant factor for cities that want to transform to a knowledge-based economy, due to the increasing demands of high-skilled workers. Rotterdam has several attraction factors for these groups of workers: a modern architectural setting, an international character and all the amenities of a large city. Rotterdam offers a multitude of leisure activities in several fields: sports (both recreational and professional, including an annual international marathon), culture (cinemas, museums, theatres, the International Film Festival, summer carnival and annual dance parade, etc.) and a vibrant nightlife. In 2005, Rotterdam ranked second behind Amsterdam in the top five Dutch cities most visited by the Dutch. In the youth segment it ranks 7th in living attractiveness (OBR, 2006).

On the other hand, for many years, Rotterdam has lacked sufficient adequate housing for medium- and high-income residents. For a few years now, the municipality has been actively tackling this problem: more high-grade apartments and houses have been made available, e.g. in the 'Kop van Zuid' area and through large-scale neighbourhood renovation. Two Rotterdam neighbourhoods are amongst the ten best in the country (Bulterman and Van der Vet, 2006).

The city of Rotterdam has a long record of improving its quality of life and its image by investing in cultural events. Rotterdam organises the successful annual 'International Film Festival Rotterdam', and the city was a 'City of Architecture' in 2007. The number of excursions to Rotterdam has grown steadily thanks to major events and a varied catering and cultural supply; events like the 'Red Bull Air Race', 'Monaco aan de Maas' are mainly credited for increasing the visits by one million between 2004 and 2005. The aforementioned, food-orientated Markthal project fits in this effort (Bulterman and Van der Vet, 2006). Rotterdam has suffered from image problems in the area of security. In this field, Rotterdam has made – and is still making – significant progress; in the safety index (scale; 0–10) commissioned by the Rotterdam municipality, the city has succeeded in moving from an average of 5.6 in 2001 to 7.2 in 2006.[4] To integrate the city marketing activities, the city founded the Rotterdam Marketing Agency (van den Berg *et al.*, 2000).

Factor conditions

Factor conditions (prices and availability of land and labour) are highly important for the development of any manufacturing activity. Concerning the labour market, we have already seen that Rotterdam faces an oversupply of unskilled labour and a shortage of skilled workers and knowledge workers. Rotterdam is facing several additional labour market imbalances and challenges. First, like the rest of the country, Rotterdam's active population is ageing. The CWI (Dutch Centre for Work and Income) is expecting a growth of 50,000 jobs (8% of total) in employment in the Rijnmond area within the next five years while the active population is expected to grow by just 1%. A particular challenge for the next decade is the replacement of retiring workers in the industrial and port sectors, as these sectors lack appeal amongst the young people. Second, there is a mismatch between local skills and jobs. In 2005, Rotterdam had the highest unemployment level of the four largest Dutch cities, and at the same time the highest number of job vacancies. This gap is further aggravated by a large outflow of high-skilled locals towards jobs outside Rotterdam. Unemployment is highest amongst the lower-skilled workers. Third, the region's labour market is characterised by a spatial imbalance. The city of Rotterdam receives more incoming workers (138,000) than inhabitants working locally (118,000). The large inflow does not fully cover the job vacancies, while the large outflow does nothing to allay the unemployment levels. Also at the level of the Rijnmond area (greater Rotterdam), we observe a sizeable daily inflow (98,000) and outflow (94,000) of commuters. This mismatch has proven to be structural in nature. Moreover, large flows of commuters lead to an increased mobility and a decline in accessibility.

Approximately 50% of the city's population is of foreign origin, and some 167 origins and nationalities are represented. A large segment stems from recent migration waves, and is marked by a weak socio-economic position expressed in educational and income levels. Despite its disadvantages, it might prove beneficial in the long term. This segment could make up for the total ageing population. The upcoming generation, if adequately educated, could help bridge the labour market mismatch. Already, the city is witnessing a rise in entrepreneurship among minorities.

A key aspect of factor conditions is the link between labour costs and productivity: high labour costs are not a problem as long as workers are very productive. In some segments of the Rotterdam economy, however – and significant parts of the industry, the port sector among them – labour costs are relatively high due to high levels of unionisation. At the same time, productivity gains in Rotterdam's industrial sector have lagged behind the national average. This worsens Rotterdam's attractiveness for industrial companies.

The factor 'land' matters as well, in particular for the development of Rotterdam's agro-food cluster. In the last decades, the rising costs of land have pushed many space-intensive manufacturing industries and distribution centres out of the region, towards the hinterland. A similar process now occurs in the horticultural sector. In the highly urbanised southern part of the Randstad, there is an increasing pressure to move space-intensive horticultural production (greenhouses) to other parts of the Netherlands where land is more abundant. Indeed, many greenhouses have moved to North Holland, South Brabant and the northern part of Limburg. This tendency is reinforced by an ongoing consolidation in the industry, resulting in larger companies that demand more space.

9.3 The food industry in the Rotterdam region

This section describes the development of one prominent industrial branch in Rotterdam: the agro-food industry. Before turning to the Rotterdam region, we will first give a bird's-eye view on the national context.

The Dutch agro-food industry: a bird's eye view

The agro-food industry is an important sector for the country. In 2002, it generated around 10% of the Dutch GNP, 20% of exports, €15 billion in added value, and employed 600,000 people (of a 7 million active population). The Netherlands has a world reputation in food safety and certification, and Dutch companies are successfully targeting specialty and niche markets (Ministry of Economic Affairs, 2004).

An overwhelming majority of the enterprises in the Dutch agro-food sector are small and medium sized. Large firms make up some 5% of the total, but they generate 57% of employment and 78% of sales. Some of the big names are Unilever, DSM, Campina, Royal Cosun, CSM and Heineken. Among the top 25 EU food manufacturing companies, five are Dutch. Worldwide, seven Dutch companies figure in the top 100 ranking (M-team, 2006). The Dutch agro-food industry manifests itself in many segments and activities. There is much primary production in the western, southern and eastern parts of the country. A significant component of this production is horticultural.

The Dutch agro-food industry is highly innovative. Wageningen has an extensive knowledge infrastructure which centres on the Wageningen University and Research Centre. The Wageningen region is endowed with significant manufacturing and research assets, and, as we shall see later, plays a key role in innovation networks. The Rotterdam area fulfils a pivotal function for food logistics on the Northern European level.

Food manufacturing is still a significant industrial activity in the Netherlands. Firms in this subsector offer 149,000 jobs or 15% of total Dutch industrial jobs. In 2002, the industry included 4,800 firms (10% of the total industrial enterprises). Dutch food manufacturing is internationally strong. From 1997 to 2001, it generated an annual growth of 5.2%, the highest rate in Europe. In the period 2000–2001 the Netherlands was the number one in the world in terms of value added per employee, and it had the highest profit as a percentage of turnover. Dutch food manufacturing realizes approximately 47% of its turnover in foreign markets. Domestic sales run mostly through retail, in which supermarkets are the largest segment.

The agro-food industry faces a number of challenges. It needs to respond adequately to rapid technological development, changes in tastes and regulation. Quality, security and environmental regulations have become critical issues for the industry. Consumers increasingly demand healthier products, but there is also a trend towards more ease and convenience. There are new revolutionary technological developments, too (see Box 9.2 for some examples). To be successful in this complex and changing environment, value chain efficiency is increasingly important, together with a focus on innovation, efficiency and branding.

For the national government, food is a priority domain. The 'Food & Nutrition Delta' (FND) is a national governmental initiative to enhance innovation in the Dutch food industry. An important target of FND is to promote top-quality research through the Top Institute Food and Nutrition (TIFN). The government provides half of TIFN's budget, and the other half is paid by the knowledge institutions and private firms (Unilever is among the most important ones). But the FND also targets SMEs and aims to help them in their innovation process.

The 'Peaks in the Delta' is a national programme, operated by the Ministry of Economic Affairs, to support regions in their development of economic strengths. The food sector is identified as an important sector in the east of the Netherlands and the region around Wageningen, but also for the southern part of the Randstad region, the programme recognises the food cluster as promising for regional development. Authorities in this region are invited to submit plans in this domain, which – when approved – may receive national support.

In the economic priorities of the city of Rotterdam, the food industry does not play a significant role, although in its recent economic vision, the Economic Development Board Rotterdam (EDBR, 2005) noted the food/horticulture top 'cluster' as a strength of the city.

Box 9.2 Some prominent research areas of TIFN

Nutrigenomics. The aim of research in this field is to discover and validate molecular biomarkers for the early detection of metabolic stress. This will enable the development of novel food components for dietary management and prevention of metabolic stress, and ultimately metabolic syndrome. An application in this field is the so-called 'biomerkenset' which can instantly test a product for allergens. This could yield substantial results for companies as it helps them avert claims from harmed customers, and also because in packaging, allergy warnings will not be needed.

Research on food sensory and structural properties. This type of research focuses on gaining a better understanding of food sensory attributes and their origin in the interplay between food properties and mastication and on gaining knowledge of how to engineer foods to generate the required sensations. This knowledge is essential in developing excellent tasting food products that meet increasing complex and specific nutritional requirements, such as low fat, slow carb, low salt, or high protein.

Research in Microbial Functionality and Safety. The aim is to develop fermented foods with improved functionality. Such foods may also target microbial activity in the gastrointestinal tract. The research also focuses on new approaches to control food safety and preservation. The research makes use of gram-positive bacteria with known genomes as model systems.

Source: www.tifn.nl

Horticultural production in the Rotterdam region

The Rotterdam region includes a large horticultural production area (see Figure 9.3) that produces around 43% of the total Dutch horticultural production (Bulterman *et al.*, 2005). The graph shows that employment in this sector is declining, but only slightly. As described in the previous section, land scarcity and consolidation in the sector drive producers towards other regions.

An important player in this sector is The Greenery B.V., a cooperative auctioning and wholesale company whose main activity is the supply of vegetables, fruits, potatoes and mushrooms to supermarket chains, wholesale businesses, catering companies and food manufacturing companies. It employs 2,100 people, and it imports from and exports to some 60 countries. The Greenery B.V. was established in 1996, following the merger of nine fruit and vegetable auction houses and the Central Bureau of the Dutch Auctions. The Greenery

B.V. is owned by 2,000 producers, members of the horticultural cooperative 'VoedingsTuinbow Nederland' (NTV). Operating from Barendrecht, Maasland and Bleiswijk, The Greenery B.V. supplies its customers throughout the year. For this, it relies on production bases throughout the world, but 60% of its sales take place in the Netherlands, Germany and the United Kingdom.

The horticultural production that takes place in the greenhouses around Rotterdam can be considered a form of 'mass manufacturing' of fruits and vegetables. The horticultural sector is extremely productive and very innovative, not least thanks to the presence of highly innovative suppliers such as seed houses, greenhouse constructors, specialised machine builders and ICT systems suppliers. Margins are thin, and the industry is characterised by increasing concentration (fewer but larger players). Therefore, producers increasingly try to compete on innovation.

Food logistics

The strength of Rotterdam as a food logistics hub thrives on the combination of a strong local production base (primary production), imports overseas – from as far as the Far East – and imports on land – from as far as Morocco and Turkey. The imported volumes are substantial and do not solely serve the Dutch market but are channelled towards the rest of the continent. This combination offers Rotterdam great economies of scale and the ability to supply a full range of products throughout the year. Other ports in the Le Havre-Hamburg range do not have access to a local production base and thus have to reduce their operations at the end of import seasons. The outflow from Rotterdam towards the rest of the continent, and as far as St. Petersburg and Finland, is transported almost solely by road. The main markets are the Netherlands, the United Kingdom and Germany.

The port of Rotterdam has state-of-the-art facilities and reaches world-standard levels in terms of scale, efficiency and security. Containerisation is increasingly prevalent in the transport of food products. Rotterdam, as Europe's main container

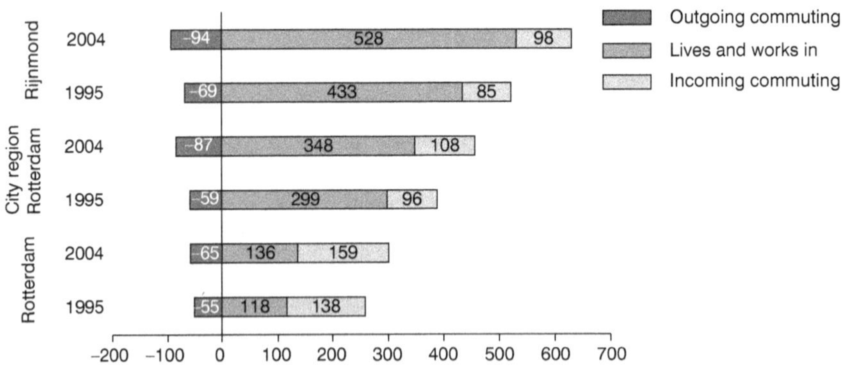

Figure 9.3 Commuting to and from Rotterdam and Rijnmond, 1995 and 2004 (source: CBS, 2006)

hub, is well placed to benefit from this development. Reefer containers can be transported by trucks but also by barge or train, thus offering scope for intermodal transport.

The Fruitport area is a focal point in Rotterdam's food logistics industry. It is one of the largest clusters of handling, storage, transhipment, distribution, repackaging, sales and trade of perishables (fruit and juices) in the world. Juice concentrates are stored, mixed and blended for the manufacturing industry. Its primary activities are supplemented by specialised transport firms, quality surveyors, inspection authorities, customs, etc. The throughput volume of the Fruitport has quadrupled since it was established in 1991, to 2 million tons a year. Within the Le Havre-Hamburg range, the Fruitport has a market share of 50% in juices and about 30% in fruits. It employs between 500 and 800 employees, depending on the season (interview). The Fruitport is dominated by fresh logistics firm Seabrex (400 employees) which specialises in the handling of fruit and vegetables.

Packaging activities (often referred to as 'value added manufacturing') are emerging as an attractive market. The Greenery B.V. conducts these activities on its three main locations. There, packaged produce is made ready for sale in supermarkets. Producers associations are also investing in these activities in the Westland. The Greenery B.V. further directs the flow to and from the packaging installations to ensure full use of the available capacity and further dissemination into European markets. The Greenery B.V. also has its own transportation company with over 400 trucks delivering all over Europe.

Transport is an issue of concern. Road transport is very efficient and highly flexible, but it causes pollution and suffers from chronic and increasing congestion. Furthermore, European countries are increasingly taking measures to restrict transit (Germany, Switzerland and Austria). At the same time, containerisation offers new possibilities for intermodal transport. That is why there is increasing talk of alternative modes of transport for food transports. One option is to use the railway for transport towards Germany; another is to use the HSL towards Paris. Our interviewees doubt whether trains and barges can outcompete the truck in the short run, but they see possibilities for the longer run. There seems to be scope for an increased use of inland shipping between the Rotterdam region and the increasingly pivotal area around Venlo (in the east of the country) that could further strengthen Venlo as a centre for food-related value added logistics.

For some products, air transport is still an important mode. This is true for light-weight highly valuable goods such as seeds and genetically modified organisms. Also in the transport of flowers, air transport is still significant, although also in this domain the container is gaining ground due to the rapid development of conservation technologies.

Food and beverages

The Rotterdam region counts a substantial but highly heterogeneous group of firms active in the food and beverages sector. They are highly fragmented in terms of size, products, production methods, etc. We have not discovered any particular niches or spatial concentrations in the Rotterdam region. Employment in this subsector is significant (almost 5,900 people in the region in 2006, see Figure 9.4). Though food manufacturing enterprises are generally small sized and locally or regionally oriented, there are instances of larger establishments, such as those of Unilever, Coca-Cola, Campina and Quaker Oats B.V. DSM Food Specialities in Delft is a prominent business group within the life science cluster of DSM. It is a leading supplier to the food manufacturing industry. Since 2004 the manufacturing industry has resurged in terms of employment. Employment in the food manufacturing sector grew by some 700 jobs.

Unilever is by far the most important actor in the Rotterdam region, with its headquarters in Rotterdam, its research facility in Vlaardingen, and some production activity in the region. There are also a number of small food manufacturing firms, but this segment is highly fragmented and weakly organised compared to other regions, where we found examples of concerted action of small food manufacturing firms to improve their image, in order to gain better access to innovation funds and institutes and to lobby more effectively for improved professional education. All over the country, 'food regions' are being created in which SMEs organise themselves (see Box 9.3).

9.4 Local and global networks

In order to analyse the linkage between manufacturing activities and knowledge activities, this section analyses the relevant networks in which the key players of the food industry in Rotterdam operate, with a special focus on the spatial dimension of these networks. We analyse the networks in the three subclusters that we have distinguished in section 9.1: horticulture, food logistics, and food and beverages. For the latter, we focus on networks of Unilever, the key player in the food and beverages industry.

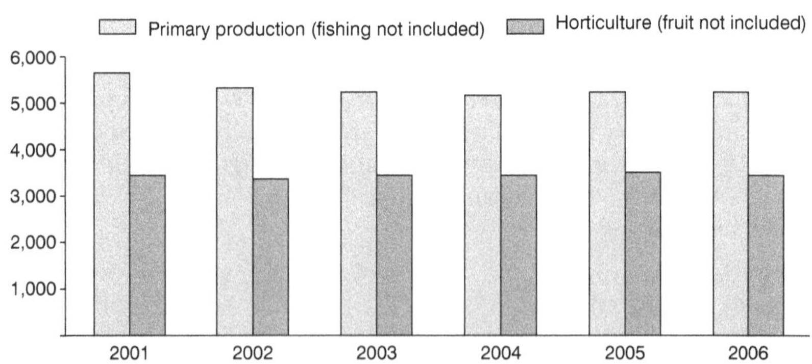

Figure 9.4 Employment in the Rotterdam region (Rijnmond) in primary production and horticulture (source: own elaboration, data from COS, 2006)

Box 9.3 The Food Regions and Food Connection Point

The Food Regions and Food Connection Point are networking platforms for entrepreneurs. The first Food Region was set up in 2001 in the east of the Netherlands, as an 'umbrella' organization for food manufacturing entrepreneurs. Its aims are to improve the image of the food sector, to realize a better fit between labour needs and education, and to represent the interests of said entrepreneurs. Food Region Helmond was the inspiration for Food Region Veghel and four other Food Regions that are currently being set up. It also lies at the basis of the Food Connection Point (top-down), which was established early 2006 and which aims to function as a central organization that operates above the Food Regions and the national level. It focuses on the fields of knowledge development, education, labour market, image and product and process innovation. The Food Region model is especially esay to follow due to its simplicity and potential in bringing together food manufacturing SMEs.

Source: own elaboration

Networks in horticulture

There are few strategic partnerships among horticultural producers (apart from 'study clubs'), but The Greenery B.V. cooperative plays a central organising role in this sector. It performs sales, marketing and logistics functions for its members (although large producers increasingly sell directly to supermarket chains). Furthermore, it defends the interests of the organised producers vis-à-vis third parties, e.g. policy makers, supermarket chains, etc. The Greenery B.V. also plays a role in the fostering of innovation among its members: for instance, it coordinates activities of producers and seed houses as the latter try to develop better seeds for producers. Recently, The Greenery B.V. has become involved in the exploitation of energy that is released by greenhouses. TNO Rijswijk and some knowledge institutes from Wageningen participate in this highly innovative project. There are already several instances of greenhouses supplying energy. The Horticultural Cooperative (WKK) and AgroEnergy are planning to jointly supply a capacity equalling that of a large electrical energy plant.

The horticultural sector is considered a strategic sector by the national government, and receives national support through the 'Pieken in de Delta' programme (Peaks in the Delta, PID). The PID programme for the southern part of the Randstad aims to boost the knowledge-intensive segments of the horticulture cluster (robotics, mechatronics, ICT, sorting systems, packaging, binding, monitoring control, etc.).

Networks in food logistics

By nature, logistic activities are network orientated, but the network 'morphology' and power relations strongly vary by subsector. It is beyond the scope of this study to go in depth here, although some trends can be mentioned. Large supermarkets play a leading role in these logistics networks; they want high quality, speed and flexibility, and put pressure on their logistic service providers to deliver this at the lowest possible costs. In the lower segments of the logistics industry (truckers, barge operators, handling activities), margins are thin, competition is fierce and knowledge intensity is relatively low. More knowledge-intensive activities take place in companies that organise networks and develop logistical concepts or IT systems for customers. There are also some partnerships between logistics firms and knowledge institutes TU Delft, Erasmus University and other partners (one example is a project of TU Delft with Seabrex). In Wageningen, a research group is developing new technologies to improve the conservation of flowers, vegetables and fruit, which may have a substantial impact on future good flows and the modal split. Another group develops monitoring technologies to make sure that imported tropical fruit enters the supermarket at exactly the right degree of 'ripeness'.

In the food industry, packaging and other value added manufacturing activities are becoming more important, as an instrument of producers (or producers' organisations such as The Greenery B.V.) to escape 'commodification' and create more value and distinction. Value added activities occur in logistical hubs. These comprise such activities as packaging in the Westland and the mixing of juices in the Fruitport. The cluster formation in these hubs is actively sought after. The Greenery B.V. has recently facilitated the installation by producer associations of new packaging installations in the area of the Westland.

Networks in food and beverages: the case of Unilever

Within Unilever, we observe an increasing locational and functional separation between innovation activity (knowledge networks; 'exploration') and manufacturing (production networks; 'exploitation'). Production takes place in plants all over the world, and is constantly being rationalised. Typically, production cannot be located too far from the consumer market, but the company increasingly considers Europe as one market and rationalises its production facilities accordingly. For some products, Unilever has its own plants, but for others it uses contract manufacturers. By 2007, 50% of Unilever's production is outsourced to contract manufacturers. The latter option brings more flexibility and lower capital investment, but also induces the risk of leakage of valuable knowledge to the contract manufacturer. Property rights protection is becoming an issue of growing concern. Unilever has moved its central purchasing and supply chain management operations to Schaffhausen (Switzerland) for tax reasons but also because of its central location in Europe and the quality of research facilities in that region.

Within Unilever, R&D activities are crucial and highly strategic. It spends €415 million on innovation (on a total turnover of €40 billion). The firm is looking to strike the right balance between in-house R&D and networks with outside partners. Its own global R&D network consists of six leading R&D centres focusing on basic research (see Figure 9.5), consisting of 13 global product development centres that focus on new product development for a particular category or technology area, and 37 regional development centres for adapting and implementing innovations and renovations in the regions. In addition, all plants have small R&D and implementation teams. The latter proves that manufacturing activities, although on a small scale, always need physical proximity of some R&D activities in order to implement new products.

So, Unilever's research network has a global reach, with a strong regional dimension in order to adapt products for regional consumer tastes and/or to fulfil government requirements. Recently, Unilever has invested strongly in new growth markets[5] via the extension of R&D centres in China (Shanghai), India (Bangalore and Mumbai), Latin America (São Paolo) and South Africa (Durban) and it aims to link these new centres with the established centres in Western Europe and the United States. The European R&D organisation employs 1,160 people in over 60 locations across Europe. From 2007 on, the company aims to reduce this number to 29 locations, with Vlaardingen as one of the leading centres. A main goal of its innovation activity is to drive down the levels of sugar, salt, trans-fats and saturated fats in its products.

In the innovation process, Unilever identifies four stages: define, discover, design and deploy. Unilever increasingly adopts an 'open innovation' model, with strategic partnerships, recognising that the company cannot do everything in-house and needs strategic partners to remain competitive. From this perspective, the company engages in several types of networks. First, in the 'define' stage, it cooperates with a multitude of academic research groups at Dutch and foreign universities (but not with Rotterdam-based Erasmus University) to conduct 'pre-competitive research'. The main partner is the TIFN in Wageningen (see Box 9.1). The outcomes of this research are uncertain, but to maximise the impact, Unilever

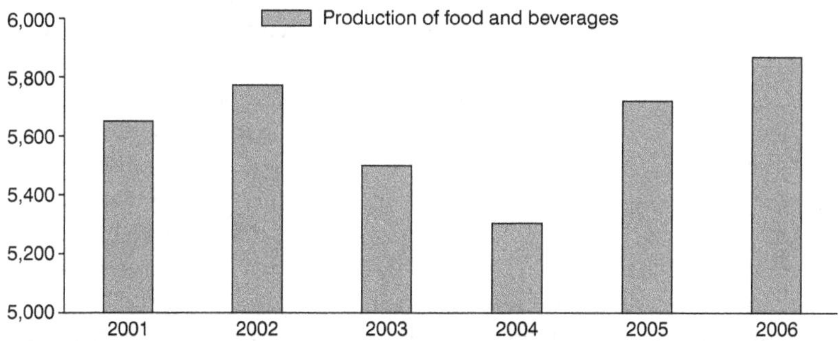

Figure 9.5 Employment in the food and beverages production in the Rotterdam region (Rijnmond) (source: own elaboration, data from COS, 2006)

Figure 9.6 Unilever' s principal R&D centres (source: www.unilever.com)

has created 'shadow research teams' in-house, that link research outcomes to company objectives. In the 'discover' stage, Unilever develops fundamental innovations of food products. This activity is mainly conducted in-house, in Vlaardingen, but also in this stage, there are cooperations. An important concept in this respect is 'technology venturing': Unilever participates in promising and innovative (small) firms (most of them in the USA), in order to stay current with new trends. Unilever has considered investing more in European small companies but could not find enough promising ones.

In the 'design' stage, fundamental innovations are translated in new products that are commercially viable. In this stage, Unilever increasingly cooperates with suppliers, for example with suppliers of functional ingredients, and the 'flavour' industry, to improve the quality of its products. In this type of product development network, research is more practical and leads to commercially viable products. The network partners are scattered all over Europe, with a focus on Switzerland. The 'deploy' stage is the actual manufacturing and sales of the new product.

9.5 Conclusions

Rotterdam's food industry is not a growth industry. Our data suggest that for each of the three subindustries (primary production/horticulture, logistics, and food and beverages), employment growth is low or even slightly negative. Turnover and value added, on the other hand, have been growing substantially in the last decade, indicating a strong rise in productivity. A common issue in each of the subsectors is the increasing demand for skilled professionals and the lower demand for unskilled labour (with horticulture as an exception: this industry needs 'hands' during peak seasons). This trend once again underlines the increasing importance of a qualified regional labour force. Overall, in each of the three subsectors, there are indications that the knowledge component is increasing. In the food and beverages industry, upgrading occurs because of increasing demands on food quality and rapid technological change, forcing companies to invest in product development. For the food-related logistics industry, the trend of containerisation of (fresh) food might imply that (like many other containerised commodities in the recent past) some activities in the food chain will relocate to the hinterland, where space is cheaper and congestion is less. Especially distribution and value added logistics are sensitive to relocation. The Venlo region (close to the German border), with its already strong food cluster and its excellent intermodal facilities, is particularly well placed to benefit from this trend. The food and beverages sector in the Rotterdam region, finally, is dominated by Unilever, making it relatively sensitive to developments within that company. One recent example is the announcement of Unilever to merge the sales and marketing divisions of Belgium and the Netherlands, with negative repercussions for employment at Unilever's Rotterdam headquarters.

In this book, one of the central questions is how manufacturing and other 'operational' activities are functionally and spatially related to other activities in the value chain, notably knowledge based activities. In the case of the agro-food industry of Rotterdam, a mixed picture emerges, with different realities in various subsectors. The *logistics subsector* is one of Rotterdam's flagship sectors. Companies in 'operational' parts of this industry (transport, storage, transhipment firms) typically have thin margins. They are located in the region because of the port and its infrastructure. They face a continuing pressure to cut costs and shorten lead times, and they have little power in the supply chain. Innovations are incremental, and mostly related to optimisation of processes. Most innovations are pushed by powerful players in the chain: the supermarkets. New conservation techniques for fresh products will have a profound impact on the sector. Firms in this branch have few if any 'strategic' partnerships for innovation. Nevertheless, this sector is in a qualitative upgrading process. Due to automation processes, the demand for unskilled manual labour is decreasing, and firms in this sector increasingly need people with professional education, for instance experts in quality control and ICT.

In the *food and beverages* sector, the situation is different. In this case study, we have focused on Unilever, one of the leading players in the region. This firm is active both in 'standardised' manufacturing activities (although it outsources some 50% of it to contract manufacturers) and knowledge based activities. We can conclude that for Unilever, there is a functional and physical separation between the manufacturing function on the one hand, and R&D/services on the other. However, the presence of small R&D teams in all production plants suggests that production always needs some R&D activities to implement new products. For innovation, Unilever increasingly acts as a 'meta-national firm' (Doz *et al.*, 2001) that strategically sources knowledge from partners all over the world via a global R&D network, the use of open innovation and technology venturing strategies. The firm's key research centre is located in the Rotterdam region, but innovation networks focus on knowledge partners in other regions (mainly Wageningen, where the agricultural university is located), and increasingly abroad. Manufacturing, on the other hand, is rationalised on a European scale, and the management of Unilever's manufacturing and purchasing activities has moved to Switzerland. Unilever is not representative of the entire industry, however. Compared to other manufacturing industries, food production will remain relatively 'local', first because consumer tastes vary per country, second for logistical reasons, and third because the sector is dominated by SMEs.

The *horticultural production sector* (the horticultural sector to the north and north-west of the city) is often cited as a 'classic' example of an upgrading cluster, where production has become ever more efficient thanks to excellent cooperation between greenhouses and innovative supporting industries (greenhouse construction companies, seed houses, developers of fertilisers and pesticides, etc.). Indeed, several of the so-called 'supporting industries' no longer depend solely on the 'local demand' and sell their products and knowledge all over the world. This may mean that gradually, the link between R&D and manufacturing will become weaker.

Notes

1 www.tudelft.nl.
2 Largely based on www.portofrotterdam.nl.
3 G4 refers to the four largest cities in the Netherlands: Amsterdam, Rotterdam, The Hague and Utrecht.
4 www.Rotterdamveilig.nl.
5 For instance, in 2008, the sales growth in Asia, Africa and Central and Eastern Europe (the 'new markets') was 14.2%, against 6.5% in North, Central and South America and only 1.3% in Western Europe. This gave the new markets a share of 36% of Unilever's total turnover. (Unilever, 2009).

Bibliography

van den Berg, L., E. Braun and A.H.J. Otgaar (2000) *Sports and city marketing in European cities*, Rotterdam: Euricur.

—— L., P.M.J. Pol, W. Van Winden and W. Woets (2005) *European cities in the knowledge economy*, Aldershot: Ashgate.

Bulterman, S. and J. Van der Vet (2006) *Economische verkenning Rotterdam 2006*, Ontwikkelingsbedrijf Rotterdam (OBR) and Rabobank.

—— Stoffels, and G. Bal (2005) *De Kracht van het Westland, Economische thermometer Westland*, Kamer van Koophandel Haaglanden and Rabobank.

Cadilhon, J.J. (2003) *Wholesale markets and food distribution in Europe: new strategies for old functions – centre for food chain research*, Department of Agricultural Sciences, Imperial College London, Discussion paper No. 2, January.

Doz, Y., J. Santos and P. Williamson (2001) *From global to metanational: how companies win in the knowledge economy*, Boston, MA: Harvard Business School Press.

EDBR (Economic Development Board Rotterdam) (2005) *Economische visie 2020, Rotterdam stad van de toekomst. Sterke stad in een welvarende regio*, Rotterdam: EDBR.

Erasmus MC (2006), annual report 2006, Erasmus University Rotterdam.

Gemeente Rotterdam (2009) *Kerncijfers & trends Rotterdam 2009*, Rotterdam: Centrum voor Onderzoek en Statistiek (COS).

Kuijpers, M. (2006) *Naar Topprestaties in de Rijndelta: Innovaties in de kennisinfrastructuur. voor haven, transport en industrie*, Rotterdam: KMR.

Ministry of Economic Affairs (2004) Industriebrief: Hart voor de industrie, Ministry of Economic Affairs.

M-team (2006) *Food and nutrition delta fase 2-business plan 2*, available online at: www.senternovem.nl/mmfiles/FND-Fase2_tcm24–227262.pdf.

OBR (2003a) *Economische Verkenning*, Rotterdam: OBR.

—— (2003b) *Programma Kenniseconomie*, Rotterdam: OBR.

—— (2004) *Vrijetijdseconomie in cijfers,* Rotterdam Festivals, Rotterdam Marketing and OBR.

—— (2005) *Economische Verkenning Rotterdam 2005*, Rotterdam: OBR.

—— (2006) *Plezierige zaken in Rotterdam 2006. Vrijetijdseconomie in cijfers*, Rotterdam: OBR.

Unilever (2009) Annual report 2008, Rotterdam/London: Unilever.

Websites used

www.eur.nl The Erasmus University
www.efmi.nl Erasmus Food Management Institute
www.portofrotterdam.nl The Port of Rotterdam
www.foodvalley.nl Food Valley
www.foodconnectionpoint.nl FoodConnectionPoint
www.duvo.nl Foundation for Sustainable Food Chains
www.ez.nl Ministry of Economic Affairs
www.thegreenery.nl The Greenery B.V.
www.unilever.nl Unilever
www.wur.nl Wageningen University and Research Centre
www.seabrex.nl Seabrex
www.tifn.nl Top Institute
www.schipholgroup.nl Schiphol Group
http://unstats.un.org United Nations Statistics Division
www.food.gov.uk/foodindustry UK Food Standards Agency
www.ers.usda.gov Economic Research Service (USDA)
www.nugo.org/facts/24023 The European Nutrigenomics Organisation
www.technopolisinnovationpark.nl Technopolis Innovation Park
www.wb-online.nl Wageningen University and Research Centre
www.actransport.nl Academic Centre TransPort (ACTP)
www.agriholland.nl AgriHolland
www.rabobank.com Rabobank
www.edbr.nl The Economic Development Board of Rotterdam
www.hightechconnections.org High Tech Connections
www.portofrotterdam.com Port of Rotterdam
www.Rotterdamveilig.nl Safe Rotterdam
www.markthal.nl Markthal Project
www.spaansepolder2015.nl Industrial park Spaanse Polder

10 São Paulo: automotive industry

10.1 Introduction

One hundred years ago, São Paulo's economy was still dominated by the coffee crop and trade. The development of infrastructures like roads and ports, together with an influx of immigrants required by the coffee business, set the foundations for the industrial development of what is today the biggest and most influential economic centre of Brazil. Nowadays, São Paulo is South America's most important headquarters and advanced services location, while keeping a strong and diversified economy. Besides hosting many high-tech manufacturing activities, like aeronautics (with Brazilian flagship corporations, like Embraer), the state of São Paulo also has an important share of Brazil's rich and abundant natural resources – it is a net exporter of commodities like orange juice, soya, ethanol, oil and many others.

For almost 100 years, the automotive industry has been a flagship of São Paulo's industrial growth. The first assembly line was set up by Henry Ford in 1919, for the classic Ford Model-T; the first Chevrolet from General Motors followed 10 years later. Since then, the sector has gone through different political and economic orientations, ranging from governmental policies of import substitution in the 1960s to the progressive economic liberalisation of the 1990s. The sector's evolution has also been linked with the production and use of ethanol (a bio-fuel derived from sugar cane). First supported by government subsidies as a response to the oil shocks of the 1970s, ethanol became the prime fuel of Brazilian cars until the late 1980s, when the production subsidies ceased and the largest share of the Brazilian fleet become oil-driven again.

Nowadays, benefiting from its fast growing market and from stable political conditions, São Paulo hosts plants of the major world OEMs and their suppliers. Moreover, and associated with environmental and climate change pressures, bio-fuels became strongly intertwined with the automotive industry. New technologies made the use of ethanol stable and cheaper; sugar cane harvesting became more mechanical and environmentally friendly. Recently, a new set of technologies – *flex-fuel systems* – mainly developed in São Paulo, allowed vehicles to run on conventional petrol, ethanol or a mixture of both. Presently, the demand for this type of vehicle is soaring. Public and private R&D expenditures in bio-fuel related issues make Brazil and São Paulo a centre of fuel research and applications. Economy is booming and wealth is increasing – many people are buying cars and other consumer goods for the first time – but, in the medium run, social disparities and growing congestion may threat the sustainability of this economic performance.

This chapter analyses the development of the automotive industry in São Paulo's urban region, with a special focus on its links with the bio-fuel related technologies and activities. The chapter is structured as follows: in section 10.2 we present a brief background analysis of the main trends in the Brazilian automotive industry and the dynamics of bio-fuels in Brazil. Section 10.3 depicts the foundations of the automotive industry in São Paulo and section 10.4 presents the characteristics of its main players. Section 10.5 illustrates production and knowledge networks taking place at different geographical scales, and their main governance modes. Section 10.6 analyses the role of public policy and section 10.7 concludes.

10.2 Brazil and automotive: background and contextual features

Automotive industry in Brazil

The automotive industry in Brazil has almost one century of history and is presently one of its most dynamic industrial sectors. Not claiming to be comprehensive, this section depicts some of the most relevant milestones of the industry, as well as main trends shaping the sector in the country.

The first automotive plant in Brazil was founded by Henry Ford in 1919. General Motors followed, also in São Paulo, to assemble the first Chevrolet in Brazil. In the same decade, General Motors opened a second factory in São Caetano do Sul, today São Paulo's metropolitan area. The number of cars produced in Brazil rose fast between 1920 and 1939, coming to a halt during and after World War II. Until then, automotive components were almost exclusively imported to be assembled in Brazil, and thus the fleet was outdated quickly due to war market restrictions.

In order to develop the industry and improve Brazil's balance of payments, the policy response was to ban imports of assembled cars and impose strong restrictions on the imports of car components. This policy meant the sector had to develop internally: most of the components of a car needed to be produced in Brazil. By the early 1950s, automotive became a 'leading sector in a broad import-substituting industrialization push' (Shapiro, 1996). Until 1960, new car producers invested in Brazil, both in Brazilian-owned and joint ventures with foreign players, strongly supported by government subsidies. From the 1960s until the late 1980s, many political and economic events (military dictatorship in Brazil, oil shocks in the 1970s) shaped the automotive sector evolution, which faced periods of sharp growth and investments, followed by many downturns and market contractions.

The beginning of economic liberalization in the early 1990s set a new milestone: the government progressively banned import restrictions, compelling automotive multinationals to modernise plants, processes and products through investments from incumbent and new entrant firms. These processes took place simultaneously with the global restructuring of the industry's value chains and globalisation trends (Shapiro, 1996). During this period, many mergers, take-overs and other restructuring processes driven by external investment took place in Brazil, shaping today's sector structure.

As Figure 10.1 depicts, the process of economic stabilisation, together with favourable economic growth conditions, has been leading to a steady rise of the Brazilian vehicle fleet. In 2007 most of the world's OEMs produced and assembled cars in Brazil (see Figure 10.2), namely Volkswagen, Ford, General Motors and Fiat. French and Asian OEMs like Renault, Peugeot-Citroen, Toyota or Honda built facilities and are expanding production in Brazil. Brazil is also an important automotive exporter: during the last decade, as Figure 10.3 depicts, the ratio 'export of vehicles as a share of total Brazilian production' has been increasing. The external markets of Brazilian cars have mainly been other South American countries, Mexico and Germany (ANFAVEA, 2007). The industry of automotive components is also a strong net exporter (see Figure 10.4) to in geographically broader and more demanding markets, exporting also to the USA, Germany and the United Kingdom (Sindipeças, 2007).

Although absolute employment levels have been decreasing since 1990 due to automation and increased efficiency, the recent sales and exports growth has led to the net creation of jobs in the Brazilian supplier sector (Figure 10.5).

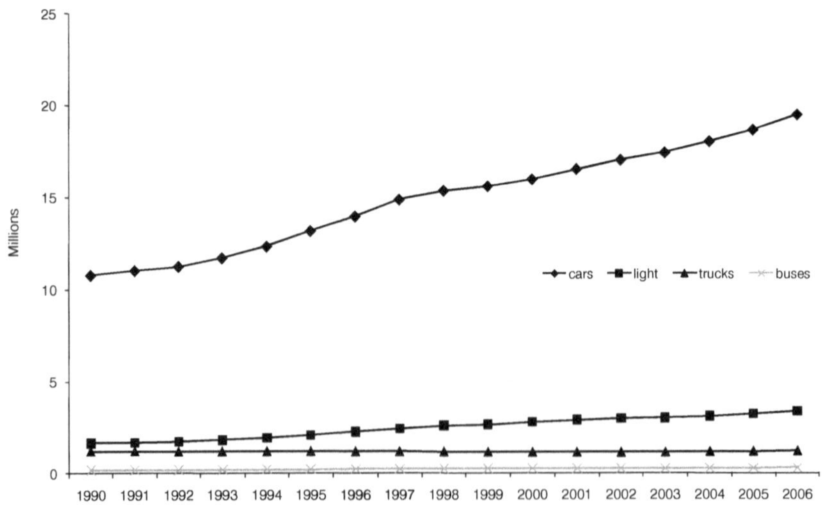

Figure 10.1 Brazilian fleet (source: Sindipeças, 2007)

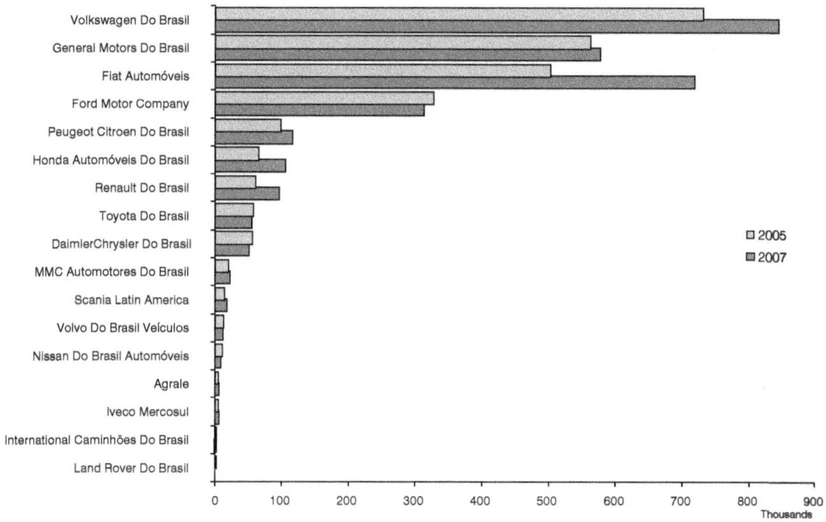

Figure 10.2 Brazilian automotive production, OEMs, number of vehicles per year (source: ANFAVEA (n.d.), own calculations)

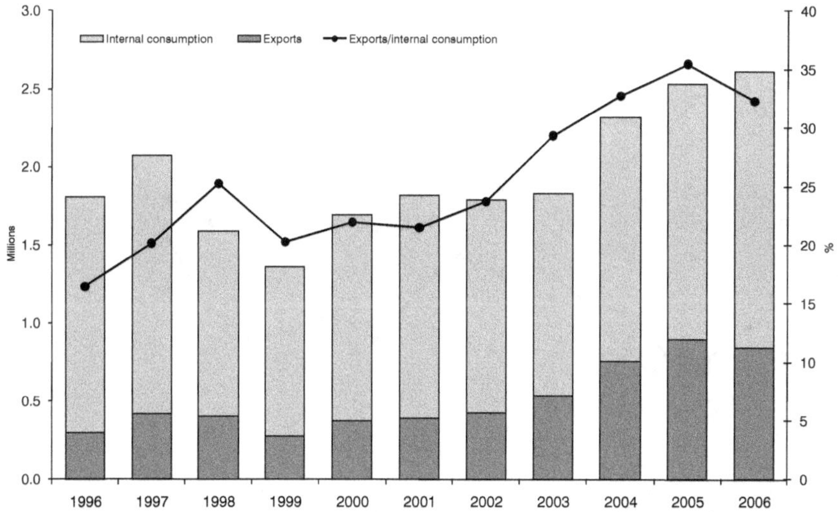

Figure 10.3 Brazilian automotive production, millions of vehicles and % (source: ANFAVEA (n.d.), own calculations)

Note: Figure includes cars, light commercials, trucks and buses; excludes agricultural machinery.

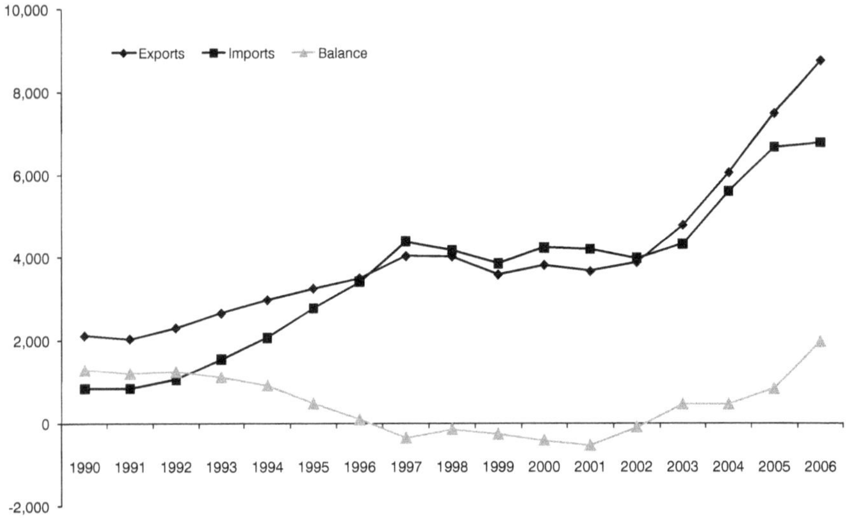

Figure 10.4 Auto parts production in Brazil, 1990–2006, imports and exports, $US million (source: Sindipeças, 2007). Note: values FOB (free on board).

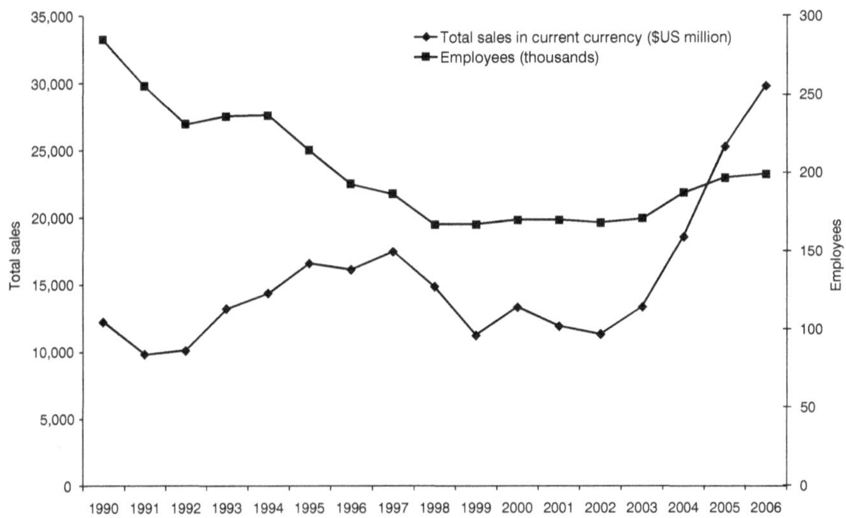

Figure 10.5 Autoparts production in Brazil, total sales and employment, 1990–2006 (source: Sindipeças, 2007)

The evolution of ethanol and bio-fuel use in Brazil

Brazil is one of the world's pioneer producers and users of bio-fuels as an alternative to conventional petrol. The most well known is *ethanol*, a result of the fermentation process of glucose from sugar cane.[1] In Brazil, the first experiments of alcohol ethanol as a fuel date from the 1920s, when one Ford Model-T travelled from São Paulo to Rio de Janeiro running on ethanol. During the 1930s, the first ethanol standards for car use and petrol mixes were established by the government.

Yet, the boom of ethanol production for car use in Brazil started in the 1970s, through a package of government regulations and subsidies to foster ethanol production and consumption – the 'PROALCOOL' programme. The objectives were twofold: 1) reduce the oil and energy dependency of the country, and 2) foster the automotive industry, shaken by the rising oil prices. The PROALCOOL set ethanol-buying guarantees from the government; controlled prices between sugar, ethanol and petrol; set incentives to buy cars running on ethanol; and ordered regulations for the percentage of ethanol that should be mixed with petrol. In order to cope with the new mixture requirements (by that time around 20/80 for ethanol/petrol), in 1979 Volkswagen developed one fully nationally designed car running on ethanol. In the mid-1980s, around 96% of the car sales were ethanol cars (Teixeira, 2005).

However, during late 1980s, Brazil faced an 'alcohol crisis'. Oil prices fell, the federal government cut many of the former ethanol-producing subsidies while sugar prices rose, making it less attractive to produce alcohol instead of sugar. As a result, ethanol became scarce in filling stations and ethanol car sales dropped sharply. Moreover, by this time, the use of ethanol in cars was not yet fully efficient and trustworthy – for example, cold engines had difficulty starting with ethanol and the car performance was lower. As a result, petrol cars got back in the market.

Nevertheless, recent technological developments in ethanol production, rising oil prices and climate change pressures brought bio-fuels and ethanol production strongly back in Brazil. This set of factors set attractive market conditions to the development of a new automotive technology in Brazil – flex-fuel engine systems – capable of running on petrol, ethanol or a combination of both.

Moreover, nowadays Brazil is developing sound bio-fuel research efforts beyond ethanol. The state of São Paulo has recently announced a large fund for interdisciplinary research in many fields related to bio-fuel technologies. Besides on-going investments to stay as the front-runner in ethanol and sugar cane production, one of the main research streams is associated with new raw materials and processes to produce biodiesel (using soya, sunflower oil or chicken fat). Meanwhile the Brazilian automotive industry is working on complementary engine solutions and systems for biodiesel mixtures.

Brazil has so far countered the main critiques pointed at bio-fuels as a cleaner and cheaper alternative to petrol, namely 1) the indirect CO_2 generated by the harvest of sugar cane, and 2) the rise in food prices and deforestation caused by the use of cattle fields and forest to produce sugar cane or other raw materials. As highlighted by many commentators and recent research (for an overview see *The*

Economist, 2008a), the automation of sugar cane harvest avoids burning plantation fields; moreover, Brazil still has a huge land potential, estimating that the rise in land use for bio-fuel production will not endanger other plantations and food prices (UNICA, n.d.). Moreover, compared with the bio-fuels produced in the United States (namely through corn), ethanol from Brazilian sugar cane appears to be much more energy-efficient, producing the same energy with much fewer units of sugar cane/land, namely due to Brazil's specific climate characteristics. Furthermore, production costs have been decreasing as a result of technological developments (CTC, n.d.).

All in all, internal production of bio-fuels is growing fast, as are its exports, despite the still high imports quota imposed by the United States. Brazil is presently a net and growing exporter of diverse types of high quality bio-fuels (namely ethanol and biodiesel), whose demand has been growing in the last years (e.g. in Sweden). Additionally, Brazil (and namely São Paulo, see Box 10.1) enjoys a front-running position in bio-fuels associated knowledge (biotechnology, engineering, engine and systems technology), as well as many economic-related activities, ranging from medium-sized bio-fuel producers in all the stages of the value chain to foreign transnational companies and high-tech university spin-offs in bio-fuel technologies and 'green innovation' (*The Economist*, 2008a).

10.3 Foundations of São Paulo

São Paulo, located on the south-eastern Brazilian coast, is one of the world's largest metropolises. According to recent statistics, the state of São Paulo has roughly 40 million inhabitants. More than 20 million are concentrated in the urban agglomeration of São Paulo (and growing) and roughly 10 million in the central city.

After many decades of urban de-concentration and sprawl, São Paulo's metropolis can be broadly decomposed in six different, though interlinked, urban areas: 1) the city of São Paulo, nowadays specialised in advanced services to firms (e.g. ICT, finance, marketing) and creative industries (like media or design); 2) its first urban sprawl and former area of strong industrial growth before the industrial de-concentration of late 1980s – known as *ABC Paulista*[2] – the urban area between the city of São Paulo and Baixada Santista, where the Port of Santos is located; 3) the Vale do Paraiba's urban agglomeration, on the eastern extension of São Paulo towards Rio de Janeiro – encompassing cities like São José dos Campos and Taubaté – where new greenfield industrial units benefit from good accessibility and lower congestion; 4) the western extension of São Paulo in the direction of the state of Paraná, namely the growing industrial agglomerations around the cities of Sorocaba and Jundiai; 5) The metropolitan region of Campinas, 100km north of the city of São Paulo, hosting an important university with interfaces with the industry, as well as high-tech start-ups, automotive suppliers and the technological centre for ethanol and sugar cane (municipality of Piracicaba); and 6) a number of small and medium sized non-contiguous cities in the north and north-western part of the state, like Ribeirão Preto, São Carlos or São Jose do Rio Preto, with

Box 10.1 State of São Paulo: main arena of Brazilian bio-fuel development

The State of São Paulo is by far the most important bio-fuels 'hot spot' of Brazil. According to the Sugar Cane Technological Centre – (CTC, (n.d.), in 2007, São Paulo was responsible for more than 70% of all the ethanol Brazilian exports. The production and export of other bio-fuels from São Paulo, like biodiesel, is growing fast. The production process of bio-fuels is concentrated in the rural and inner parts of the State, but the concentration extends itself towards the main urban areas, as well as towards Rio de Janeiro and other states like Minas Gerais and Paraná.

Not only are the most important crop fields of sugar cane concentrated in São Paulo, but also the refineries and the core of the transformation and storage processes, making it by far the largest bio-fuel 'cluster' in the country. The state hosts large economic corporations, exploring value chain synergies (raw materials and transformation), as well as smaller producers selling the fuel to the large distributors, namely Petrobras, Brazil's biggest fuel company, regulator and distributor. São Paulo is the Brazilian leader in testing new raw materials to be applied to biodiesel production.

Moreover, research and intermediate institutions (such as firm associations) related to bio-fuels are located in São Paulo, like UNICA (Union of São Paulo sugar cane producers), Abiodiesel (Association of the Brazilian Industry of Biodiesel), CTC (Technological Centre of Sugar Cane) and many other university based research institutes. CTC, located in the municipality of Piracicaba, has a large number of research units focused on the process of ethanol production, with industrial patents, and several associated firms and related institutions. São Paulo (Piracicaba) is presently investing more than US$40 million on an integrated technological and science park focusing on bio-fuels, with incubation facilities, to host firms and academic research units.

Source: own elaboration

good universities, much less congestion and growing economic activity in small productive systems, like medical equipment, jewellery and leather products.

For the purpose of this study, and bearing in mind this spatial organisation and all the criss-cross fluxes within it, we considered in the analysis a loosely defined 'urban São Paulo', without strict boundaries, due to the fast urban growth of the last decades, but mainly because we are interested in observing phenomena that do not stop at the boundaries of smaller delimitations – although there might be, as we will see, intra-regional specialisations.

Economic base

The metropolis of São Paulo as a whole has the strongest and most diversified economic base of Brazil and South America, with a highly favourable mix and interactions between industry, services and even agriculture. In 2005, the State of São Paulo represented roughly 33% of the Brazilian added value (see Figure 10.6). However, this share was higher than 40% in sectors like advanced services to firms (finance, insurance, legal assistance, ICT), media, advertising, education and health services, as well as in manufacturing industries. Also agriculture, farming and forest activities have a relevant share in the state (13%), rich in natural resources and commodities (e.g. sugar cane, orange, soya, coffee, cattle, etc.).

All economic sectors and subsectors are well represented and count with large national and multinational players, namely Brazilian transnational companies like Petrobras (energy) or Embraer (aeronautics). Together with several production plants, the metropolis of São Paulo concentrates the highest number of regional headquarters of US transnational corporations. In this favourable setting, the potential for new combinations and innovation across related sectors is extremely strong, due to the very diverse types of knowledge and capabilities present in the economic base.

Knowledge base

According to OECD (2006), the Brazilian working population has on average 4.1 school years. This is less than China, which has on average 6 school years.

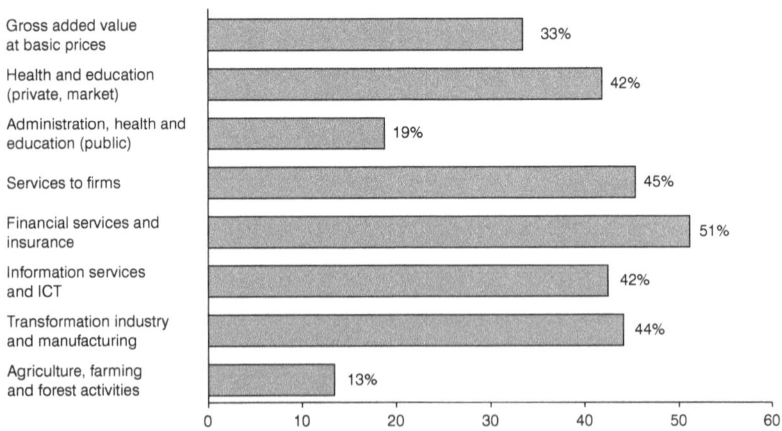

Figure 10.6 Gross value added of São Paulo (state) as a share of the Brazilian economy, %, 2005 (source: IBGE (n.d.))

Qualifications are growing in Brazil, but the general drop-out rates are still very high – this deficit yields difficulties and extra costs to cope with the increasingly demanding industrial quality standards, namely in the automotive industry. The qualification circumstances in São Paulo are however relatively better. Generally school years are on average higher – 42% of the population has more than 4 school years. At the other extreme, most of the high qualifications of the country are concentrated in São Paulo: 12% of the active population has a university degree, and São Paulo is endowed with a very large pool of universities, R&D institutes and polytechnics, covering all scientific fields and research areas, constituting the most robust scientific and innovation pole of Brazil.

Brazil's R&D efforts remain relatively low (around 1% of total GDP) and are still mostly supported by public funds; however, in 2005, São Paulo accounted for 53% of the total public resources for R&D in Brazil, as well as 60% of the total private R&D investments (Investments São Paulo, n.d.). Due to strict legislation, services rendered by academic researchers to private firms cannot be directly funded and the academic career regulations penalise any interaction and joint work between universities and the private sector. As a result, there are very few examples of stable cooperation, and the most visible ones exist mainly outside the main agglomeration of São Paulo, namely through the spin-offs of the University of Campinas and the Technological Institute of Aeronautics, in São José dos Campos.

Polytechnic schools are more open to private cooperation, joint applied research and teaching programmes (like the ones developed by the Mauá Institute of Technology São Paulo in the automotive field), but have much fewer resources. São Paulo also has many specialised training institutes for industrial workers, like SENAI National Industrial Learning School, present in different municipalities, and pilot certification and skills programmes, increasingly required to meet industrial standards.

Accessibility

São Paulo has the best accessibility and logistical infrastructure of Brazil, but faces strong pressures and bottlenecks. International accessibility is assured by two large international airports (although there are more than 30 secondary airports within the state), as well as by the two largest ports of the country – Santos and São Sebastião. The Port of Santos is endowed with a pipeline network for the transport of different types of fuels, as well as a gas duct. The region has the densest and most modern network of roads and highways in the country, linking it to the contiguous and dynamic states of Paraná, Minas Gerais and Rio de Janeiro, and then to other South American economies. The large majority of passenger and cargo traffic within the region is made by road, with the environmental and congestion consequences. Long traffic jams are almost permanent in the metropolitan area, with implications for just-in-time logistics, as well as for passenger mobility; the city of São Paulo has one of the biggest heliport networks in the world, extensively used by top executives to escape the permanent road congestion.

Presently, a large ring road is being constructed (Rodoanel Mario Covas) to ease the dense traffic flows of the metropolitan region, to reduce heavy traffic within the city of São Paulo and allow for faster and better connections between the busy port and other parts of the state. However, the pace of economic growth (leading to the rise of cars and cargo on the roads) is likely to provoke further pressures on the road infrastructure. The new urban growth areas (outside metropolitan São Paulo, like São José dos Campos, Taubaté, Campinas, Sorocaba) still have moderate congestion and good accessibility to other important regions, like Rio de Janeiro, Curitiba or Minas Gerais.

Quality of life

São Paulo is endowed with all the high-level amenities of a metropolitan region of its size: top leisure amenities, cultural and diversified events, several quality museums, international schools, public and private hospitals, urban parks and nature, etc. Diversity is strong, and the city is a magnet for highly qualified immigrants and ex pats. São Paulo breeds vibrancy and a 'business city' image, linked with a corporate atmosphere, mirrored by its many skyscrapers.

However, it is still considered one of the most unsafe cities in the world. Social segregation and sharp income disparities between the haves and have nots (many migrants from other Brazilian states) are a well known reality. Crime rates are high, though decreasing (*The Economist*, 2008b), including organised crime and homicides. Associated with high congestion costs and high prices, the attractiveness of the city faces sharp tensions. Firms face higher surveillance and robbery prevention costs than in other states. The federal government has been heavily investing in poverty reduction and education enhancement schemes (like Bolsa Familia), contributing to increased purchasing power for poor families. Despite the sharp social and territorial disparities within, the state of São Paulo presents the best living conditions in Brazil, namely in smaller urban regions outside metropolitan São Paulo, which cater for less congestion and good amenities.

Factor conditions

Like in the rest of Brazil, the relative cost of capital for investment and interest rates has been rather high. The recent depreciation of the US dollar together with the Brazilian economic growth has been leading to an unfavourable exchange rate for Real/US dollar, reducing the relative attractiveness of Brazil for industrial FDI and production of tradable goods like cars and automotive components.

The city of São Paulo and its first sprawl is further harmed due to the high prices of land and labour. Congestion and growth led to a rise in land prices, and scarcity. On the other hand, labour costs are higher in São Paulo compared to the rest of Brazil. The metropolitan region of São Paulo, namely its industrial core (like in the ABC region), has a tradition of active labour unions, increasing the relative global costs of labour in the region. However, in close urban agglomerations, but outside the metropolitan area, factor costs are still much lower, explaining recent large greenfield industrial investments in these locations.

The political context of the country is stable: a mature federative democracy, where different states have strong political power and responsibilities, powerful institutions, tax collection and budget management. Moreover, many companies face extra costs of operating in Brazil, in relation to other possible locations ('Brazil cost', see, e.g. *The Economist*, 2007), like diverse and overwhelming taxation – a company can pay around 50 different taxes to different government tiers – and bureaucracy.

10.4 The automotive industry of São Paulo

Despite the industrial de-concentration trends of the last two decades towards other urban regions and Brazilian states, São Paulo is still and by far the core of automotive production in Brazil. It has the largest concentration of OEM plants and suppliers in Brazil; in 2006, according to Sindipeças (2007), roughly 70% of total sales, employees and plants of the auto parts industry in Brazil were based in the state of São Paulo, as were 60% of total exports.

Not claiming to be comprehensive, in this section we present the main players of the automotive industry in São Paulo, as well as their main production strategies, technology trends (with focus on the recently developed flex-fuel engine technology) and geographical location.

OEMs

São Paulo houses many assembly plants as well as the South American headquarters of several transnational OEMs, specifically for cars, buses and trucks. Overall, the main rationale of São Paulo's transnational OEMs is the supply of the fast growing Brazilian and Latin American markets. The average car produced in Brazil is of medium/small size corresponding to the demand characteristics of the market. As a response, Brazilian car exports (see ANFAVEA, 2007) are mainly directed towards Latin America and South Africa.

In order to supply these markets while reducing extra capacity and financial risks, most of the OEMs historically present in São Paulo adopted, during the last two decades, 'convergent business strategies mixing compromise and flexibility' (Mariotto, 2003). OEMs invested new capacity in different productive units to enlarge flexibility of operations (e.g. rapidly changing the production lines and productive specificities, through 'neat and lean' factories) but assuming a compromise and long-term commitment by investments in specific resources and fixed costs. This strategy implied investments in other countries, like Argentina (tapping the benefits of the MERCOSUL free trading area), but also investments outside and within São Paulo's state.

In the last two decades, territorial competition for new automotive investments rose sharply in Brazil, steered by different states and municipalities, which offered incentives and tax cuts to attract foreign OEM plants and their suppliers.

This territorial 'fiscal war' (Rodriguez-Pose and Arbix, 2001), associated with increasing congestion, prices and safety problems in the city and urban sprawl of São Paulo, led to the emergence of new automotive location trends (e.g. outside the traditional automotive region in ABC Paulista) within and outside the state of São Paulo (see Figure 10.7).

The largest share of the new FDI took place indeed outside São Paulo, by the incumbent OEMs and new entrants (see Figure 10.7). The contiguous states of Paraná (location of Renault-Nissan in Brazil, which also received new investments from Volkswagen) and Minas Gerais (plants and headquarters of Fiat in Brazil, its second largest world market) got the biggest share, followed by Rio Grande do Sul (investments by General Motors, Toyota), Rio de Janeiro (location of PSA), and Bahia (recent largest Ford plant in Camaçari). Many of these investments involved not only production and assembling facilities, but also investments in R&D units and design centres, in order to adapt and design new models for the Brazilian and Latin American markets. Moreover, these new investments implied large labour training programmes to move closer to the higher qualifications levels of the automotive job market in São Paulo, by far the most qualified in the country.

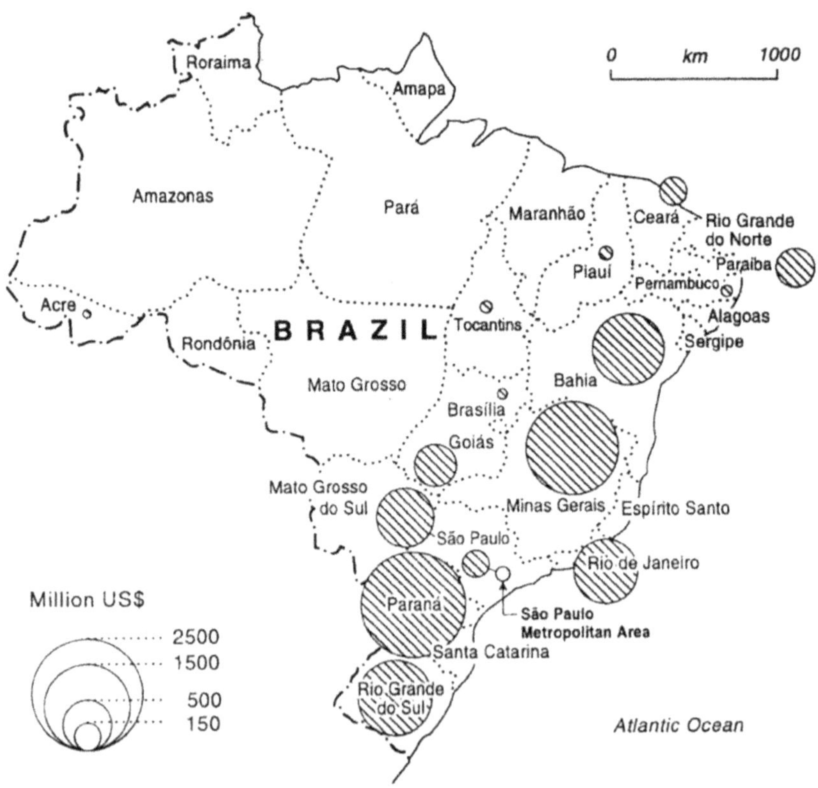

Figure 10.7 Distribution of planned FDI in new Brazilian automotive plants (1996–2001), by state (source: Rodriguez-Pose and Arbix, 2001)

To sum up, and despite the de-concentration of OEM production plants to new locations inside and outside the state of São Paulo, the metropolitan agglomeration continues to be one of the most important production sites in Brazil. Even when incumbent OEMs de-concentrated production, a significant part of it was kept in the traditional location. Nowadays, different plants of the same OEM tend to be specialised in certain functions (e.g. engine production, or assembling of different car models) under a 'Latin-American corporate network' in which São Paulo, besides assuming important production functions, also concentrates the 'brains' and main engineering functions.

Automotive suppliers

Just like with OEMs, the state of São Paulo concentrates by far the largest share of automotive suppliers in Brazil. However, as Figure 10.8 depicts, during the last decade, there was a de-concentration towards other inner regions of the state, as well as to other states. This does not mean that the number of suppliers located in metropolitan São Paulo diminished sharply in absolute terms, but that the new investments of the last decade took place mainly outside São Paulo. Nowadays, the largest share of automotive suppliers is located in the interior of São Paulo (e.g. close to the agglomerations of São José dos Campos and Campinas), benefiting from the presence of a qualified labour force – including qualified engineers – very good accessibility to the majority of Brazilian OEMs, quality of life, and lower land and labour costs.

The economic liberalisation in the 1990s brought a significant 'denationalisation' and change in capital ownership of automotive suppliers: the number of foreign suppliers operating in Brazil increased, while fully owned Brazilian suppliers decreased in the process (see Figure 10.9), through a combination of bankruptcies, mergers and takeovers. Moreover, there was a division of labour between suppliers in Brazil and other points of the global network, although Brazilian-located suppliers generally increased their production and assembling functions for the large internal and South American markets. Today, almost all the components needed to build a car, bus, truck or motorcycle are produced in São Paulo. For example, Volkswagen estimates that a very large share of their suppliers are located in São Paulo, though 70% are international companies and only 30% are nationally owned.

However, in regard to R&D activities after liberalisation, the reality is contrasted. Through the insertion in global corporate networks (namely by means of mergers and takeovers), R&D efforts tended to be optimised within the organisational chain, and according to the most efficient solutions, competences and specific corporate strategies. According to Cassiolato and Lastres (2000) there was a trend towards the 'triadisation' of R&D departments and competences towards Europe, the USA and Japan, shrinking many of the innovative efforts and R&D developed in Brazil, except for smaller adaptations for local markets and tastes. Many of the platforms for new model development started to be fully global.

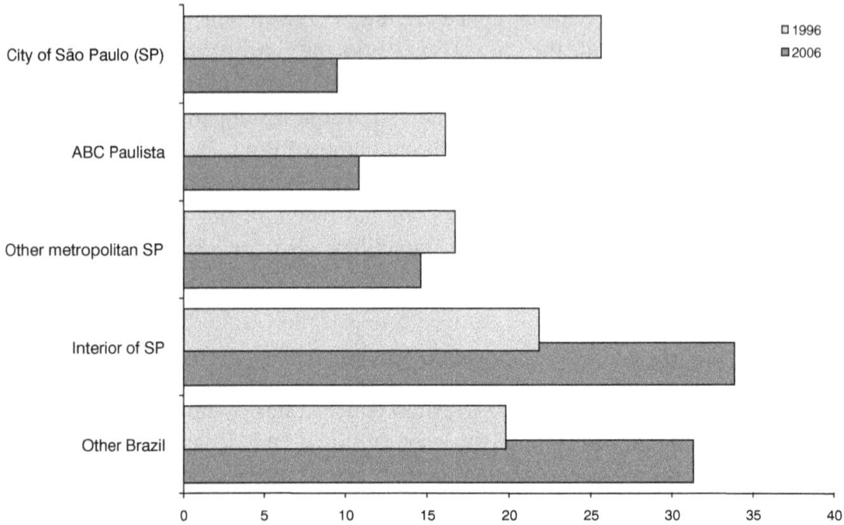

Figure 10.8 Auto parts suppliers: location trends in São Paulo (source: Sindipeças, 2007)

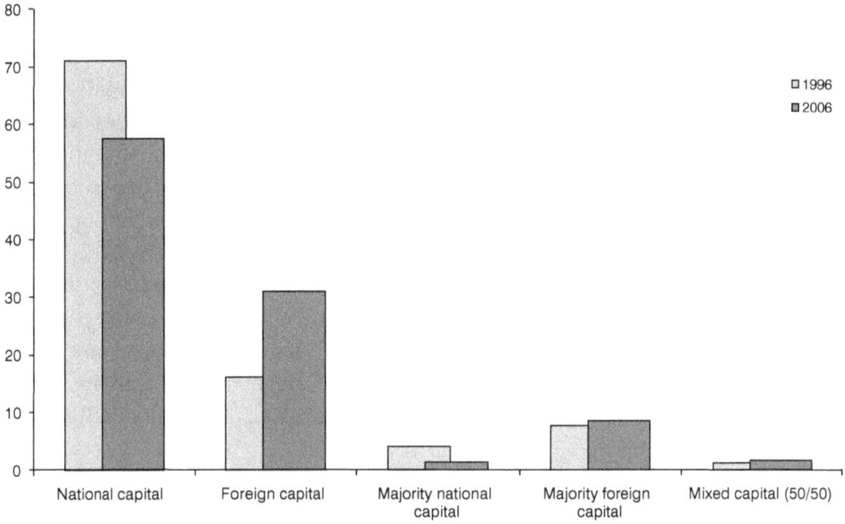

Figure 10.9 Automotive suppliers in Brazil, by origin of capital, % (source: Sindipeças) Note: the Figure depicts the capital ownership by number of suppliers, not by dimension, employment or turnover. If measured by those indicators, the gap between national and foreign suppliers would illustrate stronger foreign capital dominance.

Moreover, R&D and innovative capabilities of Brazilian-located automotive suppliers tended to evolve towards a concentration in specific niches with competitive advantages, as is the case with fuel-related technologies (engines, ignition systems, special materials, filters, specific software, etc.). In the next section we focus on one of these specific technologies – the flex-fuel engine systems.

The case of flex-fuel technologies

The portfolio of new products and technologies related to the flex-fuel car systems is strongly linked with regional past expertise in bio-fuels and engine adaptation technology. Developed by the corporate R&D departments of transnational systems producers in São Paulo, the core of the flex-fuel engine technology is an articulated group of sensors that read and compare the different types of fuel in a car tank, adjusting the engine combustion parameters accordingly. The first models with flex-fuel technology were launched in 2003 (the first one was Volkswagen Golf), and there are presently 10 car brands in Brazil with flex-fuel technology in 63 models (ANFAVEA, n.d.). As Figure 10.10 illustrates, while cars running exclusively on ethanol almost disappeared from the market during the 1990s, flex-fuel car production and internal sales have been growing fast since 2003.

On the other hand, the evolution of auto parts and flex-fuel engine exports during the last years have been remarkable. Some specific components, namely related to bio-fuel engines and systems (for ethanol and biodiesel) have increased more than 400% in value, pointing towards a very strong and growing presence of

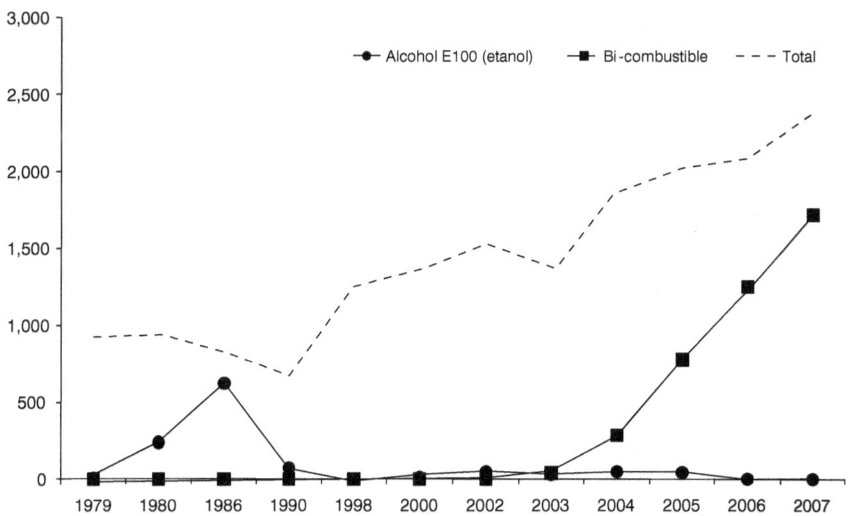

Figure 10.10 Ethanol car manufacturing in Brazil, 1979–2007 (selected years) (source: ANFAVEA, n.d.)

Brazilian-based companies in global auto component markets (Table 10.1). Unlike assembled cars, the exports of auto parts have as main destinations developed and mature markets, like the USA (number one destination) and the EU (Sindipeças, 2007). Exports of biodiesel-related parts, whose technology is not so widespread in Brazil as ethanol (at least for cars, it is much more for buses and trucks), is growing fast, indicating that technological developments in Brazil may have been absorbed first in other countries where (bio) diesel is relatively more important, like in the USA, namely for trucks, buses and agriculture machinery.

Additionally, with the use of flex-fuel engines running on ethanol or petrol, besides new software for the car injection system, many other changes are required (ignition systems, carburettors, fuel pumps, fuel pressure regulators, fuel tanks, suspension systems, filters, convertors, escape systems, cold ignition engines, etc.). This has been leading to new technological developments and enlargement of the portfolio of many auto suppliers operating in Brazil in the last years. These new components are also being designed, engineered and produced in Brazil, as a result of applied research of many firms located in São Paulo, isolated or in partnership with other corporative R&D departments (in the case of the many transnational suppliers located in São Paulo).

Knowledge institutes

Many knowledge institutions in São Paulo (universities, research centres and polytechnic schools) have relevant competences in automotive-related fields, producing a large number of engineers and qualified technicians. The state of São Paulo still concentrates the 'brains' of automotive knowledge in Brazil, with good

Table 10.1 Main auto parts exported by Brazil, 2000–2006, USD million

Rank	*Exported auto parts*	*2000*	*2006*	*Variation 2000–2006*
1	Other sub assemblers and accessories for automotive	443	1003	126 %
2	Spark ignition and internal combustion engines	106	599	452 %
3	Other diesel and shaft engines of vehicles	146	521	257 %
4	Cylinder heads and sumps	123	462	276 %
10	Fuel nozzles	55	198	261 %
11	Fuel injectors for compression engines	129	187	46 %
13	Pistons	73	132	82 %
16	Other compression-ignition internal combustion piston engines	20	108	447 %
19	Piston engines, compression ignitions (diesel or semi-diesel)	12	100	822 %

Source: own elaboration, based on Sindipeças (2007)

and available qualifications at different levels (medium, intermediate and higher education), and research potential.

In metropolitan São Paulo, automotive skills are widely available, but also growing in other close agglomerations, with good educational institutions. Some examples are the University of São Paulo (USP) and the University of Campinas (UNICAMP), both with strong engineering departments, and undergraduate and post-graduate courses. However, if these institutions produce highly qualified staff, the interaction with the private sector in general, and the automotive sector in particular, is recognised to be very low and only occasional, and driven by personal contacts. There are however some exceptions. The University of Campinas has a joint lab and some cooperation projects with Magneti Marelli in the fields of bio-fuel-related technologies, and some joint platforms of research partnerships are growing, although they are still in their infancy. Recently, the state of São Paulo announced new funding for different automotive and bio-fuels research lines, mainly for public and academic research institutes.

São Paulo also has applied technology institutes and polytechnics working closely with the automotive private sector. Two cases are FEI–University Centre and the Mauá Institute of Technology, located in the ABC region of metropolitan São Paulo. The institutes develop applied research with OEMs and suppliers, specifically for testing solutions, product development and incremental applications. Their undergraduate and post-graduate courses use this accumulated knowledge, and the institute provides tailor-made 'in-company' courses. For industrial intermediate workers, São Paulo hosts several facilities of SENAI–National Industrial Learning School, the largest in Latin America for professional industrial training and industrial knowledge diffusion. Some training specialities are particularly relevant for the automotive sector, e.g. automation, automotive, metal, polymers, and for bio-fuels-related technology, e.g. energy, oil and chemicals.

10.5 Global and local networks

In this section we pay attention to the positioning of São Paulo within the global automotive value chains and to the involvement of local players in these networks, alone or through cooperation platforms. What is the geographical span of different types of networks? In which types of networks are the knowledge developments (like flex-fuel technology) taking place? Are there other relevant examples of interaction between automotive production and bio-fuel-related technologies? How are these networks managed?

Production networks

Production is still by far the key reason why many automotive OEMs and suppliers locate in São Paulo and in Brazil. Location factors are similar to many other automotive investments worldwide: by producing in Brazil, OEMs access the large and fast growing Brazilian and Latin American markets, producing with lower factor costs compared to their home locations (in the case of foreign firms). On the other hand, automotive suppliers follow the assembling OEM to their locations in order to facilitate interaction, gain market share and reduce logistic and transport costs, many times invited by the OEM itself.

Automotive production in São Paulo takes place through many global and local networks, as well as at other 'intermediate' geographical scales, like between Brazilian states and Latin America as a whole. Production networks are frequent between different OEM plants of the same corporation. Within the state of São Paulo, one example is Volkswagen, which illustrates the operations model of many of their competitors. Volkswagen has three different production facilities within the state of São Paulo: one in ABC Paulista (called Anchieta), another in Vale do Paraiba and another in São Carlos, close to the agglomeration of Campinas. The engines produced and assembled in São Carlos are inserted into different car models assembled in the other two locations; different plants specialise in certain components, but with some room for flexibility and adaptation. Moreover, some components come from Volkswagen units in Argentina, like diesel engines and gearboxes. Another example is the one of Ford: some components produced in São Paulo are assembled hundreds of kilometres away, in a new Ford plant in Bahia.

Networks between São Paulo's OEMs and their suppliers also take place at different spatial scales. Procurement and buying departments of OEMs and their main suppliers are generally based in São Paulo, but they search globally. For customised pieces and materials (basic tools, plastics, etc.) the pressure to reduce costs implies that suppliers vary according to the best offered conditions worldwide – usually there is no 'spatial rule' concerning this type of component. Nevertheless, around 95% of Volkswagen's components are produced in Brazil. This is the case for many suppliers as well.

We found cases where components were bought through other global corporate procurement departments. For example, a leading American supplier in exhaust and fuel systems imports more than 700 simple components from global suppliers, but some come from other corporate branches, namely in the USA, Mexico and Argentina. However, when it comes to supplying OEM production lines, there are spatial constraints. Some big parts, like fuel tanks, are usually produced close to the OEM line, in order to reduce transport and logistical costs. Other production networks are usually local, specifically when related to characteristics of local markets (e.g. the production of dark window glass and thicker metal for the doors).

The networks between São Paulo's suppliers and new OEM locations (e.g. Paraná) are strong – many suppliers did not immediately follow the OEM to the new location, relying instead on good road connections. However, the growing just-in-time requirements and 'zero stock' policy of many OEMs means that critical suppliers increasingly relocate in order to avoid delays and production disruption (see Box 10.2). Being that São Paulo is an extremely congested metropolis, components previously produced in the metropolitan area and assembled in other Brazilian OEM locations are now produced, or at least finished, close to the OEM production line.

Knowledge networks

In 1972, the automotive model 'Brasilia' was the first to be fully designed and built in São Paulo, outside the German headquarters. The design and engineering efforts were developed between Brazilian Volkswagen and their main (few) external suppliers. Presently, technological innovation efforts and Brazilian-based automotive R&D is moderated: product development tends to be adapted for local specificities and tastes. The core of product development and breakthrough innovation efforts usually takes place in corporate headquarters and R&D departments elsewhere, or through temporary teams composed of engineers of different R&D corporate departments.

One example of incremental development for taste adaptations is Renault's first design centre in Latin America. Renault, with operations in the Brazilian

Box 10.2 Integrated auto parks for OEM suppliers – an expanding model

Many OEMs in São Paulo outsource the majority of the vehicle physical production, keeping mainly design, assembling and quality control functions. In order to maximize the production line efficiency, avoid stocks and disruption risks, many OEMs started to promote, together with their suppliers, integrated 'auto park' facilities. These facilities consist of a specific supplier's park within the OEMs' plant, in order to bring the critical (usually 'tier one') suppliers closer to the production line.

The case of Volkswagen is illustrative. When the plant reduced their production capacity in metropolitan São Paulo, the vacant space was used to install suppliers within the plant. More supplier facilities are being developed in other Volkswagen plants within São Paulo (e.g. Taubaté). Some suppliers are already inside this park (like LEAR and other metal parts suppliers) and others are moving to ensure physical proximity. For example, Robert Bosch, based in Campinas (100 km from metropolitan São Paulo) will have its own finishing facility within the Volkswagen suppliers park in Taubaté, to process new orders directly and avoid the risk of stock failures.

These supplier facilities are usually logistic and storage locations, but in some cases host small temporary engineering centres. In other situations, the supplier is fully integrated within the production line. Starting by being an efficiency-related facility, it generates trust between the supplier and the OEM, leading to more frequent interactions, problem solving activities and learning.

Source: own elaboration

state of Paraná, recently opened its first design centre in São Paulo (city), with designers from Brazil, Argentina, Colombia and France. The location in the inner city was chosen for its vibrancy, artistic atmosphere and top design schools. The centre constitutes a platform with Latin America (adapting to local tastes, namely the external 'look' of the car), but the main design and engineering functions will stay in France.

Likewise, while some knowledge and technology networks exist between Brazilian-based OEMs and their main suppliers (for product adaptations and development in joint platforms), the spill-over of these knowledge networks to other (2nd and 3rd) level national suppliers is relatively low – there is a big 'knowledge intensity' and 'power' gap between different types of suppliers, resulting in low technological interactions and joint learning.

However, there are important exceptions related to the recent development of a vast portfolio of bio-fuel and automotive-related technologies, presently shaping some of the most relevant knowledge networks in São Paulo. The first efforts towards the development of flex-fuel technologies were primarily made through the corporate knowledge networks of three big system producers located in São Paulo (Robert Bosch, Magneti Marelli and Delphi), close to the agglomeration of Campinas (see Box 10.3). Moreover, the diffusion and further developments meant further development networks between system producers, OEMs, other suppliers, bio-fuel producers and research institutes.

Today, system suppliers concentrate R&D efforts in adapting and developing new bio-fuel-related technologies, representing important nodes of their respective corporate R&D networks. In the coming years, R&D investments in Brazil (namely in the state of São Paulo) will increase, reinforcing their bio-fuel specialisation and engineering departments. Magneti Marelli has the largest R&D centre of fuels outside Europe, and estimates a 30% growth until 2010. The establishment has more than 200 engineers, and more than 100 specialise in flex-fuel technology (namely fuel-related software and systems). Robert Bosch, in Campinas, with more than 400 engineers working on flex-fuel technologies, hosts the 'World Competence Centre for the Development of Conventional and Oxygen Fuel Systems', with world patents in fuel combustion materials, hydraulic systems and fuel software. Their production and R&D investments will also keep growing in the near future, just like Delphi: the corporation has several quality awards in Brazil, where it also develops and produces several fuel and engine-related systems.

São Paulo is becoming a reference and learning centre in terms of bio-fuel technology, specifically automotive-related applications. Many countries, including the USA, China and India, are sending engineers to Brazil to study and buy related technologies. Likewise, there is evidence that technology produced in Brazil is being absorbed by other countries, not only through exports, but through the networks of global companies operating in Brazil. For example, General Motors invites Brazilian engineers to live in the USA for a period, in order to work with their R&D centres in Detroit; Swedish subsidiaries of General Motors

Box 10.3 The development and innovation process of flex-fuel automotive technology

Some of the most relevant developments in 'flex-fuel' automotive technologies took place in the last ten years through the R&D corporate networks of the big automotive system producers located in Brazil. The first systems were patented by Robert Bosch and Magneti Marelli, followed by Delphi. Although the patents of the systems were registered in the name of the global corporation, the core of the know-how and development process took place in São Paulo, more precisely close to the agglomeration of Campinas (Piracicaba – Delphi; Hortolândia – Magneti Marelli; Campinas – Robert Bosch).

The initial R&D efforts and follow-up

Robert Bosch in Brazil pioneered the development of a technology to sense different types of fuel inside an automotive tank. Benefiting from previous research done in its USA branches during the 80s, a first patent was registered in Brazil (1988) in the name of the global corporation. In the next years, Brazilian engineers of Robert Bosch developed and tested new sensors for fuel mixtures, in cooperation with German R&D departments, whose intermediate results were presented in a seminar in Brazil (1994) with the reference parameters that allowed developing the system. However, by this time, the new sensors were still very expensive to produce, making its generalized use not profitable for OEM, who didn't show enough interest for joint research towards further calibration.

In the mid 90s, independently from Robert Bosch, Magneti Marelli R&D in Brazil launched efforts to reduce the high cost of the sensor, eliminating most of its physical components (like infra red technology) and basing the system in a software configuration directly inserted in the injection system. To do so, a multidisciplinary team was set up in Brazil, with the participation of engineers from the Italian R&D centres – although the know-how was in Brazil, the corporation wanted to get in touch with the new technological developments.

Until the early 21th century, system suppliers in Brazil invested heavily (but in isolation) in R&D, calibration, prototyping and patenting of flex-fuel system technologies, as well as in reducing its production costs. These knowledge developments were done 'in-house' (São Paulo), in collaboration with other corporate R&D departments. By this time Delphi started to develop its own sensors.

The institutional and networking turn

In 1998, a seminar organized by IPT – Institute of Technological Research of São Paulo – set the basis for the diffusion of the technology, and represented a turning point in the OEM´s interest in its adoption. By 2000, OEMs and system producers in Brazil started the first partnerships for the joint development of the first flex-fuel technologies in OEM's engines. During this time, the Brazilian Association of Sugar Cane Producers (UNICA), the national energy company Petrobras, the IPT and the state of São Paulo analysed the relevance and impacts of the technology and, in 2002, the federal government issued a law putting flex-fuel engines in the same category of alcohol-run cars for incentives and tax-deduction schemes. After this legal move, OEMs were definitely convinced, and the first flex-fuel cars went on the market in 2003, thanks to Volkswagen (with Magneti Marelli), General Motors (with Delphi) and Ford (with Robert Bosch).

Source: own elaboration

(namely Saab) are developing an alcohol vehicle in partnership with the Brazilian centres.

As previously mentioned, the developments relating to bio-fuels have been generating new local and regional R&D links between OEMs, suppliers, research institutes and biodiesel producers. For example, first-tier suppliers like Magneti Marelli developed a joint platform with 13 Brazilian suppliers to supply its production line of fuel-injection nozzles, one of its most high-tech (and fast-growing) products. Moreover, these suppliers were offered the opportunity to supply other corporate plants of Magneti Marelli in the USA and Europe. Other first-tier suppliers are now working closely within a Ford-led platform to adjust their products (pumps, fuel injection systems, tanks) to new biodiesel engine requirements.

Research links between universities, research centres and private companies, despite being considered still low are also growing. Universities were not involved in the initial development of flex-fuel technology; one of the few, between Robert Bosch and University of Campinas didn't go through. However, cooperation efforts are catching up – research incentives and funding by the state of São Paulo have been facilitating consortia and cooperation. For example, auto suppliers are increasingly working with EMBRAPA (Research Centre of the Agriculture Ministry) for the development of new bio-plastics for anti-fuel corrosion; Ford tests new bio-fuel combinations with Petrobras and the University of Bahia; Delphi showed interest in new joint projects with the University of São Paulo and Campinas for flex-fuel innovations.

Sometimes, bio-fuel producers take the lead in promoting new applied research projects and tests. The main rationale is to improve the raw materials and mixtures towards more energy-efficient solutions, which is presently the case of Brazilian biodiesel. The Brazilian group Bertin (agro-food) is presently trying to optimise biodiesel production using animal skin and fat, as well as vegetable oils. The group developed the project 'Mistura 20' (2007–2008) to test new biodiesel mixtures. To do so, a partnership was established with other firms in São Paulo – Volkswagen Bus (trucks), Robert Bosch (injection systems and filters) and Cummins Engine Company (US diesel engine producer located in São Paulo). In this partnership, Bertin, with the support of the regulator and buyer Petrobras, tests new biodiesel components and mixtures (B20), while Volkswagen tests modified trucks with electric engines from Cummins and fuel injection systems from Robert Bosch.

Biotechnology research related to bio-fuels, namely ethanol and biodiesel, is also growing fast. This has so far been done through public funding, but there are examples of private spin-offs from the University of Campinas, like CanaVialis and Alelyx. Both are working on new varieties of sugar cane and ethanol, as well as on genetic modifications of raw materials. In universities and other research institutes, many other groups are working on similar, though more fundamental research. In some research consortia, like the ones organised by the Association of Sugar Cane Producers (UNICA), many universities and research centres are involved. One example is the project for 'Ethanol Production via Enzymatic Hydrolysis', which involves many Brazilian universities and research centres from different states, the National Centre for Biomass Development (São Paulo), and the Centre for Sugar Cane Technology Development (Piracicaba, São Paulo); moreover, it has the international collaboration of universities in Sweden, Spain and Argentina. This 'hub' character for bio-fuel-related technologies has been attracting transnational high-tech firms in multiple sectors to source and establish knowledge networks with São Paulo. Bayer, already in São Paulo for decades, is now developing new process technologies to transform and produce biodiesel. Amyris, a leading California biotechnology firm in renewable fuel, recently established a joint venture with Crystalsev, a large ethanol distributor in São Paulo, to work on the joint distribution of renewable fuels like ethanol and biodiesel.

Governance of the networks

Production and knowledge networks have differentiated governance modes, ranging from stable equity participations and corporate control to volatile market contracts.

As stated before, production networks between different OEM facilities usually happen within the same corporation, whether the links are global, regional (Latin America, Brazil) or local. Many times, privileged relations also exist between OEMs and suppliers belonging to the same corporate group; for example, Magneti Marelli belongs to the Fiat group and thus the supplying networks are managed within the same group. Many foreign suppliers located in São Paulo also have privileged relationships with corporative establishments in other regions and countries – they specialise in certain production categories and exchange through the network. Overall, this means that a very large number of strategic producing decisions which impact on São Paulo are indeed taking place in global headquarters elsewhere, namely in Europe and in the USA.

On the other hand, many inter-firm production links may also be fully market-based. Procurement departments in Brazil buy diverse components and raw materials through specified and competitive market contracts. OEMs and tier-one suppliers usually have strong power when contracting and choosing their partners. However, initial market-based governance forms sometimes evolve towards other more stable cooperation modes – many OEMs wish to reduce the number of suppliers, and establish more privileged and trust-based links with the more efficient, capable and trustworthy ones. The new integrated auto parks demonstrate the physical outcome of more stable partnerships and development platforms between OEMs and suppliers.

When it comes to knowledge and innovation networks, equity and corporate belonging is also a key determinant of network building. As for the development of the flex-fuel core technologies, many R&D networks take the form of temporary local/global platforms, where different R&D corporative departments interact, and spread knowledge. These types of networks allow multidisciplinary groups to gather together and to cross the competences of the network. In some cases, and due to strict confidentiality, interaction with external players (to the group) is rare. There exist however examples of other governance modes for joint technological developments and innovation. After the successful adoption of the core flex-fuel systems by Brazilian OEMs, research consortiums – with risk sharing – become common governance modes between OEMs and specialised suppliers, e.g. to adapt new engine systems to new models, to test new bio-fuels in engines, or to develop new bio-fuel pumping systems. Research platforms in São Paulo are even emerging with fierce competitors, like Robert Bosch and Delphi, namely to produce new engine motorcycle systems.

In São Paulo there are no specific cluster platforms or cluster-based policies focusing on the automotive sector, or in the automotive industry and its links with bio-fuels production (although they exist for other sectors, like plastics). Market is the main and almost exclusive driver of the production and research networks of the industry. On the other hand, cooperation between firms and universities is very low – the governance process of this type of network is complex, legal incentives do not yet fit to the sector's needs and former experience in managing joint research is almost non-existent, despite the exceptions of some knowledge transfer academic agencies, like INOVA, from the University of Campinas. However, the Research Foundation of the state of São Paulo (FAPESP) recently launched incentives towards joint research platforms and research contracts between private automotive players, not only through some funding for research within firms but also through the organisation of meeting platforms for firms and universities, called 'Tech Days'. During these specific days, an interested firm (usually large) meets with research institutes of different universities of São Paulo to find possible complementarities. For example, the American system supplier Delphi met with research institutes and knowledge transfer agencies of the Universities of São Paulo and Campinas to find possible complementarities and research cooperation in the fields of flex-fuel injection and electric systems.

Automotive sector associations exist at the national level and are based in São Paulo. On the side of automotive, the most important ones are ANFAVEA (for OEMs) and Sindipeças (for auto parts producers); UNICA aggregates sugar cane and ethanol producers while Abiodiesel focus on the producers of biodiesel. They all foster links between their members by means of seminars, information sharing and sector intelligence. Moreover, they are key institutions in lobbying with the state and federal governments on fiscal issues and diverse legislation. However, interactions between bio-fuel and automotive associations have so far been relatively low, and joint projects between automotive and bio-fuel associations have been occasional.

10.6 The role of public policy

In Brazil in general and in São Paulo in particular, despite fiscal and regulatory issues, public policy has a very limited intervention in economic development and innovation issues. In regards to the development of the automotive industry, inspired in the USA model, policies promote 'laissez-faire' and give almost full sovereignty to private players, once framed by legislation and regulatory schemes.

However, at different administrative levels, policies do have impact on the development of the automotive industry and innovation in São Paulo. The federal government, besides controlling interest rates, provides tax breaks for firms' R&D; moreover, it decides many standards for bio-fuel mixtures and quality, as well as flex-fuel automotive regulation. For example, only after the legislation by the federal government in 2002, determining fiscal incentives, OEMs started to invest in flex-fuel technology. Moreover, the federal government collects many of the more than 50 different taxes on automotive firms, totalling around 40% of the value of a new car.

The state of São Paulo promotes different policies and funding to support both automotive and bio-fuels developments. Besides marketing the state externally to foreign investitures and fostering access of local industries to external markets (in association with the industrial association of São Paulo – FIESP), the state has announced a large fund to support specific research in bio-fuels and related technologies, in order to strengthen the position of the state in bio-fuels and bio-energy-related technology (BIONEN). Besides basic research, some research focuses on cooperation between academia and firms. The main research topics focus on processes to produce ethanol and biodiesel, as well as application of bio-fuels in automotive engines, internal combustion engines and fuel cells, bio-fuel genomics and alcohol chemistry. Moreover, the state has its own research institutes, some of them working in automotive and bio-fuel-related fields, like the IPT – Institute of Technological Research, responsible for the 'Competitiveness Agenda of São Paulo', a recent strategic plan for competitiveness, innovation and local development within the state.

At the local level, municipalities deploy a vast array of incentives to attract (any kind of) economic investments and some of them have very active and aggressive investment attraction policies. The responsibilities of São Paulo's municipalities are usually focused on zoning, infrastructure and land provision. Usually, they all have a package of tax cuts for a number of years for new industrial investments – this is the illustrative case of the municipality of Taubaté, in Vale do Paraíba, which has been receiving many new greenfield industrial investments in the last decades (in automotive and other industries). The municipality of São Bernardo do Campo (ABC Paulista) recently renovated its incentive scheme. It benefits not only industrial activities, but also commerce and services and new activities of former employees of the automotive industry.

In some subregions of São Paulo, like ABC Paulista, inter-municipal cooperation for economic development issues already has a tradition. It counts a development agency, the oldest in the whole state, gathering not only the mayors of the seven municipalities but also other public and private players, like sector associations, universities and large firms. However, with a limited budget and a small technical staff, the impact of the initiatives has so far been modest. Despite fostering joint discussions on economic development issues, the agency's activities focus on marketing the region and developing consultancy for the different municipalities.

All in all, the state of São Paulo has sound and focused incentives to promote effective links between automotive players, and connecting those with strong competences and emergent players in bio-fuel production. The programmed competitiveness pole of bio-fuels (close to the agglomeration of Campinas), if acting as a think tank and facilitator of initiatives and networks, has potential to foster new value chain synergies with automotive and other actors. Moreover, it is intended to develop brokerage activities, fund research and gain support for public policy design.

10.7 Conclusions and perspectives

With roughly one century of automotive industrial tradition, São Paulo is presently the most important Latin American automotive hub. Despite the trends of de-concentration to other Brazilian states, the corporate strategies of the most important global OEMs and automotive suppliers still keep São Paulo as a key node of their global networks. Although their main rationale in São Paulo is to produce for the fast-growing Latin American market, R&D competences are becoming more and more relevant, linked with the development of bio-fuel-related technologies. As a result, many transnational automotive corporations are enlarging their competence centres in São Paulo and its vicinities.

Industrial optimism is high and it is hoped that Brazil will attract new investments in the coming years (both by the expansion of incumbent players and by new entrants) in order to meet growing demands. Due to the high internal costs of capital, a large share of these new investments is likely to be FDI from transnational companies exploring growing markets, but also increasingly willing to access new knowledge related to alternative fuel technologies.

However, like in the last decades, most of the new investments will probably not be located in the more traditional automotive locations of São Paulo, like ABC Paulista. Competition from other states and municipalities is fierce, offering fiscal benefits, but, mainly, less congestion, good accessibility and lower factor costs. In the recent past, OEM's de-concentration from the state of São Paulo took the shape of two different 'types' of firm location strategies, towards very different types of regions. On one hand the 'Bahia strategy', e.g. when Ford moved a large part of its production from São Paulo to the peripheral Brazilian state of Bahia, benefiting from big tax cuts, low costs of land and labour, but also having more difficulties in finding qualified staff. On the other hand, other OEMs opted by the 'MERCOSUL Strategy', e.g., when OEMs formerly present in São Paulo moved part of their operation to states like Paraná or Rio Grande do Sul, benefiting from good logistic conditions, proximity to Argentina, quality of life and qualified labour at moderate costs. Other investments in the states of Rio de Janeiro and Minas Gerais followed a similar rationale, benefiting from the proximity to São Paulo, but reducing the main congestion and factor cost constraints. Most of these relocation and investment strategies from OEMs implied that the main suppliers also moved close to these new locations, generating spill-over and multiplier effects in the local and regional economies.

Nevertheless, in the last decades, supported by favourable internal and external conditions, the sector showed significant growth in São Paulo as a whole, measured, for example by turnover and exports. If employment decreased in the more traditional locations, it increased in other poles of the state. Moreover, automotive production was upgraded and showed technological dynamism, mainly through the portfolio of new flex-fuel technologies and associated R&D efforts, but also through new advanced functions like design. At the same time, the automotive sector in São Paulo deepened its dependence from the powerful transnational players – OEMs and first-tier transnational suppliers – whose strategic decisions on production, but also on R&D, depend on decisions of global headquarters located in Europe and in the USA. Even though recent knowledge and R&D efforts in São Paulo are associated with a very specific natural resource (bio-fuels) and the networks between transnational firms and local players are getting deeper (research institutes, national firms), the dependence of São Paulo on global corporations is (and is likely to remain) very high.

The foundations of São Paulo are remarkably strong. The *economic base* is diversified and vibrant, offering many opportunities for new productive combinations, by crossing the competences of different sectors. Automotive is indeed one of the sectors benefiting from that, since a car is a composite effort of many different types of industries, technologies and services, all present in São Paulo. The production in São Paulo of different types of bio-fuels (mainly ethanol) facilitated and steered new automotive engines and system tests, as well as an important share of recent technological developments in the sector, reinforcing this through the attractiveness of the region to foreign players.

The *knowledge base* is very strong, encompassing all qualification levels. Automotive tradition produces qualified and experienced industrial and automotive workers. The many educational facilities of São Paulo produce qualifications at all educational levels, including top engineers in their very good universities and top research departments. R&D is likely to increase in the coming years. Many public funds are being channelled to fundamental and applied research, namely in engineering and biotechnology related to alternative fuels in top research centres; moreover, automotive corporations are expanding their R&D capacity in São Paulo, seeking more bio-fuel-related patents and breakthroughs. However, the potential of interactions between firms and universities remains largely unused – history of cooperation is very low and institutional arrangements (law, regulations and values) are not yet favourable to this kind of cooperation.

At the same time, relatively high *factor costs* of land and labour in São Paulo are not attractive to automotive production functions, intensive in labour and space. *Accessibility* and infrastructure are the best in the country, but nowadays extremely congested (although less polluted since flex-fuel vehicles), making just-in-time logistics risky. Moreover, regarding *quality of life*, despite the presence of top amenities for the well off, social disparities are resilient and endanger the sustainability of the region. Companies have higher safety costs than in other places. The combined effect of these disadvantages threatens the long-term attractiveness and sustainability of São Paulo as a place to do business, work and live. Many automotive suppliers' and OEMs' recent investments within São Paulo are located in new urban agglomerations away from the central city, like Campinas or São José dos Campos (including R&D departments), benefiting from lower congestion, quality of life and, still, the presence of highly qualified workers. At the same time, congestion and crime have been pushing many high-qualified workers away from the metropolis, looking for a better quality of life in new economic poles. Furthermore, high prices in the metropolis tend to force out intermediate qualified workers.

These trends, in the medium term, are likely to change the 'automotive geography' inside the state of São Paulo. Some signs are already visible. The congestion, lack of space and high prices in the city centre will keep pushing productive functions away; it will likely specialise in related functions like design, high finance, legal services and regional headquarters. The metropolitan area of São Paulo, especially former automotive strongholds like ABC, although keeping an important automotive production share from the incumbent players, may see a further reduction in the coming years, due to further de-concentration. Also here, firms are specialising in automotive-related services, distribution and applied development. On the other hand, the agglomerations close to Campinas and São José dos Campos will likely concentrate increasingly more automotive functions. These areas still have plenty of greenfield space, good accessibility, quality of life, and the required qualified people, including graduates and post-graduate engineers. Production from OEMs and their suppliers is already being made there. Proximity to good universities, as well as to the core of bio-fuel new developments, will contribute to the re-enforcement of bio-fuel-related R&D functions in these new locations. In this context, the networks and automotive interactions within the urban agglomerations of São Paulo are likely to increase, making it an even more integrated region, at least for automotive issues.

The automotive industrial structure in São Paulo is led and dominated by the market-based decisions of large and powerful transnational corporations – OEMs and tier-one suppliers – who interact with other local component and raw material suppliers mainly through semi-hierarchical governance modes. This is the typical implementation model of automotive OEMs in Latin America (Cassiolato and Lastres, 2000). The role of institutions and government in shaping this structure is moderated. Institutions set the 'rules of the game', through legislation, (high) taxation and general investment incentives, but let companies work in a 'laissez-faire' fashion.

The transnational insertion and the global networks of the main automotive players present in São Paulo have influenced the shape and character of its local interactions. Local suppliers benefit from the presence of the global OEM and supplier corporations – they get in touch with their quality standards, and sometimes get training to cope with it. In some cases, once they supply the transnational company in Brazil, they get the chance to supply other nodes of the corporation worldwide. However, in other cases, the spill-over is more moderate. Some components and raw materials are bought and accessed through corporate global procurement departments and the technological link between 'tier-one' and other tier suppliers is very weak, and fully market-based.

In São Paulo, automotive production is separate from research. R&D of transnational corporations based in Brazil is done mainly internally and in multidisciplinary teams with other global corporate departments, usually in the company's European or American headquarters. Despite adaptations to the local market, the design and engineering of a new car is mainly done in other locations. The liberalisation of the 1990s reinforced this situation; take-over and mergers between Brazilian players and global corporations led to a shrinkage of R&D functions in Brazil and growth in other points of the global network. However, there are notable exceptions. The design, engineering and production of the new flex-fuel engine systems took place in São Paulo, due to the proximity of a very specific natural resource (sugar cane) and local know-how. Nowadays, regarding bio-fuel-related technologies (in the fields of engineering, ICT, biotechnology and chemistry), São Paulo has R&D excellence centres, exporting technology to mature markets. This is one of the few areas where the R&D networks of transnational companies spill and interact with local suppliers and other players, like bio-fuel producers and research institutes.

Nowadays, there is a lot of R&D taking place in São Paulo in automotive technologies. These efforts are both public and private, and with several links to external sources of knowledge (R&D corporative departments worldwide, universities outside São Paulo, etc.). However, there is still plenty of room for further integration between public and private research efforts and firm–university research consortia are still in their infancy. Nevertheless, academic spin-offs are growing (namely in biotechnology and related 'green' innovation) and cooperative efforts in firm–university consortia are also developing, namely through a new generation of joint research incentives from the state of São Paulo.

In this situation, it is likely that the state of São Paulo will keep production capacity, due to market proximity and installed capacity, but also the most important 'brains' of the automotive sector, namely in bio-fuel-related engineering, reinforcing its position *vis-à-vis* world-level centres of excellence in the field. Moreover, these changes in the automotive R&D and production structure are likely to have spatial effects. Metropolitan São Paulo will share its knowledge and R&D dominance with other agglomerations. It is the case of the agglomeration of Campinas, where front running companies are expanding their R&D centres. The University of Campinas produces state-of-the-art knowledge in related fields, will host a science park, and new technological facilities are being planned in the municipality of Piracicaba, the main ethanol production and research hub in the state of São Paulo.

Notes

1 Sugar cane is a Brazilian (and other countries) specific commodity. In countries like the USA the fermentation of maize (corn) is more popular, though the energetic power and quality is much lower (e.g. OECD, 2008).

2 Urban–industrial area comprising the municipalities of Santo André, São Bernardo do Campo and São Caetano, recently expanding to the municipalities of Diadema, Mauá, Ribeirão Pires and Rio Grande da Serra.

Bibliography

ANFAVEA (2007) *Anuário da Industria Automobilistica Brasileira*, São Paulo: ANFAVEA

—— (n.d.), available online at: www.anfavea.com.br (accessed 28 May 2008).

Cassiolato, J. and H. Lastres (2000) 'Local systems of innovation in Mercosul countries', *Industry & Innovation* 7, 33–53.

CTC (Tecnological Centre for Sugar Cane Technology) (n.d.) available online at: www.ctcanavieira.com.br (accessed 28 May 2008).

IBGE (Brazilian Institute of Geography and Statistics) (n.d.) available online at: www.igbe.gov.br/home (accessed 28 May 2008).

Investments São Paulo (n.d.) available online at: http://english.investimentos.sp.gov.br/portal.php/information (accessed 17 July 2008).

Mariotto, F. (2003) 'Flexibility and compromise in the Brazilian automotive industry', *Annals of the Third International Conference of the Iberoamerican Academy of Management*, 1, 1–20.

OECD (2006) Economic survey of Brazil 2006 – fostering long-term growth: the challenges ahead, Paris: OECD.

—— (2008) Economic assessment of bio-fuel support policies, Directorate for Trade and Agriculture, Paris: OECD.

Rodriguez-Pose, A. and G. Arbix (2001) 'Strategies of waste: bidding wars in the Brazilian automotive sector', *International Journal of Urban and Regional Research*, 25:1, 134–54.

Shapiro, H. (1996) 'The mechanics of Brazil's auto industry', *NACLA Report on the Americas*, Jan/Feb. 1996.

Sindipeças (2007) *Brazilian autoparts industry performance 2007*, São Paulo: Sindipeças.
Teixeira, E.T. (2005) *Flex fuel tecnology in Brasil*, São Paulo: DNA Institute.
The Economist (2007) *Special edition on Brazil,* 12April 2007.
—— (2008a) 'Lean, green and not mean', 28 June 2008, pp. 59–60.
—— (2008b) *Getting less violent*, 23 August 2008, p. 43.
UNICA (n.d.), available online at: www.unica.com.br (accessed 21 June 2008).

11 Shanghai: automotive industry

11.1 Introduction

In the last decades, China has emerged as an economic dragon on the world stage. The country is now often referred to as the 'workshop of the world': in many industries, China has proved to be highly competitive as a location for production. The industrial development has entailed a massive migration from the countryside to cities, and has lifted millions of people out of poverty. The major economic achievements of China are ascribed to the country's vast human resources potential, its massive investments in infrastructure and its carefully planned process of integration into the global economy. In the last few years, China is entering a new stage and actively seeks to raise its profile as a leader in research and development. The Chinese government has designed a variety of measures to achieve these goals. Among other things, it has committed itself to raise expenditures on education and R&D, boost innovation in specific sectors, attract knowledge-intensive investments from abroad and, last but not least, learn from foreign knowledge and expertise. One of the industries targeted by the Chinese government is the automotive industry. The Chinese government has formulated the ambition to develop a full-fledged Chinese automotive industry, including Chinese brands, design and technology, and has committed substantial resources to realise this ambition. These ambitions and policies are formulated by the national government, but their implementation has an important regional dimension. There are just a handful of regions where the automotive industry is significant. This chapter describes and analyses the dynamics of the automotive industry in the most important one: Shanghai. Over the last decades, Shanghai has emerged as the largest and most advanced automotive cluster in China. Using our conceptual framework, we describe and analyse the development of this remarkable cluster.

This chapter is organised as follows. First, we provide a brief overview of the development of China's automotive industry since the 1980s (section 11.2). In section 11.3, we move to the regional dimension, and describe the urban economic context in which Shanghai's automotive cluster has developed. Section 11.4 describes the firms and knowledge institutes that make up the automotive industry in Shanghai. Section 11.5 provides an analysis of the local, national and international networks of the automotive cluster. Section 11.6 concludes, and puts the findings in the perspective of the framework of analysis of this study.

11.2 The automotive industry in China: an overview

The development of the modern Chinese automotive industry started in the beginning of the 1980s when the government selected the automotive industry as a prime target of industrial policy. In the 1980s and early 1990s, the Chinese market was dominated by Volkswagen, which was one of the first foreign OEMs that entered the Chinese market via a joint venture (JV) with Shanghai Automotive Industry Co. (SAIC). In Shanghai, it produced outdated models for the Chinese market. There were also indigenous car producers but they were far less efficient than global firms, and products were inferior.

In the 1980s, the government (state and local governments) deployed three main tools to promote the local automotive industry: 1) the obligation for foreign investors to engage in joint ventures with Chinese companies; 2) local content requirements, and 3) import quotas and tariffs (200% in the 1980s; 80–100% in the 1990s). These measures aimed to make the indigenous companies stronger, and to tap the knowledge and resources of foreign investors. The JV strategy implied that foreign automotive firms could only invest in China if they would team up with a state-owned enterprise (SOE). One of the goals of this measure was to learn from the foreign expertise: valuable knowledge (on production and technology, but also organisation and logistics) would be transferred from the foreign company to the Chinese partner. Gradually, the learning process shifted from transferring knowledge about basic manufacturing and project management skills to more sophisticated knowledge on engineering, design and branding. For foreign suppliers, the rules were less strict: they could also enter the Chinese market without a JV partner, as wholly foreign-owned enterprises[1] (WFOEs) (KPMG, 2003; IMF, 2004).

To promote the indigenous industry, the government also put local content requirements to foreign producers. OEMs were obliged to buy a particular percentage of their inputs from local suppliers. This policy aimed not only to increase the direct local 'spin-off' of foreign direct investment, but also to develop manufacturing competences within a region by replacing imports with supplies of Chinese firms (Depner and Bathelt, 2005). In practice, the local content was defined as domestic content, which means OEMs could also buy their inputs from other regions in China. Only Shanghai-Volkswagen (SVW), and Shanghai-General Motors in a later stage, were able to reach a high local content rate, thanks to the high quality of local suppliers in Shanghai. The quality of the local supplier network was no coincidence. As we will see later, it was the result of a conscious and sustained policy effort of the Shanghai municipality. Other regions have not succeeded in creating such a supplier network and were increasingly relying on Shanghai (Thun, 2004).

By the late 1990s, the automotive industry entered a new stage. The private market for vehicles expanded strongly (see Figure 11.1), competition increased

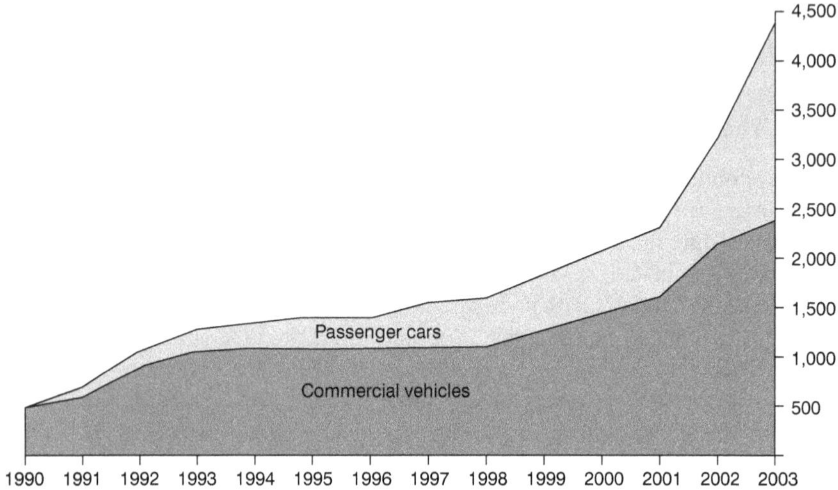

Figure 11.1 Vehicle production, 1990–2003, in China (1,000 units) (source: IMF, 2004)

and there was an increasing level of global integration. Many foreign car manufacturers entered the Chinese market and invested in production facilities. This led to rapidly falling prices and rising demands on supply firms. Chinese suppliers continued their process of upgrading, and several indigenous Chinese OEMs emerged as significant producers.

An important catalyst for change was China's entrance to the World Trade Organization (WTO) in December 2001. This implied that local content rules and import quotas and tariffs would be phased out. The regulatory differences between the period 'before' and 'after' the WTO entrance are depicted in Table 11.1.

The retail prices of domestically produced passenger vehicles dropped significantly by 2002 (KPMG, 2003). Demand surged (see Figure 11.2), thanks to lower prices of vehicles in combination with strong economic growth (Weider, 2004). The growth of vehicles sold in China was a staggering 69% in the first nine months of 2003 (Heneric *et al.*, 2005). In 2005, nearly 4 million passenger cars were sold in China.

There were also qualitative changes in the market. Vehicle demand in China used to be concentrated in the coastal 'mega-urban areas', such as Shanghai, Guangzhou and the capital city Beijing (IMF, 2004), but since 2004 demand also surged in the heavily populated inland areas of the country, where people now could afford to buy cars. While sales increased in the second-tier cities, the demand in first-tier cities, such as Beijing and Shanghai, was shifting to newer and more modern vehicles (*Financial Times* (*FT*), 2006d). There, the increasingly wealthy and demanding customers were no longer content with older, re-worked models, which are launched in China months or even years later than in other parts of the world (*FT*, 2006c). Another qualitative change was the shift in demand from institutions to private individuals. In the 1990s, the largest share of cars was sold to fleet owners, such as taxi companies and the government. Since the

Table 11.1 Changes of boundary conditions after WTO entrance

	Before WTO entrance	*After WTO entrance*
Import tariffs	200% in the 1980s; 80–100% in the 1990s	25% from 2006
Import quota for foreign OEMs	30,000 vehicles per year	Increase of 20% per year; from 2006 no more restrictions
'Local content' rules	40% in the 1st year of production; rising to 60% in the 2nd and 80% in the 3rd year	Dissolution of the local content quota
Foreign participation at sales and distribution	Limited wholesale through JVs; prohibition of common sales organisations for import and JV vehicles	For both wholesale as for retail of own cars allowed; integrated sales organisations allowed from 2006
Vehicle finance for Chinese customers	Prohibition of the finance	Finance by foreign, non-bank financial institutes, first in selected cities, later in the whole country

Source: Weidner (2004)

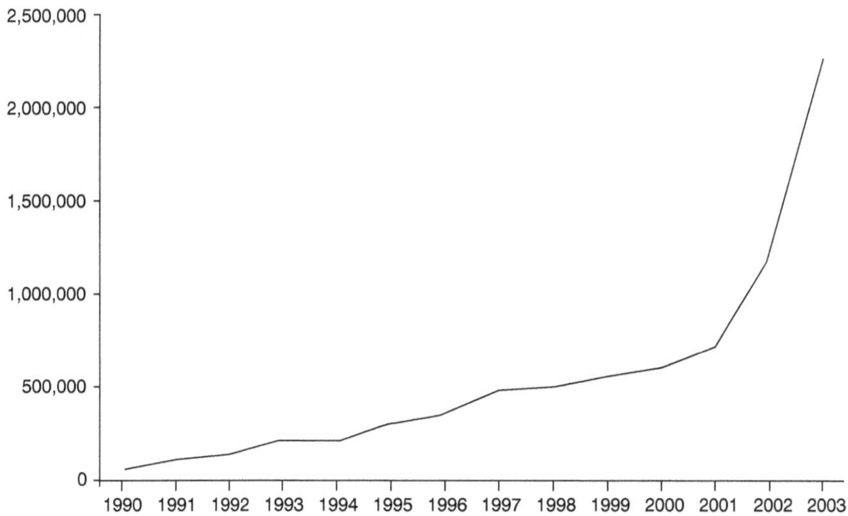

Figure 11.2 Passenger car sales in China (in numbers) for the period 1990–2003 (source: Weider (2004)

beginning of the twenty-first century, this has changed radically (see Figure 11.1 for a change in the production as a consequence of the change in demand). This shift in demand is important, as private customers are more sensitive to changes in price and are more concerned with quality and style (IMF, 2004; Weider, 2004; Heneric *et al.*, 2005; Thun, 2006). Also, the number of available models exploded: in 2000, there were fewer than 10 models available on the market; in 2003, the Chinese customer had more choice, with over 60 models available (Becker, 2006).

Somewhat surprisingly, the lower import tariffs induced by the WTO entrance did not significantly increase the number of imported vehicles. The rise in demand was mainly met by domestic production of foreign OEMs (in JVs), and more recently, by indigenous Chinese OEMs. Besides the OEMs, the automotive suppliers have also been heavily investing in China: most of the large multinational first-tier suppliers are now active in China. The four largest global automotive suppliers (Delphi, Robert Bosch, Visteon and Denso) together had a market share of about 9% in 2002 and are still expanding their presence (KPMG, 2003).

Most car plants in China are located in the urbanised and most developed eastern part of the country. The plants are located in six major clusters, and include plants of domestic firms as well as JV plants (such as Nanjing-Fiat, a JV between Yujing Motor Group and Fiat). The provinces of Shanghai, Jilin and Guangdong together accounted for 71% of the production in 2001. Shanghai has the largest share, with 40% of the total passenger car production, and Jilin and Guangdong have shares of 20% and 11% respectively. The remaining 29% of the passenger car production is fragmented between 19 other provinces and municipalities. The supplier industry is concentrated in the provinces and the municipalities around the Yangtze River Delta, accounting for 44% of the total production in 2001. The leading centre is Shanghai with a market share of 20%, followed by Zhejiang (14%) and Jiangsu (10%). Other large production bases can be found in Hubei (10%) and Jilin (7%). Other main production areas are Guangdong, Liaoning and Shandong, with a combined share of 15%. The remaining 24% of the production of the suppliers is divided between 21 other provinces and municipalities (KPMG, 2003).

Policy

In the 1980s, the Chinese government assigned the automotive industry the role of ‘pillar industry’. Since that time, the state has supported the industry heavily and has played a major role in the industry’s development. Within the general framework set by the national government, local and provincial governments have substantial freedoms in the implementation of national policy. Furthermore, many provinces (with Shanghai as the most important one) are themselves players in the automotive industry: they own automotive companies, and/or are involved in JVs with foreign partners.

In the Five-Year Plan for the 2005–2010 period, the state has set several major development goals for the automotive industry. The first is the *development of an indigenous Chinese industry*, which is defined as ‘own Chinese brand, own design and own R&D'. This reflects the ambition to make the Chinese car industry less dependent on foreign technology and skills. Although the country has made great progress as a manufacturing hub, it still depends heavily on foreign expertise in the fields of engineering, marketing/branding and organisational concepts. The target is to increase the share of 'Chinese nameplates' from 20% in 2006 to 60% by 2008 (*Asia Times*, 2006b). To reach this target, the government has developed a subsidy programme for OEMs and research institutes including universities. Not only are national state-owned enterprises (SOEs) eligible for subsidies, but also OEMs that are owned by provinces, and private OEMs. Furthermore, OEMs receive subsidies directly from provinces (or the municipality in the case of Shanghai). Each of the companies has developed its own strategy to reach the goal of creating an own brand, design and technology. For instance, SAIC, the major OEM in Shanghai, is developing an own model that is based on a Rover model, which SAIC has bought. However, there are some doubts to what extent this will be a real Chinese model and brand. Moreover, there are doubts about the development of Chinese technology since many Chinese companies still depend heavily on their foreign JV partners.

The second development goal is to *make the car industry cleaner and less dependent on fossil fuels*. To reach these goals, the government spends large amounts of money on R&D in OEMs and research centres. For the 2005–2008 period, a budget of €1.5 billion was reserved for R&D on new fuels.

A next goal is *to develop China as a car exporting nation*. To that end, the government has created free trade zones, some of them dedicated to the car industry. The district of Jiading (a district in Shanghai) is one of them. In these free trade zones, companies related to a certain product get certain privileges, such as lower taxes. The export zones are not only supported by the state but also by lower levels of the government.

Some other policies are relevant for the automotive industry as well. One of them is the policy to redress the spatial imbalances by favouring investments in inland locations rather than the already prosperous and well-developed coastal areas. The government tries to favour inland locations for the establishment of manufacturing plants through tax incentives (lower taxes for inland locations). Lower labour costs can be another incentive for companies in favour of inland locations.

The central government is worried about the overcapacity in the Chinese automotive industry and promotes a consolidation in the highly fragmented industry. For political reasons, this is far from easy, however. Many Chinese provinces are heavily involved in the car industry via SOEs, in JVs with foreign OEMs, and have few incentives to limit their ambitions. There is a strong competition between provinces, which forms a barrier for cooperation and consolidation (Huang, 2002; Weider, 2004; Heneric *et al*., 2005; Thun, 2004).

11.3 Foundations of Shanghai

Shanghai, which literally means 'on the sea', was originally a fishing town. Before 1840 the city was a medium-sized city with 200,000 inhabitants. The modernisation of the city started after the Treaty of Nanking in 1842, following the First Opium War 1840–1842. The treaty stipulated that China had to open five commercial ports, of which Shanghai was one. The city grew rapidly after the treaty, and it became the largest city in China. In 1936, Shanghai was the seventh largest city in the world, with 3.8 million inhabitants (Yeung, 1996; Gamble, 2003; Miaoyang, 2004). In 2005, the city had a population of 13.6 million inhabitants (Shanghai Municipal Statistics Bureau, 2006). This is the number of officially registered inhabitants, but the city also counts a large, illegal 'floating population' of several millions of people.[2]

In the last decades, Shanghai has developed as the main automotive hub in China. To understand the current dynamics of the cluster, it is important to understand the local context in which the cluster has developed. Therefore, this section provides a description of the 'foundations' of Shanghai's automotive cluster: the economic base, the knowledge base, accessibility, quality of life and factor conditions.

Economic base

The economic development of Shanghai can be divided into different periods. Thun (2006) identifies three periods based on political events. The first is *pre-1949 Shanghai,* the time period before the victory of the Chinese Communist Party. In that time, the city was running large-scale 'modern' enterprises and had a strong industrial base. In the mid-1930s the city accounted for half of the total production of the country. The city had internationally competing shipyards and was an international textile manufacturing centre. The second is *post-1949 Shanghai*, during the communist regime in which Shanghai was appointed as the industrial core for the entire country. It played a key role in the central planning system. The city was under strict control of the national government, and was the cash cow for the national economy in financial terms. During the 1970s, the city provided on average 25% of the national revenues (Wu, 1999). Besides this, Shanghai was a major supplier of physical and human resources. For instance, when First Automotive Works (FAW) was created in Changchun in 1953, over 100,000 engineers and workers from Shanghai travelled to Manchuria to build it. *In the reform area* (after 1978), the city still heavily depended on the state. The relationship between the city and the state was one of co-dependence: Shanghai firms were able to demand preferential treatment of the state because the central government depended heavily on Shanghai firms. In the period between 1979 and 1990, Shanghai's growth was substantial but lagged behind Guangdong and other southern areas in China. Since 1992, Shanghai has become a national growth engine, with annual growth rates above 10% (Shanghai Academy of Environmental Sciences, 2005).

The entrance of China to the WTO (December 2001) marked the beginning of a new era. This event has had consequences for the economy of China as whole, but in particular for the internationally orientated city of Shanghai. The WTO entrance has increased the internationalisation of the economy, reflected, among other things, in increasing FDI inflows. The dynamics of Shanghai's economy is reflected in Figure 11.3. In a decade, the GDP per capita increased by a factor of 3.5, from ¥20,000 in 1995 to almost ¥70,000 in 2005.

The economy of Shanghai can be characterised as internationally orientated, diverse and knowledge intensive. The international orientation is reflected in the large number of foreign companies that have chosen Shanghai as their base for Chinese (or sometimes even Asian) operations. Many foreigners live and work in the city. The international orientation dates back centuries, and relates to the city's strategic geographical location. The integration of China into the global economy in the last few decades has reinforced Shanghai's standing as an international business centre. Currently, major foreign companies have their headquarters and production plants in Shanghai, and increasingly, their R&D units. More than 140 R&D units have already been established by foreign companies (Sigurdson, 2005).

The economy of Shanghai is highly diverse, and no single sector dominates. Besides the automotive sector, key industries of the Shanghai economy are finance, trade (trading and circulation), ICT, machinery and real estate business. Taken together, in 2005 these six industries provided 43.7% of the total value added (against 40.4% in 2000). Shanghai has the largest petrochemical complex in China, and produces the biggest steel output. Figure 11.4 shows the sectoral employment shares of the Shanghai economy. The manufacturing industry is by far the most important industry in terms of employment: 32% of all workers are active in the manufacturing industry. This is twice as much as the second most import industry, retail and wholesale (15%). The relatively low percentage of workers in modern service industries, such as ICT services, hotel and catering,

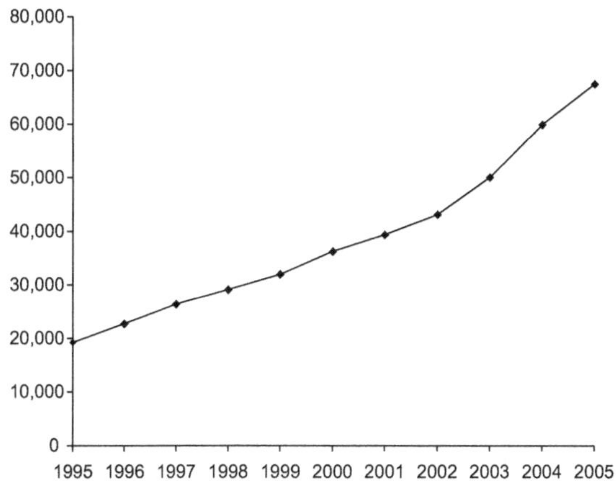

Figure 11.3 GDP per capita, Shanghai, in yuan, 1995–2005 (source: Shanghai Municipal Statistics Bureau, 2006)

finance, and real estate is remarkable. Less than 5% of all workers work in these industries. Nevertheless, the tertiary sector has rapidly gained importance in the employment structure. The share of workers in the tertiary sector grew from 21.6% in 1978 to 44.9% in 2000 to 55.4% in 2005. In terms of value added, the picture looks different: in the period 2000–2005, the share of the service sector decreased from 52.1% to 50.9%. This reflects the enormous productivity gains realised in the manufacturing sector. The significance of the manufacturing industry is also reflected in the sales of the largest companies: nine of the ten largest firms are in the manufacturing industry (see Table 11.2).

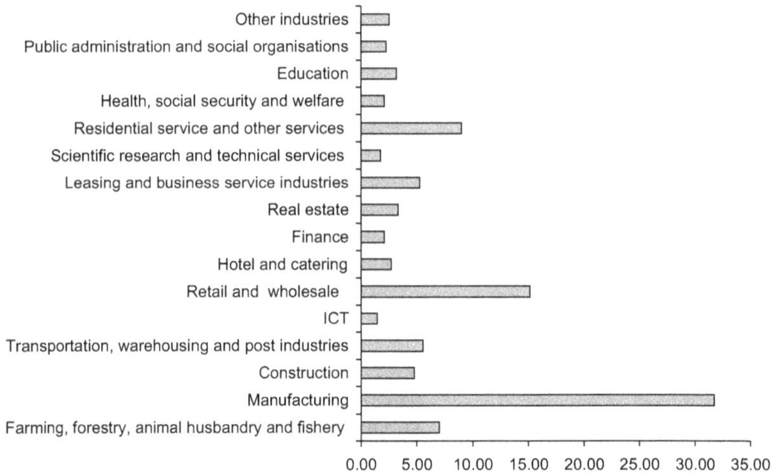

Figure 11.4 Economic structure, Shanghai (2006, based on employment) (source: own elaboration, data from Shanghai Municipal Statistics Bureau, 2006)

Table 11.2 Top 10 largest enterprises (based on sales)

Rank	*Enterprise*
1	Baoshan Iron and Steel Co Ltd
2	Tech-Front (Shanghai) Computer Co Ltd
3	Shanghai General Motor Co Ltd
4	Shanghai Volkswagen Automobile Co Ltd
5	Sinopec Shanghai Petrochemical Co Ltd
6	Shanghai Municipal Power Company
7	Lianhua Supermarket Co Ltd
8	China Petrochemical Co Ltd
9	Shanghai Gaoqiao Subcompany
10	Tech–com (Shanghai) Computer Co Ltd

Source: *Asia Times*, 10 January 2006

The Shanghai economy is becoming more knowledge intensive. Within the secondary (industrial) sector, there is a gradual shift towards high-end products (*Asia Times*, 2006a). According to Leman (2002) some major successes are achieved in the restructuring of the manufacturing industry in the 1990s. Low value-added manufacturing industries, especially textile and heavy-equipment industries, have decreased significantly, while Shanghai makes progress in some high and medium value-added industries, such as motor vehicle production, computer manufacturing, telecommunication equipment, integrated circuit manufacturing and packaging. Since the early 1990s, large FDI inflows have propelled Shanghai's economy as a national leader in sectors such as automotive, chemicals, micro-electronics and advanced transportation equipment. The process of the restructuring has consequences for the urban structure of Shanghai. Many labour- and space-intensive manufacturing activities have relocated outside of Shanghai. A number of plants have shifted to outlying cities and towns in the Yangtze River Delta megalopolis, and even more inland to low-cost locations in central and western provinces.

Medium-tech manufacturing industries are developing in suburban towns. These industries have transformed the suburban towns within the Outer Ring Road (which used to be 'rural' towns) into important extensions of the urban core. All counties in Shanghai (except Chongming Island) have recently been upgraded to administrative districts, and have been given the same powers and responsibilities as traditional urban districts within the Outer Ring Road. In this way, Shanghai is involved in a poly-nucleated structure, similar to other global cities such as Tokyo and New York.

The Shanghai municipality has an explicit policy to create clusters of high-tech industries. The micro-electronics industry is concentrated in Zhangjiang High-tech Park in the Pudong New Area. The automotive industry is clustered in the Jiading district. The chemical industry, finally, is concentrated in Shihua Town on Hangzhou Bay, east of Shanghai.

The shift towards high-tech manufacturing is reflected in the municipal development strategy 'Third, Second, First industry', in which technological innovation is identified as the main driver to promote industrial upgrading in all aspects. In this strategy, priority is given to the development of the knowledge-intensive service industry and the high-tech industry (Shanghai Urban Planning and Design Research Institute, 2001).

Knowledge base

Shanghai has a highly qualified workforce and strong educational institutes, some of which can compete on an international level. Several of our interviewees confirmed that Shanghai is one of the best places in China for skilled people, and it attracts skilled people from all over China. Many Chinese students who have studied abroad come back to Shanghai to find a job.

Shanghai is the home of many of China's top and oldest universities. The city boasts 59 institutions (colleges and universities) of higher education, with a total enrolment of 600,000 students in 2004. The city has 10 universities listed in the national top 100, which have been selected by the Ministry of Education to receive special treatment and additional resources. Three of the universities in Shanghai are recognised as world-famous research universities: Fudan University, Tongji University and Shanghai Jiaotong University (Sigurdson, 2005). An often-cited problem in the educational system is the heavy emphasis on theory, while insufficient attention is given to problem-solving and working in teams. Other problems concerning education are: overloaded classes and deference towards hierarchy.

Both in absolute and in relative terms, expenditures on R&D in China show a strong increase, especially since 2002. In the 2002–2005 period, expenditures doubled from ¥102 million to ¥214 million in 2005 (see Figure 11.5). China overtook Japan in 2006 in terms of R&D spending and ranks third after the US and EU-15 (FT, 2007e). Also in relative terms (as percentage of the GDP) the expenditures on R&D have risen. It increased from 1.3% in 1995 to 2.3% in 2005 (see Figure 11.6).

Firms accounted for the largest share of R&D investments (68% in 2005), followed by research institutes (21%), and institutions of higher education (11%) (Shanghai Municipal Statistics Bureau, 2006). It should be noted that these percentages do not exactly reflect the division between public and private investments, since a lot of enterprises in Shanghai are SOEs.

China is investing heavily in its education system and knowledge infrastructure, with the aim to boost innovation, but the innovation system shows several flaws. A first problem is the financial system. Small, private companies have been seen as the main engines of innovation, but the financial system in China does not support private entrepreneurs. Instead of this, most of the credit goes to large SOEs. Second, there are severe problems of plagiarism and fraudulent research.

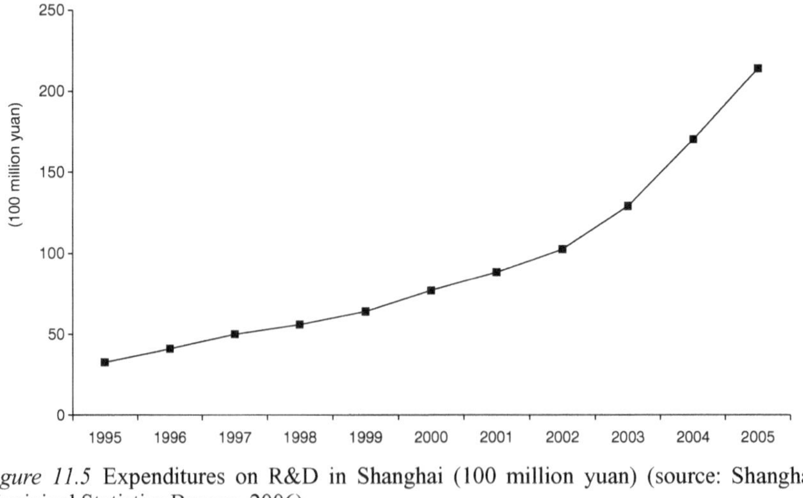

Figure 11.5 Expenditures on R&D in Shanghai (100 million yuan) (source: Shanghai Municipal Statistics Bureau, 2006)

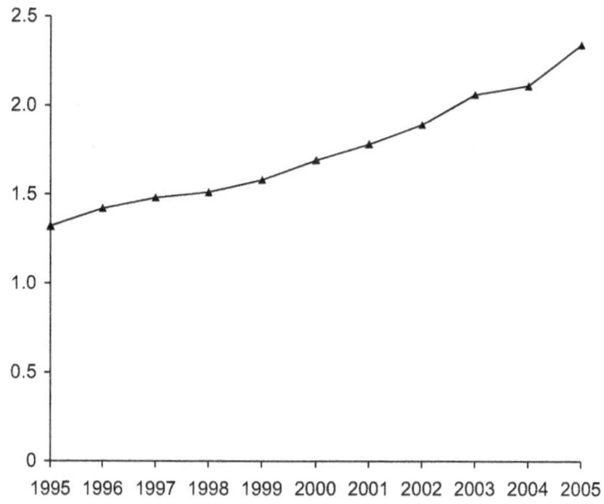

Figure 11.6 R&D expenses as percentage of GDP (source: Shanghai Municipal Statistics Bureau, 2006)

Third, there are doubts about the quality of the research, despite high investments in education and R&D by the government. An indication for this is China's high ranking in terms of published scientific papers (9th place), while the ranking in the average number of citations per paper is much lower (124th place). Some argue that research facilities (including universities) depend too much on the state and therefore miss the link with the industry (*FT*, 2007e).

To boost its knowledge base, the city of Shanghai created a number of university science parks and high-tech parks. Five universities have a university science park: Fudan University, Tongji University, Shanghai Jiaotong University, Shanghai University and Donghua University. The main purpose of university science parks is to support technology transfer and to commercialise technologies, in close relation with universities. High-tech parks differ in several ways from university science parks. They attract larger companies rather than small ones, and they are more orientated towards export markets. In total, 53 national high-tech parks have been developed in Shanghai (Sigurdson, 2005).

Accessibility

Thanks to its favourable geographical position, Shanghai is well connected to China and the rest of the world. As a major transhipment hub, it serves not only the densely populated areas in North and South China, but also cities in Japan, Korea and in the Asia-Pacific region (van Hoek *et al.*, 2005). In 2005, the city handled in total 443 million tons of cargo, making it the largest port in the world. Its connections make Shanghai attractive for export and trade-orientated industries, including the automotive sector.

Connections by air are also good and still improving. Shanghai has two airports: Hongqiao International Airport in the west and Pudong International Airport in the east. Pudong International Airport is the largest, and caters for about

60% of the total aircraft movements in Shanghai. It has connections with more than 70 international and regional destinations, and 60 domestic destinations. Hongqiao International Airport mainly serves for domestic flights, after the airport authority transferred international and regional flights from Hongqiao Airport to Pudong Airport. However, it still keeps standby capacity for international flights.[3] Together, the two airports can handle nearly 30 million passengers a year (van Hoek *et al.*, 2005).

Road transport in Shanghai is severely hampered by congestion, caused by the fast growth of Shanghai's private car fleet in a limited area with a very high population density. The number of private passenger vehicles increased from 8,100 in 1996 to 221,000 in 2003. The dense traffic negatively affects the air quality in Shanghai. Some measures are taken to control this situation: only cars with a green sticker (cars with relatively clean engines) can enter the city centre. Also, as will be discussed in more detail, the government promotes the development of environmentally friendly engines that produce fewer emissions.

In order to meet the rising traffic demand, Shanghai has sped up urban road construction since the mid-1990s, and invests heavily in the rail transit and metro system.

Quality of life[4]

Shanghai offers big-city dynamics, many job opportunities, a vibrant business life and almost all possible amenities in the fields of sports, leisure, culture, nightlife, etc. Also, the offer of facilities and amenities is highly geared to an international clientele. On the downside, however, the city faces environmental degradation (notably the poor air quality), very dense and congested traffic, and relatively high crime rates.

Shanghai offers all the facilities a modern metropolis needs, such as entertainment districts, shopping areas, museums, theatres, hotels and restaurants, etc. The most important shopping street is the pedestrian mall of Nanjing Road. In the city there is a wide variety of restaurant choices, varying from 16 different styles of Chinese food, such as Fuijan, to Japanese, Indian and Western food. The facilities do not only attract Chinese people, but also foreigners (both as visitors and as residents). There are a lot of facilities for foreigners, such as international schools. Besides this, many people in Shanghai speak English and a lot of information is available in English, such as the signposting on public transport, on the roads and on menus in restaurants. Shanghai's numerous facilities and the international atmosphere give the city a key advantage in the attraction of highly-qualified (foreign) labour.

Factor conditions

Despite strong economic growth and rising wages, China can still be considered low-wage country. As can be seen in Table 11.3, the average labour costs in China are much lower than in the Triad countries and other low-cost locations in

Table 11.3 Comparison of labour costs

Country	*Labour costs ($ per hour)*
China	0.59
Eastern Europe	4.0
Japan	19.01
USA	20.32
Germany	24.07

Source: FHA, 2004

Central and Eastern Europe. Within China, there are marked differences in factor costs, and Shanghai is one of the most expensive locations. Real estate, land, labour and management costs are significantly higher than in other locations. This explains the shift of manufacturing facilities to more inland locations with lower labour costs. This trend is reinforced by the government, that wants to push manufacturing to more inland locations to reach a more balanced development and to prevent social imbalances.

The relevance of low labour costs for the development of an export-orientated automotive industry is limited, however. Lower production costs are partly offset by high shipping costs of complete vehicles (due to much 'dead air space'). Furthermore, to realise the government's ambition to develop a full-fledged automotive industry, skilled engineers and research capabilities are needed rather than cheap labour. (*Asia Times*, 2006b). The costs of these resources are already high and rise steadily.

11.4 Shanghai's automotive industry

Over the last decades, Shanghai has emerged as the largest and most advanced automotive cluster in China, and over the years, the municipality of Shanghai has played a very active role in this process. The city is a major player in the industry: it owns a large group of automotive firms (SAIC group). Also, it controls, steers and promotes the development of the entire local automotive cluster, including independent firms and foreign firms. Finally, it directly influences the location of the automotive industry through urban planning initiatives. In the 1980s and 1990s, Shanghai acted as a 'developmental state', by leading and controlling the development of the local automotive industry. In this respect, Shanghai distinguishes itself from other Chinese cities and regions with a significant automotive cluster (Thun, 2004, 2006).

The main players

SAIC group (SAIC) is by far the largest automotive business group in Shanghai. It is one of the major SOEs that are active in the auto industry (two other large ones are FAW and Dongfeng). The majority of its shares is controlled by the municipality of Shanghai, and therefore SAIC is closely interlinked with and supported by the city's policy makers (Depner and Bathelt, 2005). SAIC has a vertically integrated organisational structure, which spans the entire value chain. Its manufacturing capabilities cover all vehicle systems, ranging from engine, power train and chassis components to vehicle body, interior and exterior parts (IMF, 2004). By the end of 2001, the company was engaged in 55 JVs and employed almost 62,000 people (Depner and Bathelt, 2005). Two JVs stand out: one with Volkswagen (SVW) and one with General Motors (SGM).

SAIC has been supported strongly by both the state as well as by the municipality of Shanghai (Depner and Bathelt, 2005) and it was able to create a strong automotive cluster. SAIC has a hierarchical structure, with the municipality of Shanghai (the owner) on top. During the first stage of automotive development in China, the hierarchical structure of SAIC had three functions. First, it maintained control over all operations within the group. Second, it served as a planning base for all enterprises in the group (of OEMs and suppliers). Third, the structure catered for learning during the development process. Information and knowledge on management and production techniques was transferred top-down via the municipality and VW, the major foreign JV partner at that time (Thun, 2004). In this period, SAIC forced its main JV firm (SVW) to make use of suppliers that also belonged to the SAIC group, even though their prices and quality did not always meet VW's standards. The position of SAIC has changed radically since the entrance of China to the WTO. Competition became more intense, and prices were under constant pressure. This urges SAIC to innovate and cut costs. In this light, the company faces the following challenges:

- *Meet the goals of the Chinese government.* In particular, the SAIC group is a key instrument of the government to realise the goals of an 'own Chinese brand based on own Chinese design and own Chinese technology'. For this, SAIC has developed a car which is based on the design of the Rover 75, a former British OEM that was bought by SAIC and Nanjing Automobile. Also, in order to gain access to technology and competence, SAIC has bought foreign firms. Until now, SAIC has bought (parts of) three foreign companies: Rover MG in England, Daewoo-GM and Ssangyong Motors in Korea. By acquiring foreign companies, SAIC buys access to foreign markets, new technology, engineering capabilities and management expertise. However, foreign companies also have problems, mainly caused by cultural differences. An example is a strike at the Ssangyong Motors plant in Korea in 2006 for 49 days, which cost SAIC US$1.3 billion. SAIC also plays a role in the consolidation goal of the government. In December 2007, SAIC bought its competitor Nanjing Automobile, another state-owned auto group, for ¥2.1 billion (€198 million) (*FT*, 2007c). Via this deal, SAIC now has access to all of Rover's facilities.
- *Develop more environmentally friendly cars*, in line with government ambitions. For this purpose, SAIC conducts its own research, and is engaged in many joint research projects with JV partners and universities.

- *Increase exports* to benefit from the opening of international markets. The total export value increased strongly from ¥17 million in 2001 to ¥70 million in 2004 (SAIC, 2005). Automotive parts have been exported for a long time and are sold to all parts of the world. Exports of complete vehicles by SAIC, on the contrary, are relatively new and limited to a few countries. In 2004, 150 SVW-Polo models were exported to Australia and some Santanas will be exported to Kazakhstan (SAIC, 2005). However, there are plans for SAIC and other Chinese OEMs to increase exports to the West (EU and USA).
- *Increase and improve service activities.* After a restructuring of SAIC in 2004, the group defines its role as 'a comprehensive investment corporation in advanced manufacturing industry and modern service' (SAIC, 2005). Besides the manufacturing activities by OEMs and suppliers, SAIC also has multiple service activities. It has an extensive sales and service network in China, consisting of 4S[5] service centres, after-sales service centres and agents. Other services that SAIC offers are: R&D, vehicle finance and insurance, vehicle and parts logistics, car rental and lease, marketing, leisure services (among others the auto museum in Auto City), and industrial real estate (for instance SAIC owns all the land in Auto City). The share of services in the total revenue was 7.5% in 2004 (SAIC, 2005).
- *Coordinate vertical and horizontal growth.* It will be more complex for SAIC to manage the group because of a strong expansion in Shanghai, other parts of China and abroad via JVs and foreign mergers and acquisitions. Conflicts between different partners can arise through cultural differences and international property rights. A major example of the latter is a conflict between SVW and SGM (two major foreign JV partners) and SAIC-Chery (a local JV partner from Anhui province). SAIC-Chery had been accused of using components from VW in Germany, and for copying two models from GM. In the end, SAIC was forced by VW and GM to give up the stake in SAIC-Chery in 2003. More conflicts between foreign OEMs and SAIC can be expected in the future since SAIC has announced that it will build cars entirely on its own through its publicly listed subsidiary Shanghai Motor Co. (Depner and Bathelt, 2005). Furthermore, conflicts may arise due to the changing regulations caused by the WTO entrance, which limits SAIC's options to protect the local automotive industry.

Volkswagen (VW) played a pioneering role when it opened its JV plant with SAIC in 1984.[6] The early entrance of VW gave the company a 'first-mover advantage' and it reached a market share above 50% in China. The cars that were built used to be old models from Germany and other Western countries. In recent years (2000–2005), VW has been losing market share to local OEMs and other foreign OEMs, which entered the Chinese market in the second development stage. In 2005, the market share of VW dropped to 15.7%, and the company made losses on its Chinese operations (*FT*, 2006a). The rising competition in the Chinese market has had major repercussions for SVW. First, there is pressure to cut costs, which is also transferred to suppliers. Second, competition increases the urge to develop new models faster. Third, where the emphasis used to be put on production quality (models were imported from Germany), today the key point, besides costs, is design and engineering to adapt the models to the Chinese market.

SVW is located in Auto City, where it has three assembly plants with a total annual production capacity of over 450,000 vehicles. Also, the company has an engine plant and a technical centre. Based on the interviews, we estimate that SVW employs between 15,000 and 20,000 people, of which 60% are blue-collar workers. The firm employs about 500 engineers.

General Motors opened a JV plant with SAIC in Shanghai in 1997: Shanghai-General Motors (SGM). SGM has one manufacturing plant and one engine plant in Shanghai, and is located in Shanghai's Pudong Jinqiao Export Processing Zone. The base in Jinqiao has a total annual production of 320,000 vehicles – 100,000 automatic transmissions and 200,000 engines. In the beginning, SGM made only luxury cars for higher segments, mainly for government officials. Nowadays, the total product line of SGM is made up of four brands (Cadillac, Buick, Chevrolet and Saab), with models for all different segments. Besides the production facilities, GM and SAIC own a research centre (a 50/50 JV), the Pan-Asia Technical Automotive Center (PATAC). SGM has been successful: it managed to expand its market share from 5% in 2002 to 11% in 2005, and the company enjoys healthy profits (*FT*, 2006b; Weider, 2004).

Besides the two SAIC JVs, other (smaller) *local OEMs* are established in Shanghai: Shanghai Wanfeng Auto, Shanghai Maple Automobile and local OEMs that are owned by SAIC. Shanghai Wanfeng Auto is a non-SOE share company. Its assembly plant has an annual capacity of 50,000 vehicles. The firm produces pick-up trucks and MPVs. Shanghai Maple Automobile (SMA) is a relatively new local firm (created in 2000); it is a subsidiary of Geely, but it operates independently. The firm produces affordable cars for the lower market segments. The capacity of the plant can be extended from 300,000 to 400,000 vehicles, a goal that SMA wants to reach in 2010. However, the production is about 50,000 vehicles and 100,000 engines. The company produces three different car models, engines for their own vehicles and for other Geely models, and it develops a hybrid vehicle. SMA has developed its own models, and its own brand (with support from the government) and develops its own design in-house. However, the company is being sued by the French PSA Group (the group that produces, among others, Peugeot and Citroën brands) for using its design and chassis. SMA has about 2,000 employees of which 900 are manual workers. SMA does not have its own research facilities, but it has set up a research centre at the plant site, in cooperation with Jiaotong University. Also, SMA uses the facilities of Jiaotong University, and has cooperation agreements with Tongji University and Jilin Technical University.

The supplier industry

In the car industry, suppliers are important players: they produce up to 80% of the total value of a car. Shanghai has the strongest supplier base in China. Suppliers of all tiers can be found in Shanghai, ranging from raw material suppliers to tier-one suppliers. Large foreign multinational suppliers and small domestic suppliers are active in Shanghai. Furthermore, all functions (manufacturing, research and office functions) can be found in Shanghai. However, it should be noted that some suppliers prefer Beijing as the location for their head office for political reasons.

The strong supplier base in Shanghai did not emerge spontaneously: it was the combined result of coordination and control of the SAIC group and strong government support from the state and the Shanghai municipality. Thun (2006) describes how the Shanghai government acted as a 'development state', through several policies aimed to raise the qualitative level of the suppliers. Also, local content rules were very helpful in this respect, and by the late 1990s, Shanghai managed to reach the highest local content percentage in China (see Table 11.4). Currently, the presence of many OEMs in Shanghai makes it an attractive location for suppliers because they are not dependent on one single OEM, but can offer their products to a range of OEMs in the region.

The strength of Shanghai's supplier base is illustrated by the fact that many Shanghai suppliers not only produce for OEMs in Shanghai, but also for OEMs in other regions. By 2003, for instance, the automotive industry in Beijing was sourcing 50% of its components from Shanghai, compared to only 10% locally and 40% from other provinces (Thun, 2004). Of the top 20 largest automotive suppliers in terms of turnover in 2001 (Heneric *et al.*, 2005), 18 are active in Shanghai (directly or via JVs); the 10 largest automotive suppliers are all present in Shanghai.

Shanghai has a number of location factors that attract suppliers. One is that Shanghai has always been the industrial centre of China. For instance, Shanghai has the largest steel complex in China (Bao Steel) and the largest chemical cluster. These industries deliver many inputs for the automotive industry. Also, there are logistical advantages; Shanghai has the largest port in the world, two major airports and good connections with the locality of Shanghai. Other location factors

Table 11.4 Percentage of components sourced locally by JV (by component not by value)

Joint venture	*1985 (%)*	*1997 (%)*	*2003 (%)*
Beijing Jeep	0	20–30	5–10
SVW	0	88–90	88–90[a]
Guangzhou Peugeot	0	20	–
SGM	–	0	>60[b]
Guangzhou Honda	–	–	22[c]

Source: Thun, 2004

Notes:

a The value for SVW is unchanged since 1997 because it is at the highest possible level.

b The percentage of SGM includes only the value of components sourced from SAIC suppliers, all of which are within the Shanghai municipality. The percentage from all Shanghai firms is higher.

c The percentage for Guangzhou Honda is for the number of suppliers in Guangzhou. Consequently, the percentage for Guangzhou Honda alone is probably lower. To give some indication of this, Honda is only using four former Peugeot suppliers (Thun, 2004).

in Shanghai are its openness to foreigners, a good quality of life for employees and a large labour pool of high-skill workers.

There are some major (global) developments in the automotive industry that influence suppliers in Shanghai. One is the increasing outsourcing of activities from OEMs to suppliers. Suppliers are increasingly involved in development and design activities. Second, suppliers in Shanghai are facing an increasing price pressure by OEMs. Finally, suppliers increasingly become exporters. In 2005 alone, the export of auto parts from China increased by about 50% (*FT*, 2006d).

Automotive-related research and education

Shanghai is home to a growing number of automotive-related research centres, owned by different actors. SAIC has its own independent research centre in the city, Shanghai Automotive Engineering Academy (SAEA).[7] This is a technology centre administered by the state, which is combined with Shanghai Automotive Industry Technology Center and Shanghai Internal Combustion Engines Academy. The Shanghai Internal Combustion Engines Academy is a centre accredited by the State Quality and Technique Supervision Bureau, which supervises and inspects domestic internal combustion engines. The centre has about 700 employees, of which about 510 are technicians. SVW has its own technical centre on a site close to the plant. With GM, SAIC set up the Pan-Asia Technology Centre (PATAC), a product development centre. This is a 50/50 JV employing over 11,000 employees, with about 35% having a masters and/or doctorate degree. PATAC has different facilities for automotive design and technology: a virtual-reality design studio, a prototype laboratory, a power train lab, a kinematics and compliance lab, a noise, vibration and harshness test lab, emission testing facilities and a short-distance test track that can simulate all types of road conditions.[8] In addition to the development centre PATAC, GM invested US$250 million in a new research centre that will also conduct basic research in Shanghai (*FT*, 2007b). Besides the research facilities in Shanghai, SAIC has overseas research facilities through the foreign take-over of Ssangyong Motors in Korea and Rover in the UK. SMA has set up a shared research centre with Jiaotong University at the site close to the plant.

Most of the R&D activities by Shanghai firms concern development in the form of adaptation of vehicles and auto parts to local tastes and legislation. Foreign companies are still afraid of doing basic research in China for fear of losing control of intellectual property developed there (*FT*, 2007b), so their major engineering system integration and R&D activity still take place in their home nation. For example, SVW's key engineering activities are located in Germany (Depner and Bathelt, 2005).

Many of VW's models are based on 'global platforms' and are sold in other continents as well. The upgrading and engineering activities for platforms take place in Germany, and only slight modifications for the Chinese market requirements are done in Shanghai. The exceptions to this rule are the models designed specially for the Chinese market, such as the VW Sedan 2000. GM goes one step further; it opened a centre in Shanghai for basic research.

The most important university for automotive research and education is Tongji University, which has set up a College of Automotive Engineering. This college is located at the Jiading Campus in the educational area of Auto City. It offers various bachelors and masters programmes for students and offers special training programmes for people who are active in the automotive industry. The college has sophisticated research facilities, including various testing tables for cars and spare parts, large-scale software for design and analysis, and a Testing and Trial Production Base. Furthermore, the college is conducting contract research for the automotive industry, and is involved in a number of state and local key research projects. Another key university for research in the field of automotive engineering is Shanghai Jiaotong University (SJTU). Various faculties, especially the faculty of mechanical engineering, provide research and educational programmes that are relevant for the automotive industry. It also has specialised automotive research centres, such as PACE (Partners for the Advanced of Collaborative Engineering Education) Center, which was established at SJTU in 2002 through a donation of hardware and software facilities by an consortium of leading companies including GM and Siemens. An important research centre is the National Center of Supervision and Inspection on Motor Vehicle Products Quality (Shanghai), located in the centre of the R&D area of Auto City. It conducts and develops technical tests, for instance on vehicle safety and emissions.

Several interviewees noted that those engineers in China have a lower status than managers, and therefore many Chinese people prefer to become managers instead of engineers. The level of engineering education can not yet meet international standards. Some of our interviewees doubt whether this will change in the short to medium term: 'as long as China can make money through manufacturing (which is still possible because of lower wages and good margins), it will be difficult for China to become a more engineering country'.

For the government, the development of cleaner energy is a research priority, for various reasons. The first is oil scarcity, which is partly caused by the rapid economic development of China and an increasing use of (private) cars in China. Second, the government wants to address global warming and concerns over environmental degradation that threatens the liveability of Chinese cities. Third, China has the ambition to compete with established car makers in the 'new game' of alternative energy use. The technology of traditional fossil fuels is matured; China cannot compete with the main Western competences in this field, but may become a leader in the development of new fuel, for two reasons. First, the country does not yet have the expensive petrol infrastructure and can build a 'new fuel' infrastructure from scratch. (*FT*, 2006c). Second, the application of new energy in cars is still in its infancy, which increases the chances of China catching up and eventually becoming the leader in this domain. Also, the development of new types of engines affects requirements of all other car parts, and thus it can revolutionise the entire industry. China hopes to benefit from this development.

The Shanghai government seeks to build competences in this domain in several ways: 1) by increasing the cooperation between companies and universities in the Yangtze River Delta, 2) by intensifying the cooperation with VW and GM and 3) by creating and funding a dedicated research institute for SAIC. There are also plans to cooperate with Japanese OEMs, which are leading the way in the development of cleaner vehicles (though this is a politically sensitive issue).

New energy research focuses on the use of three types of new energy: mixed fuel, hydrogen and representative fuels. All three types have already been developed in concept versions, or are in use on a limited scale in Shanghai. For instance, taxis in Shanghai are using representative fuel (such as LPG), and on one bus line, a hybrid bus is used. Despite these concept versions, some interviewees suggest that the plan of the government to develop cleaner energy is too ambitious and not realistic, given the lack of high-level engineering and system integration competences.

Some of our interviewees were critical about the R&D capacities of Shanghai's automotive cluster, and questioned the ambitions of the government in relation to development of the new energy. One comment is that SAIC (but also other Chinese automotive companies) would rely too much on foreign technology. Technology is 'imported' through foreign JVs (SVW and SGM) and by foreign takeovers (Rover and Ssangyong). Most of the development is done abroad, while the research in Shanghai is limited to low- and medium-tech. Another critique is that automotive R&D is driven by the government, not the industry. The government provides generous subsidies to OEMs and universities, but private investments are relatively low. Furthermore, although state investments in R&D are large, there are doubts about the outputs and results (*FT*, 2007e). Also, some interviewees suggested that the cooperation between universities and the industry is difficult, since professors typically lack industrial experience and do not understand the needs of the industry.

Related service industries

There is a growing service industry related to automotive production. Consultancy firms, global financial service providers and logistics firms serve the car industry. Moreover, many automotive suppliers and OEMs have offices in Shanghai for purchasing, sales, marketing and finance functions. Many firms choose Shanghai as their regional headquarters (GM is one of them), not because Shanghai is a production hub but rather for the excellent accessibility, talent pool and international orientation.

The downstream part of the value chain is well developed in Shanghai. Besides the logistical service providers, there is a well developed sales network of dealers and agents. The after-sales service segment is still in development, and lags behind the development of the rest of the auto industry (manufacturing and sales); for instance, there are too few points for after-sales services. Also, there is a qualitative problem in relation to the after-sales services: a lack of qualified labour and good technology.

The geography of the automotive industry in Shanghai

Automotive companies (OEMs, suppliers, R&D centres and other services) can be found in all districts of Shanghai. The supplier industry is scattered over Shanghai, and on a larger scale over the Yangtze River Delta. Nevertheless, in Shanghai, three major concentrations of automotive firms can be identified: Jiading, Pudong and Jinshan. In Jiading, the place where SAIC-Volkswagen (SVW) is located, the government has developed Shanghai International Automobile City (in short, 'Auto City'), a special industrial zone for automotive activities. In Pudong and Jinshan are the plants of Shanghai-General Motors (SGM) and Shanghai Maple Automotive (SMA), respectively. A fourth location of interest for the automotive industry is the city centre because multiple service functions and the research centre of SAIC, SAEA, can be found there.

Auto City[9] officially started in 2001 as a key urban project to develop and stimulate the automotive industry. The ambition of the government was to develop a 'modern automotive service cluster', which would combine manufacturing, R&D, education, trade and 'vehicle culture', and tourism in one area. Furthermore, the government considers Auto City as an international brand and it is referred to as the 'Detroit of the East'. It is located at Anting Town in the Jiading district in the north-east of Shanghai where the plants of SVW and the head office of SAIC are located, as are many supplier firms. Anting has a symbolic value for the government because the first vehicle produced in Shanghai was produced there 40 years ago. Auto City has a central business district (which consists of an R&D area, a theme park and a golf course), a production zone (located around the plants of Shanghai Volkswagen), an educational park, a residential area, and an international racing (F1) circuit. There is a 'dealer street' and a used-car market. Currently, there are dealers for 11 different brands in the dealer street, but the ambition is to have 26 brands. The used-car market has an auction hall, a trading hall, an evaluation system and inspection equipment. Furthermore, there are facilities for the trade of automotive parts. Besides this, the state has designated Jiading as an export zone for auto parts. In this export zone, companies related to automotive products get certain privileges, such as lower taxes.

Tongji University has a campus in the area, and some other research institutes are located in Auto City as well. There is a large exhibition centre, Shanghai Automobile Exhibition Center (SAEC), which should demonstrate the power of Shanghai as an automotive cluster. Entertainment and leisure are present at the site: an F1 circuit, an auto museum, a pedestrian boulevard, a park and a golf course. The government wants to use these attractions to develop tourism in Auto City.

Despite government ambitions and many facilities that Auto City offers, there are some critiques of the concept. One is that only SVW would benefit from the cluster. There are plans to attract more OEMs to the area,[10] but this is not very likely to happen since there is a 'golden rule' that two large OEMs can not co-exist in the same area. Also, it will be difficult to attract other suppliers, since suppliers have high relocation costs (including high land costs). Besides this, many suppliers produce for multiple OEMs and therefore they need to find an optimal location. Only the suppliers that are supplying mainly to SVW and suppliers with multiple plants may locate to Anting.

Pudong is the second 'concentration' district of the automotive industry. Pudong is the new part of the city centre on the eastern bank of the Yangtze River.[11] Large parts of Pudong are planned as locations to advanced manufacturing industries. Several car plants have located there, among which are GM and Shanghai Wangfeng Auto and several major international suppliers and domestic suppliers. A third, emerging location for the automotive industry is *Jinshan district* to the south-west of Shanghai. Here, Shanghai Maple Automotive (SMA, an indigenous OEM) has its plant, and some suppliers are located there as well.

In the *city centre*, the fourth major location, many automotive companies (especially suppliers) have (head) offices for commercial and administrative functions. The plants of the suppliers are often located in more inland locations of China, because of (labour) cost advantages.

11.5 Local and global networks in Shanghai's automotive cluster

The automotive industry is highly networked. As already explained, OEMs and suppliers cooperate closely in the production of cars. But research institutes, universities and governments are also important actors, and the quality of their interaction is assumed to have an impact on the performance of the regional cluster. Several types of networks can be distinguished: operational networks (manufacturing, logistics), R&D networks (basic research, applied research, development, design), management/organisational relations, and other relations. The networks operate on many geographical scales: local, regional, national and international. In this section, we focus on network relations between the government, OEMs, suppliers and R&D institutes, paying special attention to the geographical scope of these networks. By doing this, we aim to shed light on the evolving relations between exploitation activities on the one hand, and exploration activities on the other.

In terms of networking, the Shanghai cluster is different from European automotive clusters in two respects. First, there is an important cultural difference: communication within and between firms is often based on personal relations, the so-called 'guanxi'. Guanxi is based on joint interest, interdependency, reciprocity, trust and openness. Guanxi has a long-term character and has benefits and obligations for each side. This can cause (management) problems for foreign companies, especially JVs, since foreign companies are used to relying on authority of contract and law instead of tacit, unwritten rules and regulations that stem from guanxi (Depner and Bathelt, 2005). Second, unlike other major automotive clusters such as in Paris or Munich, in Shanghai there are no 'platform organisations' (cluster organisations, dedicated business associations, etc.) where players in the automotive industry meet, exchange knowledge and ideas or engage in joint projects; instead, knowledge diffusion takes place though the government, and within the SAIC group. A key particularity in this case is the dominant influence of the government and of the SAIC group (owned by the government) in the automotive industry, in every link of the automotive value chain.

Joint ventures as vehicles for knowledge transfer

Shanghai's strategy has been to give companies access to the Chinese market, in exchange for knowledge and expertise; the Shanghai municipality uses JV structures to realise this ambition. The JVs of SAIC with VW and GM (SVW and SGM) have been (and still are) extremely important for the region, as sources of technological knowledge, capital (FDI), management and engineering techniques.

SVW commands modern technology, and has special training programmes for Chinese managers and engineers (partly done in Germany). There are strong links with Germany, not only through import technologies, skills and techniques, but also in terms of decision-making. The same is true for many German suppliers that followed SVW to Shanghai. However, recently, many suppliers have reduced their dependence on German headquarters. They can act more independently, and also, many suppliers have replaced German managers with Chinese ones (Depner and Bathelt, 2005).

Like Volkswagen, General Motors is also an important source of new technology, management and engineering techniques for the Shanghai region. Most of GM's new technology is developed in the US, and then applied in China. SGM was the first foreign company to introduce an all-inclusive, lean manufacturing-based system in China that handles every aspect of the production process in China, including purchasing, manufacturing, sales, after-sales and quality management. The higher management ranks of SGM are taken by foreigners, while middle and lower management is increasingly made up of Chinese people.

The relations between JV partners can be delicate. For foreign partners, intellectual property rights are a concern because they want to avoid leakage of new technology. For this reason, Volkswagen's research centre in Shanghai is relatively closed to the outside world. Volkswagen's network position is complicated, in particular because it has not only a JV with SAIC, but also with FAW (in north-east of China). The company has established a coordination centre in Beijing, to align the activities of SVW and FAW-VW.

Another delicate issue in this respect concerns the relations with SAIC suppliers and JVs. Before entrance to the WTO, SVW was forced by the SAIC board to use suppliers that belonged to the SAIC group, even if their performance (in terms of price) was considered insufficient by the German company. In the early days, VW did not complain (the market was protected and VW made a healthy profit); later, when competition increased, the system made it difficult for VW to cut costs. GM entered the Chinese market later, and managed to establish completely different relations with suppliers in the SAIC group. GM applied its own purchasing method (based on a standard process of supplier evaluation[12]) to Chinese suppliers. After an assessment of 1,000 Chinese suppliers, GM could convincingly argue in which cases it made sense to buy from non-SAIC suppliers located in China or abroad. Within SGM, this led to conflict between SAIC managers and GM managers, but the prospect of falling tariffs (after WTO accession) made the position of GM stronger. The success of this approach was clear, and in 2000, SVW started to use a similar purchasing system (Thun, 2006).

Supplier networks

In the networks between OEMs and their suppliers, the OEM is the dominant partner. Typically, it specifies which types of parts it needs for a certain car model it wants to produce. Suppliers are selected according to their ability to meet quality standards at the lowest price. There are two types of assignments: the OEM gives very detailed specifications and suppliers can submit proposals; or they give more general descriptions, meaning that the supplier has to do some design and development work. The latter option is becoming more common. Typically, suppliers get a contract for the life of the car model, and are required to reduce costs by a certain percentage per year, and OEMs maintain the right to take new suppliers under some conditions. Although the selection of suppliers is largely based on objective criteria, our interviewees suggested that personal relationships still play a very important role. Furthermore, there are indications that suppliers belonging to the SAIC group get a preferential treatment by OEMs that are also part of SAIC. For instance, after negotiations with all suppliers, SAIC suppliers are given a chance to adapt their price.

In Shanghai's supplier industry, different types of suppliers can be discerned. The most knowledge-intensive suppliers are the well-known and globally active 'system suppliers' (like Bosch, Siemens or Valeo). They produce vital parts of the car, and typically, are not only involved in manufacturing but also in product development, in close cooperation with OEMs. These suppliers conduct most of their sophisticated development work in the home nation of the company. The same goes for many foreign lower-tier suppliers (see Box 11.1 for the case of Ficosa).

Their establishments in Shanghai typically focus on manufacturing (often in JV structures). Thus, the knowledge networks are international, and knowledge mainly flows within the company; the manufacturing networks are regional. Development work is restricted to slight adaptations of systems and components. Like in the foreign OEMs, the contingents of Chinese managers and engineers in these firms are growing, and the dependence on 'imported' technology is declining.

As described, OEMs in Shanghai do not only depend on foreign suppliers: they can use a strong local supplier base. In the 1980s and 1990s, the OEMs (mainly SVW, but also SGM) played a major role in developing the local supplier base, by helping suppliers to learn manufacturing capabilities. A trans-regional, inter-departmental, cross-industry and systematic project by SVW has helped many local suppliers improve their technical capabilities and raise the quality of parts and components to international standards. Nowadays, more than 400 domestic local suppliers supply locally produced parts to SVW. Other OEMs have also

Box 11.1 The case of Ficosa

Ficosa is a lower-tier supplier in the automotive industry. It is a Spanish company and it has developed rapidly in the 1980s, with the strong growth of the Spanish car industry by that time. Ficosa produces car parts such as pedals, mirrors, and gearboxes. Worldwide, the firm has 30 plants, most of them located close to OEMs. The headquarters and main R&D operations are in Barcelona, Spain.

In Shanghai, Ficosa opened its plant in 2002. It is 100% foreign-owned. Its clients are SGM, SVW, Ford China and Mazda. It generates a turnover of €10m in China (5–7% market share). The original idea was to start a JV with a Chinese firm, but Ficosa could not find a suitable partner. The firm has its Chinese headquarters in the Jiading district (it was already there: before the plant was opened, Ficosa used Shanghai as a base for import and trading). At the site, the company also conducts product development (20 people, among which 12 foreigners), and it has its plant there (97 employees). Ficosa uses China as an export base, but only for after-sales items (spare parts); this takes some 10% of total turnover. In the next 5 years, the ambition is to increase exports of spare parts substantially from China.

The design and prototyping of new products is conducted in Spain, mainly because of the availability of high-level test facilities. In China, Ficosa conducts the less complex process testing (production process), and the engineering of slight modifications to the product for the local market. Ficosa cooperates with Tongji University in two ways: 1) for training their staff and 2) for co-designing products for the Chinese market. Ficosa plans to expand its R&D department, from 20 people in 2006 to 60 in 2010. Of these 60, only 4 or 5 will be foreigners (Spanish), and the rest Chinese. This growth will not go at the expense of R&D employment in Europe: it will mainly help to service the growing Chinese market. In every continent, Ficosa conducts research, and the firm will continue this. Interestingly, Ficosa considers moving its R&D for the Japanese market to China.

Source: own elaboration

accepted SVW suppliers. Some of the suppliers have even become suppliers for global-sourcing OEMs (SAIC, 2005).

Clearly, transferring knowledge to suppliers was in SVW's own interest: it helped to reach the required local content levels and cut costs (using local suppliers can be cheaper than importing parts from abroad; this was especially relevant in the period before entrance to the WTO, when the import tariffs were very high). SVW's interest coincided with Shanghai's, which aimed to develop a strong local automotive cluster. The success of the upgrading efforts are evident: SVW has a local content percentage of about 90%, the highest possible level (Thun, 2004, 2006). It should be noted that in recent years, the relations between SVW and its suppliers has changed. SVW's increasing pressure to cut costs is transferred to suppliers.

The presence of many OEMs in Shanghai and its region makes it an attractive place for suppliers: it reduces their risks, as they can work for several clients. Furthermore, the critical mass of the cluster makes it easier to find qualified staff, and to find partners for cooperation projects.

In the operational ('manufacturing') relations between OEMs and suppliers, geographical proximity matters to varying degrees. Some OEMs prefer to work with suppliers nearby; for instance, SMA prefers to work with suppliers located relatively close to the manufacturing plant, in Shanghai, Jiangsu province or Zheijing province (mainly for logistical reasons). SMA has invited 12 major suppliers that supply parts with high logistical costs to co-locate on the same site.

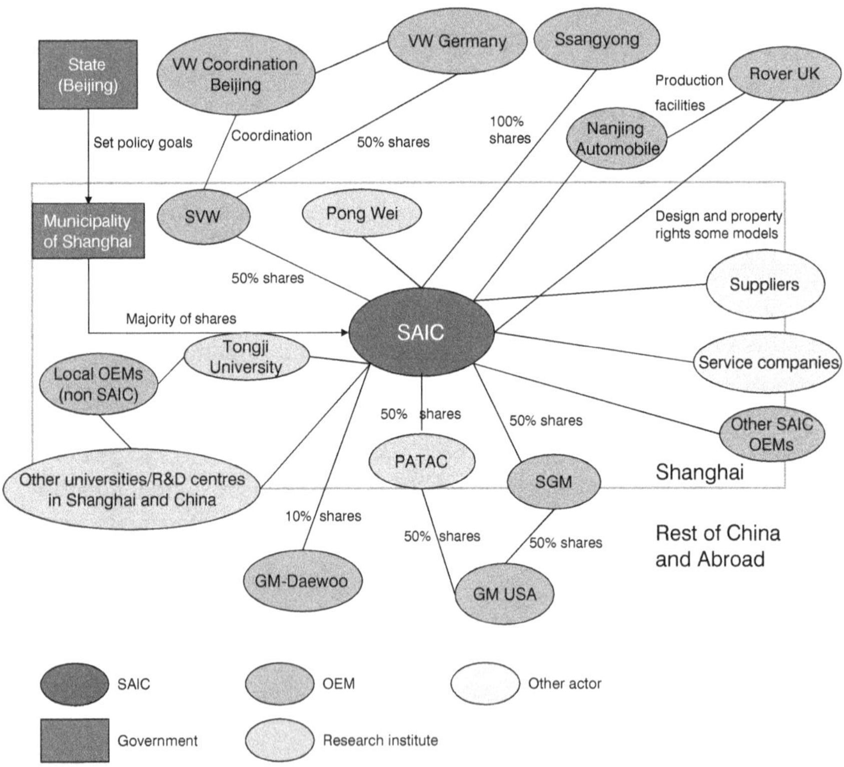

Figure 11.7 SAIC as centre of the local automotive industry (source: own elaboration, based on Depner *et al.* (2005)

The increasing price competition, however, makes cost considerations relatively more important than location, and therefore some suppliers are forced to (re)locate to cheaper inland locations farther from the OEMs. Our interviews suggested that only suppliers that are highly involved in R&D and design and suppliers with high logistical costs are located in direct proximity to OEMs.

The international trend that suppliers are increasingly involved in product development also applies to Shanghai. Many suppliers work with local knowledge institutes to train staff, to improve their development capabilities and to engage in joint projects. For instance, one interviewed supplier cooperates with Tongji University in design projects and in staff training. Suppliers prefer knowledge institutes nearby, as it makes interaction easier.

Networks for new energy

The Chinese government has set the goal to exploit cleaner energy sources and to make cars more environmentally friendly, so it invests in R&D in these directions. The funds are channelled to universities, research institutes and companies that engage in energy projects.

SAIC, being an SOE, takes up the challenge, and engages in several strategic networks to reach this objective. It has three strategy lines, each with different partners and on a different geographical scale. The first strategy line is to develop mixed-fuel applications. In cooperation with VW, SAIC is developing mixed-fuel vehicles in the SVW research centre on the company's site, near the manufacturing plant. The second strategy is to develop hydrogen vehicles. Here, SAIC cooperates with GM in PATAC, which is located in Pudong (near the large manufacturing plant).

The last strategy line is to develop representative fuels, such as LPG (liquefied petroleum gas). For this, R&D is conducted in cooperation with two local universities (Tongji University and Jiaotong University). In essence, SAIC taps different sources of knowledge from different partners.

In general, the interviewees gave the impression that the intensity of cooperation between automotive firms and knowledge institutes is increasing. The cooperation is not restricted to universities in Shanghai, but increasingly also includes knowledge centres in the Yangtze River Delta and other parts of China. For example, Shanghai Wanfeng cooperates with China Automotive Technology & Research Center, Tongji University and Jilin Industrial University. SMA works with Jiaotong University, Tongji University and Jilin Technical University. Another example is China Automotive Energy Research Center (CAARC) at Tsinghua University in Beijing. In this new government research centre, Tsinghua University, SAIC and GM will share resources and personnel to develop a strategy to reduce China's reliance on oil. CAARC will cooperate with various stakeholders in government, academia and companies. SAIC and GM will jointly invest US$5 million in CAERC (*Automotive Business Review*, 2007).

Networks between governments

The relations between governments in China have a vertical (hierarchical), and a horizontal dimension. The vertical dimension implies that national government sets the policy goals, and the provinces/municipalities take care of the implementation. As we have seen, within the national policy framework, lower-level governments have substantial freedoms to choose their own policies, and Shanghai is a prime example.

In the automotive industry, horizontal relations (mainly between provinces) are relevant as well. There is strong competition between provinces to attract firms and to develop the local automotive industry. Several provincial governments have appointed the automotive industry as a pillar industry, for its high visibility, status and linkages with many other industries. Like Shanghai, many local governments have large stakes in local/regional automotive business groups. The national government has expressed the desire to speed up the consolidation in the car industry to fight fragmentation and build fewer but stronger groups. The strong government involvement in the industry hampers the realisation of this wish, as the provinces are reluctant to give up their own ambitions.

Despite the competition between local/provincial governments, there is also cooperation. The municipality of Shanghai cooperates with the Jiangshu province to develop the automotive cluster around Auto City (which is situated close to the provincial border). Both governments realise that the automotive cluster in Anting is larger than Anting itself, and that it has many links with other locations in Shanghai and other parts of the Yangtze River Delta. Many automotive suppliers are located just across the municipal border in the Jiangshu province. For both governments it is a win-win situation, to cooperate in order to develop the cluster as whole.

Within Shanghai, there is competition between the districts of Jiading and Pudong (and to a smaller extent also with Jinshan district and other locations), to attract automotive firms. The government favours the cluster in Jiading since it is appointed as an export zone, and tries to cluster many suppliers in Jiading. However, if new suppliers prefer another location (such as Pudong), they are allowed to go elsewhere. Also, current suppliers at other locations are not forced to relocate to Jiading.

Unlike in other automotive clusters (such as the ones in Paris and Munich described in this book), there is no ‘cluster organisation’ in Shanghai, as a platform for companies to meet each other, exchange knowledge and information, or engage in joint projects. The networks are in place, but steered in a different way. The coordination mechanism in the cluster is a combination between hierarchy and market, with SAIC in a dominant position. On the one hand, SAIC is top-down driven by the government to implement policy goals. Through SAIC, the government has a lot of influence in the development of the automotive industry in Shanghai, especially since SAIC is a comprehensive business group incorporating all activities of the automotive value chain. On the other hand, SAIC is driven by international business via international business partners. This is expressed graphically in Figure 11.9. Also non-SAIC-owned firms have relations with SAIC, for instance local OEMs make use of SAIC suppliers and service companies (this is not expressed in Figure 11.9).

11.6 Conclusions

In China, the automotive industry has developed rapidly over the last decades. A strong manufacturing base was successfully built up, and China has developed as an important car-producing (but not yet exporting) nation. Foreign companies (OEMs as well as suppliers) have invested heavily in China in the form of JVs, hungry for market share in this country with tremendous potential. Through the cooperation with foreign companies in JVs, Chinese automotive companies have learned modern manufacturing skills and consistently upgraded their capabilities. The number and importance of Chinese suppliers has risen sharply, and for manufacturing, the dependence of foreign suppliers has decreased. Indigenous Chinese OEMs emerged with most of them focusing on the cheaper segments of the car market. In the last five years, developments have been even more turbulent due to accelerating demand for cars (thanks to rising incomes), the entrance of China to the WTO, and the ambitious policy objectives of the national and provincial/local governments to reduce the dependence on foreign companies and move towards a strong indigenous Chinese automotive industry, including Chinese designs and brands. Now, the Chinese car industry is making inroads into the higher parts of the value chain, and is rapidly becoming more knowledge intensive. Companies are increasingly investing in R&D centres, like GM, which invests US$250 million in a basic-research centre. Also, the Chinese government strongly promotes this development. It has launched plans for 'own brand, own design', and invests massively in R&D. However, until now, indigenous Chinese car brands have not been able to compete in the main markets in Europe and the USA, due to deficits in quality and design. Exports are limited to Middle Eastern and African countries. Competition on the home market is tough but margins are still fine, overall.

In this chapter, we have analysed the dynamics of the automotive industry in Shanghai, China's dominant automotive centre. The analysis is structured through an analytical framework. The framework assumes that the development and performance of any manufacturing industry cluster is dependent on regional economic foundations (general feature of the regional economy), as well as cluster-specific aspects (the size and composition of the cluster, the network structure and the governance of the cluster). Evidently, cluster performance and development is also heavily affected by external developments such as market trends, technological developments and (supra) national government policies. This case study of Shanghai shows the pivotal role of the city of Shanghai in the development of this cluster. Below, we will draw more conclusions, using the structure of the framework.

Conclusions on foundations

Shanghai's automotive cluster did not develop in isolation but is embedded in a regional context.

Over the years, the region has developed a highly varied set of manufacturing industries, many of which play a role as suppliers for the car producers in the region (steel, textiles, chemicals, plastics, electronics). In this respect, the car industry also benefits from the economic strength of the entire Yangtze River Delta, which has the largest manufacturing base in China and maybe even the world. But Shanghai has much more to offer than just manufacturing industries: the city has developed as one of the China's main service cities. Shanghai's maturing and increasingly complete auto cluster relies on service activities such as branding, distribution and logistics, research and development, consultancy, finance, marketing and headquarters functions. The city is making the transition from a car manufacturing cluster towards a 'full-fledged' cluster.

From this perspective, the relatively strong knowledge base of Shanghai is a crucial asset. In Shanghai, it is relatively easy to find staff with some years of professional experience at international standards (managers, engineers, sales people, etc). There are highly regarded universities that attract China's best and most ambitious students, who then find jobs in the city's dynamic industries after graduation; also the research capabilities are on a high level. The cooperation between university researchers and the private sector is not without problems, however. Professors, even in top universities, often lack industry experience and, as a result, do not understand the dynamics and needs of the industry. This makes cooperation a difficult and sometimes frustrating exercise. Improvements are being made, though. The cooperation between Jiaotong University and SMA (Shanghai Maple Automotive) is an example in which the industry partner takes the lead and invites university staff to cooperate on concrete research projects. Our research points to some weak points in the knowledge base. First, some commentators indicated that the level and depth of engineering skills in the region is insufficient to meet the ambition to have a car based only on technology developed in China. In other words, the dependence on the knowledge of foreign partners is still very high. Chinese OEMs still depend on platforms developed by foreign manufacturers, or on core components developed by world-leading foreign first-tier suppliers. Traditionally, engineering is not regarded as a high-status profession, and as long as Chinese employees prefer management jobs over engineering jobs, as some interviewees put it, the situation will be hard to improve. Another issue is the enforcement of intellectual property regulations. Foreign firms will be less likely to develop new technology in China if they continue to face the risk of copying by Chinese firms without compensation.

The automotive cluster benefits from Shanghai's excellent, logistical location. Its strategic location in the Yangtze River Delta and its excellent port facilities make the region a natural hub and facilitate the import and exports of materials, supplies and cars. This is especially relevant in light of the high export ambitions in the longer term. Also the modern and expanding international airport of Pudong is an asset. Companies in the cluster traditionally have very close relations with foreign partners (GM, Volkswagen and many suppliers) who frequently fly-in teams of experts from the USA, Europe and Japan. The importance of air connections is likely to continue and gets a new dimension when Shanghai firms sell more cars in foreign markets, and acquire foreign car producers (as SAIC has done with Ssangyong from Korea). The increased service component of Shanghai's auto cluster also reinforces the importance of good national and international connections.

Many studies have identified the importance of quality of life as a factor for cluster growth: it helps to attract and retain high-skilled workers who are the lifeline of any cluster. In many respects, Shanghai has a lot to offer: cultural amenities, leisure, restaurants, and healthcare facilities are plentiful, as are facilities for foreigners (international schools, restaurants). The city is vibrant and dynamic, and has a strong attraction power for talented Chinese people as well as foreigners. But there are problems, too. For one thing, air quality leaves much to be desired, and the rapidly growing car ownership in the city is likely to make the problem worse. Policy measures have been introduced to reduce pollution but enforcement is still a problem. There also are policy initiatives to promote the development of more green technologies in cars (hybrid and fuel cell engines), but the effects cannot be expected in the short run, and may be fully offset by the sheer growth of the number of cars. Pollution is not only caused by cars: the enormous concentration of manufacturing industries in the Yangtze River Delta brings severe environmental degradation over a large area. In the short term, Shanghai's economic growth (including the automotive cluster) seems unaffected. In the longer term, when incomes rise and people attach more value to environmental quality, Shanghai's position as a 'knowledge city' could be undermined when people prefer cleaner and greener places. The Expo, to be organised in 2010, is an important leverage factor to make progress in this respect. The theme 'better city, better life' links to many issues in the automotive industry, notably the need for reduced fuel consumption and emissions and improved traffic management systems. The Expo is an opportunity for Shanghai to present its automotive cluster to the world as progressive and leading in environmentally friendly technologies.

Cluster structure, geography, governance and performance

The Shanghai automotive industry is large and diverse, and it is making the transition from a manufacturing cluster towards a more knowledge-based industry. The region can boast four major OEMs and a large supplier base. This is highly relevant, as typically 80% of a car is produced not by the OEM but by suppliers. For suppliers, scale advantages play a role: the region is attractive for them because there are so many OEMs, which reduces their dependency and vulnerability.

In Shanghai, there is no single area where the automotive industry is concentrated. We identified several 'physical hotspots' centred around the OEMs: one in Pudong (around SGM), a small one in the Jinshan District (around SMA), and one in Jiading (around SVW and SAIC). The first two are more or less 'unplanned' and consist of just an OEM plus some suppliers and research; the last one is part of a large urban planning initiative to create Auto City, including a university, auto dealers, a racing circuit, leisure facilities and even housing. These three physical 'hotspots' are not the full picture, however. Downtown Shanghai is important as well for consultancy, marketing, finance and administrative functions. Suppliers are spread all over Shanghai and beyond, as far as 500km inland in the heavily industrialised Yangtze River Delta.

The quality of the suppliers is very high compared to the other clusters in China, thanks to sustained and consistent efforts of the local government to transfer knowledge from foreign companies to Chinese ones. In this respect, Shanghai is unique in China: it is the only city with such a pro-active approach towards the automotive industry. It is illustrative that Shanghai-based suppliers are used by OEMs in other Chinese cities. Other strengths are the strong presence of specialised automotive research institutes and testing facilities.

As an important element of this study, we have analysed networks in the Shanghai automotive sector. We can conclude that there is a functional and geographical difference between knowledge networks (R&D cooperations aimed at innovations) and production networks (aimed at the production and distribution of cars and car parts), and also, the networks are governed in a different way.

Over the years, *production networks* are very well developed, and manufacturing quality has reached international standards. These networks are predominantly local/regional, and mainly concern the relations between OEMs and suppliers. In these networks, OEMs are leading: they select suppliers on the basis of price and quality. Given the quality and quantity of suppliers located in the region, OEMs have a choice. Many suppliers export car parts to other countries.

For *knowledge networks*, the picture is different and mixed. For very advanced research and development (new platforms, systems), OEMs and first-tier suppliers in Shanghai still rely on expertise from abroad. The platforms on which the models of SVW and SGM are built are developed in Germany and the USA, respectively, and main systems within these cars are developed in the R&D labs of the main suppliers abroad. However, we observed that the cluster is clearly upgrading itself and is slowly reducing its dependence on foreign technology and experts. This process can be illustrated in several ways. First, throughout the industry, senior roles are increasingly filled by Chinese staff. Second, Chinese suppliers are ever more involved in development, and they expand their R&D activities rapidly to adapt their products to the specificities of the Chinese market. Thus, the link between OEMs and suppliers extends beyond a mere production function. Foreign suppliers also expand their R&D staff numbers in Shanghai, but unlike the fears of many in Europe, this growth comes not at the expense of R&D jobs elsewhere. A third sign of upgrading is the fact that universities are increasingly involved in research with the industry. SMA (owned by Geely) is a good example: it developed new engines and cars in cooperation with local university partners. Fourth, many companies invest heavily in R&D capacity, helped by generous government subsidies; and SGM, SVW and SAIC have opened large research centres. Finally, SAIC reduces its technological dependence by buying foreign companies and using/transferring their knowledge (it recently bought Ssangyong). All these tendencies show a decreasing dependence on external technology and expertise and an increasing 'localisation' of knowledge networks. Strong government investments help to move further in this direction.

Based on our interviews, some remarks can be made, however, concerning the different 'upgrading' strategies. First, 'buying' technology is one thing, but integrating and using this technology to develop a strong brand is a more difficult exercise, as it requires a very solid absorptive capacity in the parent firm. From this perspective, an indigenous development strategy may be more promising. Second, government policy supports investments in R&D, but there are some doubts about the quality of the outputs of these investments.

Within China, Shanghai is the largest of six big automotive clusters. It has all the ingredients to develop into a complete automotive cluster, in the long run probably on par with giants such as Paris region, Bavaria or Michigan. We found many indications of a fast upgrading process, helped by continuing efforts of the very pro-active national and local governments.

There are several critical remarks to be made, however. Based on interviews with insiders, we conclude that it will be hard to realise the government's ambitions (own brand, own design, use of new energies, export to the main world markets) in the medium term. The general level of engineering in Shanghai is not yet sufficient, and using foreign technology is still the norm. Despite strong improvements and incidental success stories, there is little design 'from scratch'. Buying technology (SAIC: Rover designs) or buying stakes in foreign companies (SAIC: Sangyong, Daewoo) are a far cry from engineering, designing and branding an 'own car'. A first step is taken by Chery, Geely and SMA, but there is a very long way to go. China is not an engineering country by tradition, and it's hard to change that. And, as long as so much money can be made by producing cars based on foreign technology, the incentive to truly design and engineer an 'own car' is limited. Universities are a crucial element for the opportunity, but their capacities are restricted and professors have no business engineering or industry experience. The curricula are theoretical, and the research is not always aligned with industry needs.

In the automotive cluster, the government plays several, sometimes conflicting roles. First, the city is a significant market player: Shanghai is the owner of a major OEM and a number of suppliers (SAIC group) and has a 50% share in two more OEMs (SVW and SGM). Second, the government is the enforcer/implementing body of national policy. Relevant policies here are the ambitions to promote the industry and make it internationally competitive, to develop clean and fuel-saving technologies, and to promote automotive research and education. And third, the government is active in spatial planning – witness the initiative to create the large 'automotive city'.

In the 1980s and 1990s, the pro-active and intervening role of the Shanghai government has been fruitful. Thanks to thoughtful and consistent policy, Shanghai firms learned engineering and manufacturing skills, preparing them to compete in a free market. In this early 'catching up' stage, the cluster was led by SAIC, helped by the local government. The question is whether the current coordination mechanisms in the cluster are adequate to guide the cluster to a higher level. The fact that SAIC is a state-owned firm used to be an asset, but in the current stage may have its drawbacks. First, SAIC suppliers, in some cases, continue to get preferential treatment, giving them fewer incentives to improve their performance. This may hurt the cluster as a whole. Second, with the success of new private-owned automotive firms in Shanghai (such as SMA), the Shanghai cluster's fate no longer coincides with the fate of SAIC alone. Perhaps the government should start to consider SAIC as a normal company.

Notes

1 However, Depner *et al.* (2005) found that a WFOE strategy of suppliers is not without risk, among reasons because WFOE suppliers are under much greater pressure to cut costs than JV suppliers.
2 Van Dijk (2007) notes that it is not difficult to find a figure citing 20 million inhabitants. This depends on where the city borders are drawn and whether the actual or the registered numbers are used. We have used the registered numbers of the Shanghai municipality.
3 http://www.shanghaiairport.com/en/index.jsp (accessed 16 August 2006).
4 http://www.shanghai.gov.cn/shanghai/node8059/BasicFacts/index.html#, http://lyw.sh.gov.cn/en/ (accessed 16 August 2006).
5 4S means Sales, Spare parts, Services and Survey.
6 Note that the negotiations for the joint venture between VW Germany and the state had already started by 1978.
7 http://www.saicmotor.com/english/yf/224.shtml+Shanghai+Automotive+Engineering+Academy+(SAEA)&hl = nl&gl = nl&ct = clnk&cd = 2 (accessed 23 March 2007).
8 http://www.gmchina.com/english/operations/patac.htm (accessed 14 November 2006) and interviews.
9 http://www.shautocity.com/english/cczl-01.htm, http://www.investment.gov.cn/2005-10-27/1130463265474.html (accessed 1 September 2006) and interviews.
10 The ambition is to invite the six largest international and the three largest national OEMs to locate in Auto City.
11 http://www.pudong.gov.cn:7001/english/common/content.jsp?sj_id = 117&ct_id = 1163, http://www.investment.gov.cn/2005-10-27/1130425544543.html. (accessed 1 September 2006).
12 GM had a handbook with fixed prices for each component, and it would not pay more to any supplier (Thun, 2006).

Bibliography

Asia Times (2005) 'Shanghai, Guangzhou become car industry bases', 20 October 2005.
—— (2006) 'Industry still king, but changes loom', 10 January, available online at: www.atimes.com/atimes/China_Business/HA10Cb04.html (accessed 20 July 2006).
—— (2006) 'Shanghai aims to be China's Detroit', 12 October, available online at: www.atimes.com/atimes/China_Business/HJ12Cb01.html (accessed 20 July 2006).
Automotive Business Review (2007) 'GM to establish $250 million research and energy centre in China', 30 October, available online at: www.automotive-business-review.com/article_news.asp?guid = 54C01E11–16BDA-46FC-85FA-493299ADB0A5.

Becker, H. (2006) *High noon in the automotive industry*, Berlin: Springer-Verlag.

Depner, H. and H. Bathelt (2005) 'Exporting the German model: the establishment of a new automobile industry cluster in Shanghai', *Economic Geography*, 81:1, 53–81.

Economist Intelligence Unit (2006) *Driven – are China's car manufacturers ready to compete with the US and Europe*, briefing paper.

FHA (Federatie Holland Automotive) (2004) *De groeipotentie van Automotive Nederland – Een visie op de internationale ontwikkelingen en de toekomstige rol van Nederland in de automotive industrie*, FHA.

FT, Financial Times (2006a) 'China – local low price options take off', 28 February 2006.

—— (2006b) 'Exports from China move up market', 4 April 2006.

—— (2006c) 'Strong growth in China sees car sales surge', 5 July 2006.

—— (2006d) 'Nissan to unveil range for China's new choosy buyers', 7 July 2006.

—— (2006e) 'GM to start selling hybrid cars in China from 2008', 7 November 2006.

—— (2007a) 'The dragon's lab – how China is rising through the innovation ranks', 5 January 2007.

—— (2007b) 'GM plans China research facility', 30 October 2007.

—— (2007c) 'China's Shanghai Auto buys rival Nanjing', 26 December 2007.

Gamble, J. (2003) *Shanghai in transition: changing perspectives and social contours of a Chinese metropolis*, London: Routledge.

Heneric, O., G. Licht and W. Sofka (2005) *Europe's automotive industry on the move: competitiveness in a changing world*, ZEW Economic studies, vol. 32, Heidelberg: Physica-Verlag.

van Hoek, R.M,P. Woets and X.F. Feng (2005) *'The Expo-tunity' Shanghai World Expo 2010: main goals for a long lasting contribution to the development of the city. A quick scan*, Euricur: Rotterdam.

Hoon-Halbauer, S.K. (1999) 'Managing relationships within Sino-foreign joint-ventures', *Journal of World Business*, 34:4, 344–71.

Huang, Y. (2002), Between two coordination failures: automotive industrial policy in China with a comparison to Korea, *Review of International Political Economy*, 9: 3, pp. 538–573.

IMF (International Metalworkers' Federation) (2004) *Auto report 2004*, Geneva: IMF.

KPMG (2003) *Chinese automotive and component market*, Hong Kong: KPMG.

Leman, E. (2002) 'Can Shanghai compete as a global city?', *The China Business Review*, Sept/Oct 2002.

Miaoyang, W. (2004) 'The culture of Shanghai and modernization', in: Xuanmeng, Y. and H. Xirong (eds.) *Shanghai: its urbanization and culture*, Washington D.C.: Council for Research in Values and Philosophy.

SAIC (2005) *Annual report 2004.*

Shanghai Academy of Environmental Sciences (2005) *Transportation situation and traffic air pollution status in Shanghai-vehicle emissions control and health benefits; technical and policy barriers to sustainable transport*, Part one report, Shanghai: Shanghai Academy of Environmental Sciences.

Shanghai Municipal Statistics Bureau (2006) *Shanghai statistical yearbook 2006*. Available online at: www.stats-sh.gov.cn/2004shtj/tjnj/tjnj2006E.html.

Shanghai Urban Planning and Design Research Institute (2001) *Summary of the comprehensive plan of Shanghai (1999–2020).*

Sigurdson, J. (2005) *Shanghai – from development to knowledge city*, Working Paper 217, August.

Swiss-Chinese Chamber of Commerce (2004) 'China automotive and vehicle industry: today's opportunities and risks', *Quarterly Information Bulletin*, 3, 20–21, available online at: www.sinoptic.ch/scccgeneva/download/bulletin/2004_03.pdf#search = %22%22China%20automotive%20and%20vehicle%20industry%22%22 (accessed 29 August 2006).

Thun, E. (2004) 'Going local: foreign investment, local development and the Chinese auto sector', in: Carrillo, J., Y. Lung and R. van Tulder (eds.) *Cars, carriers of regionalism?* Basingstoke: Palgrave Macmillan.

—— (2006) *Changing lanes in China – foreign direct investment, local governments, and auto sector development*, New York: Cambridge University Press.

Unido United Nations Industrial Development Organization (2003) *The global automotive industry value chain: what prospects for upgrading by developing countries*, Vienna: UNIDO.

Van Dijk, M.P. (2007) 'The contribution of cities to economic development: an explanation on Chinese and Indian cities', *ISS Public Lecture Series,* 2007:1.

Van Winden, W., L. van den Berg and P.M.J. Pol (2007) 'European cities in the knowledge economy: towards a typology', *Urban Studies*, 44:3, 525–550.

Weider, M. (2004) *China – Automobilmarkt der Zukunft? – Wie nachhaltig und zukunftsorientiert sind der Strategien der internationalen Automobilindustrie in China?* Discussion paper SP III 2004–2105, Wissenschaftszentrum Berlin für Sozialforschung.

Wu, W.P. (1999) 'City profile: Shanghai', *Cities*, 16:3, 207–16.

Yeung, Y.M. (1996) 'Introduction, Chapter 1', in: Yeung, Y.M., and Y. Sung, *Shanghai: transformation and modernization under China's open policy*, Hong Kong: Chinese University Press.

12 Turku: shipbuilding industry

12.1 Introduction

The maritime and shipbuilding industry in Turku (a medium-sized city in south-west Finland) has more than 300 years of history; ships have been built in Turku since 1700. Whereas many traditional shipbuilding strongholds in Europe have been driven out of the market by Asian rivals, the sector in Turku has prospered. Today, large luxury cruise ships are produced in Turku and its region. The order book for new cruise ships and vessels is responsible for a very large share of industrial employment and value added in the region. The production of ships still requires a lot of blue collar work, but we will see that the knowledge component of the production process has increased dramatically.

This chapter analyses the development of this remarkable shipbuilding cluster in Turku. Section 12.2 contextualises the main trends and developments in the global shipbuilding industry. Section 12.3 describes the economy of Turku, putting the shipbuilding cluster in its proper regional-economic setting. It shows which regional 'assets' are in place that enable the industry to flourish. Section 12.4 analyses the shipbuilding cluster itself. It shows the role of its various actors and unravels the dense relationships and interdependencies between them. We pay special attention to the role of proximity in this industry. Section 12.5 is concerned with the governance of the cluster: it discusses its coordination mechanisms (how the different companies cooperate and interact, both formally and informally) as well as the role of policy. Section 12.6 concludes and puts the findings in the perspective of the framework of analysis of this study.

12.2 The shipbuilding industry

The shipbuilding industry is an important and strategic industry in a number of European Union (EU) member states. There are more than 150 shipyards in the EU, 40 of them active in the global market for large sea-going vessels. Those shipyards employ around 100,000 people directly and more than 400,000 indirectly. Since the enlargement of the EU, the annual turnover of the EU shipbuilding industry is more than €11.5 billion (European Commission, n.d.).

Figure 12.1 Luxury cruisers by Aker Yards (source: Aker Yards)

The shipbuilding sector in the EU has long been challenged by other nations. In the 1960s and 1970s, Japan and South Korea emerged as strong competitors, and, more recently, China is making inroads in the sector by means of large state-supported investments. Today, the shipbuilding sector is mostly concentrated in those three Asian players plus the EU (constituting together roughly 90% of the whole world industry). As Figure 12.2 depicts, there have been noticeable ups and downs in a sector highly sensitive to demand, order book fluctuations and usually suffering from lack of flexibility in the installed production capacity.

During the last years, a number of important trends shaped the sector's development, namely:

- The consolidation and emergence of low-cost producers in Asia and Eastern Europe, especially in the mass and cheaper market segments, who proved able to attract substantial FDI flows;
- A marked decline in Europe (during the 1970s–1990s), with several shipyard closures leading to serious unemployment problems in its shipbuilding strongholds. Nevertheless, by 2000, the world order book rose again. Although emerging economies and Asian players took the lion's share, European shipyards also made important bookings, showing signs of recovery. Presently, the EU production concentrates in high-value market segments, namely specialised vessels, ice breakers, passenger ferries and luxury cruise liners;
- Growing complexity of product. Building a ship, in particular when it is a specialised and customised vessel, is increasingly a demanding and complex task; it is a joint effort of hundreds of suppliers, down a long and diversified (and more knowledge-intensive) value chain. Large yards are increasingly taking the role of planners and assemblers of the whole process, and depend on a vast array of external partners and specialised suppliers;
- International mergers and acquisitions, resulting in fewer and stronger players in shipyards and among their main suppliers (producers of engines, machinery or other complex parts). Aker Yards, part of the multinational Aker

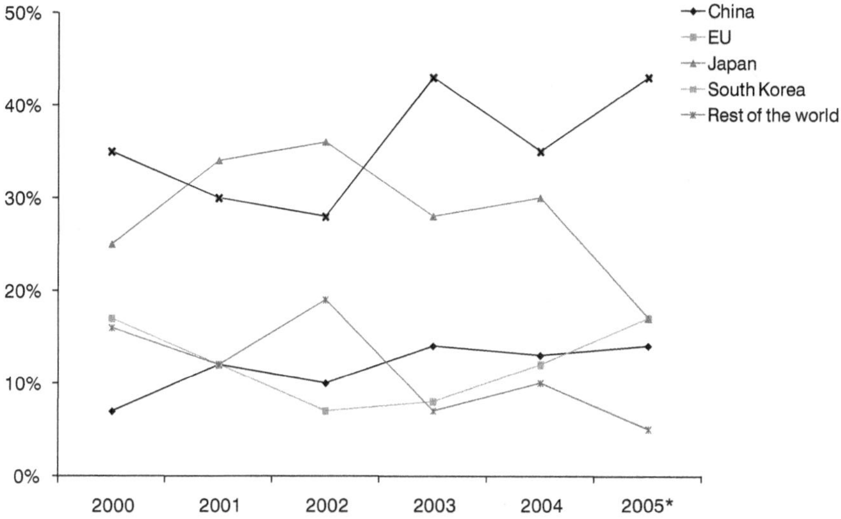

Figure 12.2 Shipbuilding sector: development of market share, % of new orders by compensated gross tons (CGT) (source: adapted from European Commission/FMI/CESA/ Lloyd's Register). Note: figures for 2005 are for the months January to June only.

Group is a well-known example of a merger (with Kvaerner Masa Yards in Finland);

- Strong demand fluctuations, sensitive to the economic conjuncture and world trade developments, putting strong demands on production flexibility and organisational engineering. Also events like 9/11, contribute to perceived safety risks and cause tourism to slow down, both of which tend to affect demand for cruises;
- Rising prices for energy and materials, putting pressure on the bottom line, as well as increasingly tight environmental and labour regulations in Western economies.

The value chain in shipbuilding

Building a ship is a rather complex process, which may take several mouths. It implies strong interaction between different actors, namely the shipyard, the ship-owner and the diverse suppliers and subcontractors involved, in different stages. Figure 12.3 represents a simplified and general overview of how the process usually works. The value chain is relatively long and encompasses three key moments:

- Contract between the client/ship-owner and the shipyard: before that moment, some suppliers besides the shipyard may already be involved, namely in preparing the project concept for the tender;

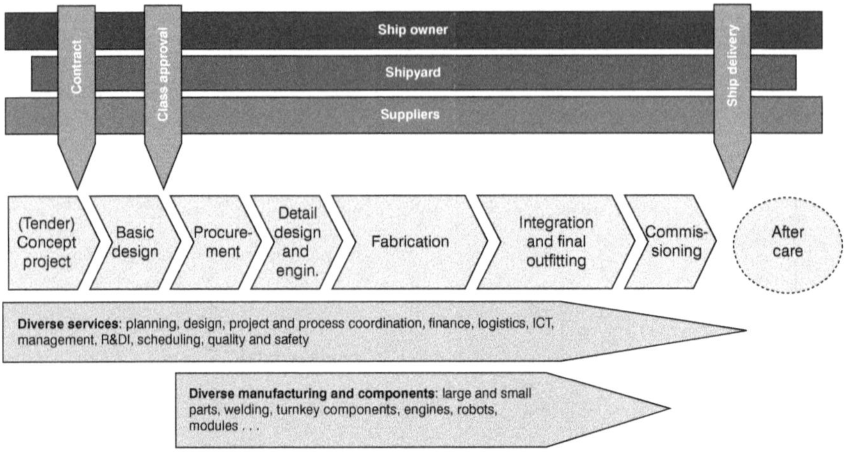

Figure 12.3 Building a ship – value chain, process and different actors involved (source: author's elaboration, adapted from Aker Yards and Deltamarin)

- Class approval: the accomplishment of this stage is preceded by the basic design of the ship, enlarging the tender specifications, including capacity, capabilities, quality, materials and other specifications;
- Ship delivery: before this moment, several key activities take place. They include the procurement for suppliers and subcontractors, the detailed design and engineering process (machinery, interiors, components, etc.), the fabrication of its different components and the final assembling and outfitting of the 'pieces', before the commissioning. This stage, despite being service-intensive, is also where the manufacturing and effective production efforts take place, both by subcontractors and by the yard's own capacity. The value chain also includes an after-care part (maintenance, etc.), in which the suppliers are also involved.

Shipbuilding in Finland

The maritime and shipbuilding industry is an important sector in Finland. The single most important player in Finland is Aker Yards (part of the Aker Group), with yards in Rauma (ferry building), Helsinki and Turku (luxury cruise and passenger ships). The Turku yard is the largest one. A large share of the sector's production is exported. The Finnish exports of ships are mainly directed to non-EU countries (see Figure 12.4) – the USA is a key market for luxury cruise ships.

Data on the shipbuilding sector reveal some interesting trends and developments. First, there are signs that the sector has become more profitable: in the last decade, the gap between value added on the one hand and production value on the other has widened (Figure 12.5), especially when focusing on a more specific sector (ISIC Rev. 4 – 3512: Build and repair of pleasure and sports boats). This reinforces the

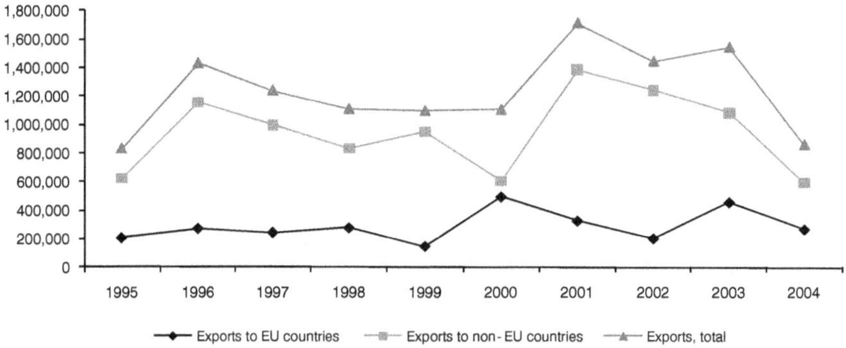

Figure 12.4 Exports, ISIC 351, Finland, Meuro (source: Statistics Finland, *Industrial Statistics on Manufacturing*). Note: ISIC Rev. 3–351: Build and repair of ships and boats.

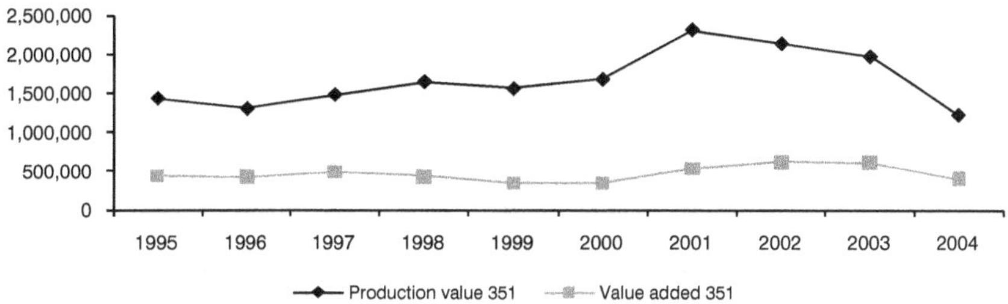

Figure 12.5 Production value and added value, 1995–2004, Finland, Meuro (ISIC 351) (source: Statistics Finland, *Industrial Statistics on Manufacturing*). Note: ISIC Rev. 3 - 351: Build. Repair. ships boats; Rev. 4–3512: Build and repair of pleasure and sports boats.

hypothesis that there is increasing profitability and higher value for some specific niche markets (Figure 12.6).

Second, we can observe a trend of outsourcing and subcontracting in the industry. The trend of increasing production value went hand in hand with a rise in subcontracting, to enhance flexibility and cater for demand fluctuations. In particular after the year 2000, there was a sharp rise of subcontracted labour, whereas contracted paid labour has been declining (Figure 12.7). However, when looking for the isolated segment of pleasure and sports boats, both the subcontracted and contracted labour has risen strongly after 2000, evidence of a more dynamic subsector (Figure 12.8). As predictable, subcontracted labour tends to vary more with the changes to the order book, which suffered a downturn in Finland at the beginning of this century.

There is also evidence that the shipbuilding sector is increasingly creating jobs and income in other business, especially in service sectors that are relevant in its value chain. Figure 12.9 shows the rising expenses of the sector on R&D

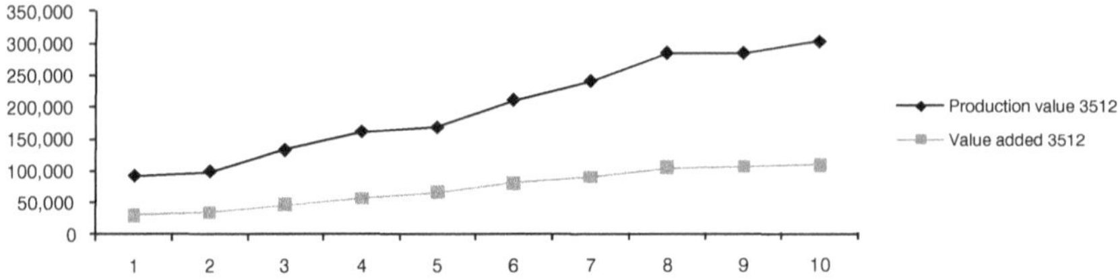

Figure 12.6 Production value and added value, 1995–2004, Finland, Meuro (ISIC 3512) (source: Statistics Finland, *Industrial Statistics on Manufacturing*). Note: ISIC Rev. 3–351: Build and repair ships and boats; Rev. 4–3512: Build and repair pleasure and sports boats.

Figure 12.7 Contracted paid labour and subcontracting, 1995–2004, Finland (ISIC 351) (source: Statistics Finland, *Industrial Statistics on Manufacturing*). Note: ISIC Rev. 3–351: Build and repair ships and boats; Rev. 4–3512: Build and repair pleasure and sports boats.

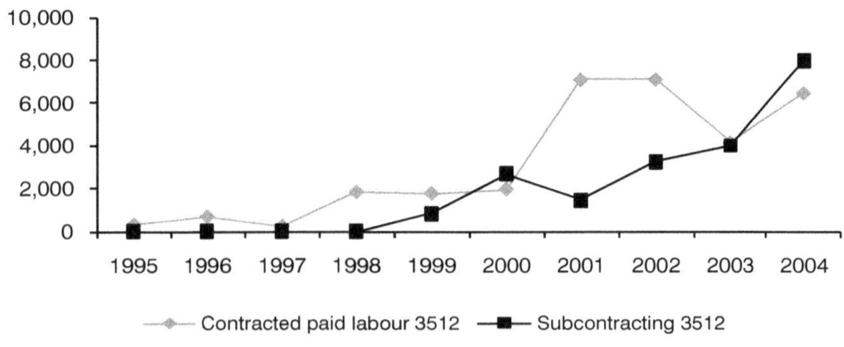

Figure 12.8 Contracted paid labour and subcontracting, 1995–2004, Finland (ISIC 3512) (source: Statistics Finland, *Industrial Statistics on Manufacturing*). Note: ISIC Rev. 3–351: Build and repair ships and boats; Rev. 4–3512: Build and repair pleasure and sports boats.

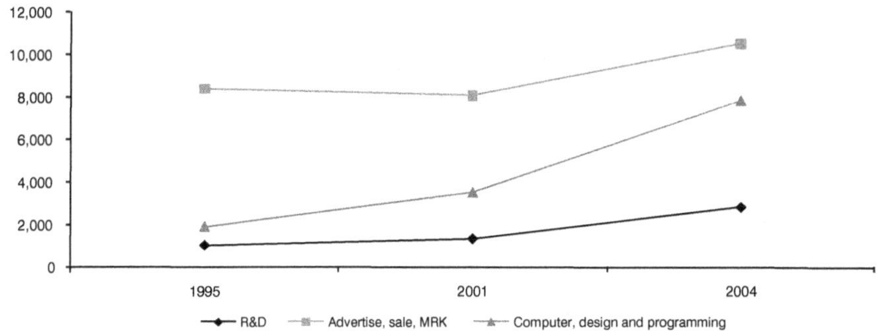

Figure 12.9 Expenses of the ISIC 351 sector, Finland, Meuro (source: Statistics Finland, *Industrial Statistics on Manufacturing)*. Note: ISIC Rev. 3–351: Build and repair of ships and boats.

(+300% in ten years!), advertising, sales and marketing (+ 25%), and computers, design, programming (+ 400%!), suggesting that the sector has become much more knowledge intensive.

12.3 Foundations of the Turku region

Turku is located in the south-western edge of Finland. It is the oldest city in Finland, and a former national capital (until 1812). The urban region counts some 300,000 inhabitants, 175,000 of which live in the city of Turku. The region is part of the dynamic urban axis of Stockholm – Turku – Helsinki – St. Petersburg. Economically and historically, the city is strongly embedded in the Baltic Sea area.

This section describes the spatial-economic 'foundations' on which the shipbuilding industry in the Turku region rests: the economic base, the knowledge base, quality of life, accessibility and factor conditions.

Economic base

Turku's urban region is the centre of one of the most dynamic provinces of Finland, south-west Finland or Varsinais-Suomi; it ranks second in Finland in terms of economic value added, after the Uusimaa province (where Helsinki is located). The economy has developed well over the last years, roughly in line with the Finnish average, showing a considerable reduction in unemployment after the large depression of the 1990s. Other cities in the province have economic relevance (like Salo, a fast growing industrial city for the electro-technical sector), but Turku is by far the largest economic centre. In fact, as Table 12.1 depicts, Turku and its urban region represent more than 60% of south-west Finland's economy.

The economic structure of the region is traditionally dominated by industrial activities. In particular, the sectors of metal and marine/shipbuilding industries used to be dominant. Although these sectors are still very important, Turku's

Table 12.1 Establishments, personnel and turnover, 2002–2004

	Establishments			*Personnel*			*Turnover (€ million)*		
	2002	*2003*	*2004*	*2002*	*2003*	*2004*	*2002*	*2003*	*2004*
City of Turku	9,119	9,044	9,178	56,727	56,656	54,764	8,944	8,992	9,056
Urban region of Turku	14,688	14,739	15,008	79,199	78,990	76,803	13,544	13,601	13,863
South-west (SW) Finland	23,746	23,884	24,368	116,633	115,575	113,082	25,205	20,865	21,342

Source: Statistics Finland

economy is rather balanced. Turku is an important service centre and transport hub, with a port and an airport, and real estate and business activities. Within the manufacturing sector, there is substantial diversity. Although two sectors have a noticeable weight – machinery and equipment (including the electro-technical industry) and other transport equipment (including ships) – other industries have an important role in the region, like food, pharmaceuticals, electronics and high-tech equipment (Figure 12.10).

In fact, a mix of traditional and emerging sectors endow Turku with a promising mix of businesses, making the region less vulnerable to the endemic fluctuations in the shipbuilding industry. Besides the metal and marine sector that employs roughly 40% of the workforce in south-west Finland, ICT industries create significant added value and employment in the region (in particular data communication), while biotech firms are growing stronger, and producing solutions for pharmaceutics and food industries (see Table 12.2). However, while the number of establishments and turnover has been growing in the beginning of the decade, the number of personnel has been declining (see Table 12.1), a structural trend in the Finnish workforce.

Compared to the national average, the share of services in this region of Finland is relatively low, and that of manufacturing relatively high. In the last decade, the sector mix has not changed dramatically in terms of value added; the share of services has grown, while that of manufacturing has declined somewhat (see Figure 12.11). Figure 12.12 shows that fluctuations in the manufacturing sector are much greater than in the service industries. However, in terms of employment, the picture is different as the manufacturing sector has been losing employment (it has become more capital and knowledge intensive), and labour productivity has increased significantly.

Within the manufacturing sector, however, there are marked differences. According to Turku's Chamber of Commerce (2006), losses are felt in the branches of textiles, clothing and shoes, building material and food, due to increased efficiency and restructuring). At the same time, the electro-technical industry

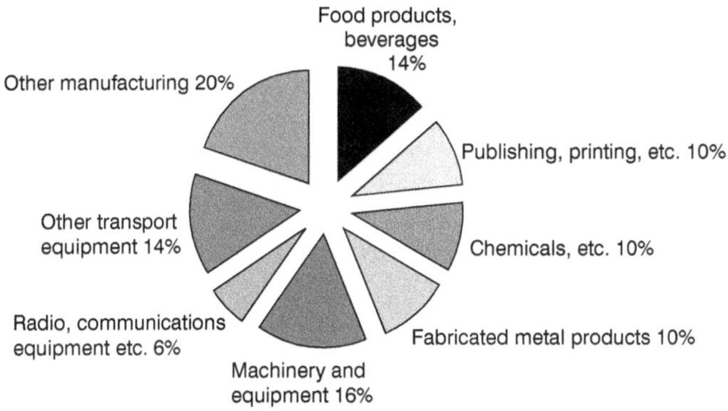

Figure 12.10 Share of personnel per manufacturing establishment, Turku's urban region 2004 (source: Statistics Finland)

Table 12.2 Traditional and emergent sectors in the south-west of Finland, 2005

Sector	*Turnover (€ million)*	*Companies*	*Personnel*
Biotech	500	60	3000
ICT	8,700	100	14,700
Metal and marine	8,800	1,200	16,300

Source: Centres of Expertise Finland

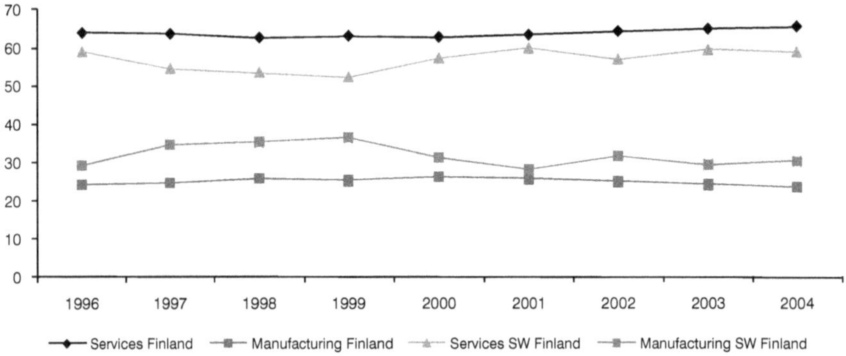

Figure 12.11 Value added share, %, 1996–2004 (source: Statistics Finland, Regional Accounts). Note: Manufacturing = ISIC sector D; Services = ISIC sectors G+H+I+J+K+L+M+N+O

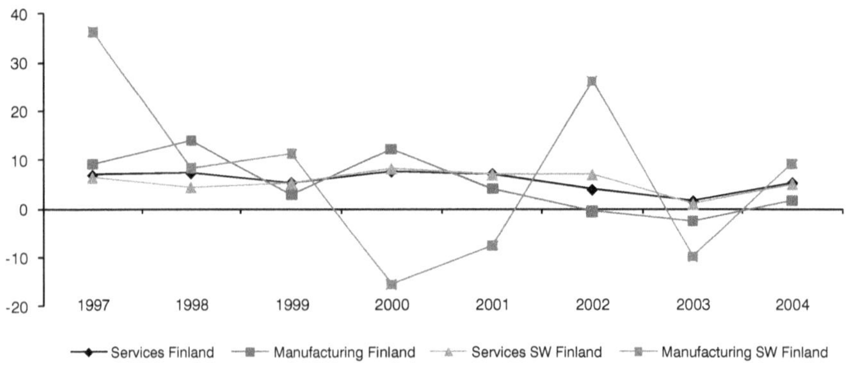

Figure 12.12 Value added, annual growth rates, %, 1996–2004: (source: Statistics Finland, Regional Accounts). Note: Manufacturing = ISIC sector D; Services = ISIC sectors G+H+I+J+K+L+M+N+O

has been creating a substantial number of jobs. It reflects the strong weight (and growth) of the metal and maritime industry, employing a very important share of the labour force of the region.

Knowledge base

Turku has a distinctive and strong knowledge base, measured by its knowledge institutions, research facilities and by the skill level of the population.

Concerning knowledge institutions, Turku's region has three universities and four polytechnics, providing relevant qualifications and skills for the region. The University of Turku is the oldest in the country, with roughly 16,000 enrolments per year, and with diverse expertise and teaching fields, throughout the social, natural and health sciences. The city's polytechnic is the largest in Finland (8,000 students), providing training in biotech, information technology, food technology, arts and media.

The science park of Turku (3 universities, 50 research centres, 13,500 experts, 750 companies, and 100 start-ups) supports innovative developments and firms in the region. Together with the educational institutions and business, it constitutes the core of the regional innovation system, developing new products for the region's diverse industrial sectors. It encompasses development facilities, aiding the survival rate of start-up firms. Other universities and polytechnics, though not in the immediate region, produce important qualifications and knowledge for the regional economic base (electro, metal and marine); the main centres are Helsinki, Tampere and Lappeenranta.

The skills of the population are aligned with other Finnish cities, but slightly better than the Finnish average (which is comparatively high from an international perspective). Over 20% of Turku's residents have a tertiary degree (Figure 12.13).

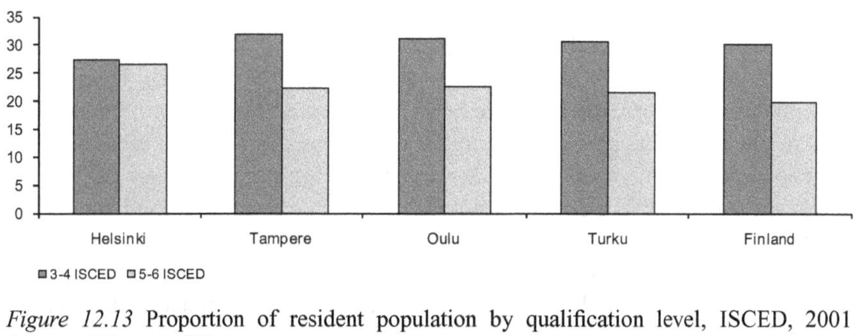

Figure 12.13 Proportion of resident population by qualification level, ISCED, 2001 (source: Urban Audit)

Accessibility

Turku is located far away from the main European markets. However, within Finland, its location is rather central: the main conglomerations in the south of Finland (where the majority of Finns live) can be easily reached by car or rail.

There are international transport connections: the city has a port and an international airport (with regular international connections to Stockholm and Copenhagen, and frequent links to Helsinki and some other Finnish cities). Moreover, the Baltic region increasingly emerges as a relevant and integrated market (Helsinki, Stockholm, Copenhagen, but also new economies to the east), and the Turku region has a strong advantage in this context. The region has emerged as an important logistics platform (top multimodal connections in Finland), which supports its industrial base. Moreover, transportation fluxes are increasing between Turku and St Petersburg.

The digital infrastructure and wireless communication service is among the best in Finland, providing the region with top digital accessibility, providing easy and efficient access for users and an innovative environment for research.

Quality of life

City attractiveness and quality of life are presently strong in the city of Turku. Although the city does not offer a metropolitan feel, it offers many urban amenities (heritage sites, events, leisure and cultural facilities). Within Finland, Turku is increasingly considered as a pleasant place to live. Lacking the vibrancy and cosmopolitan feel of other international cities, its safety and moderate costs of living are assets, and tend to contribute to lure and attract residents. Also, the surroundings of the city are attractive; Turku is close to the Archipelago, which is a strong tourist destination. Unlike many other European cities with a strong industrial base, Turku does not have the negative image associated with pollution and lack of quality of life.

Factor conditions

In this section we describe the state in Turku of two important regional structural conditions: factor prices such as land and labour, and 'soft factors', such as cooperative endeavours and social capital.

Finland has relatively high labour costs although increases have been moderate in the last decade and have been compensated by gains in productivity. Considering the Baltic area (see Figure 12.14), Finnish labour costs are relatively low compared to Denmark or Sweden, but much higher than newer Baltic EU members (like Lithuania, Poland, Estonia and Latvia). In terms of land and office space costs, although Helsinki is an expensive location, cities like Turku are very well positioned in terms of prime office rents, showing rather moderated values for this indicator and 'value for money' in terms of office space in the Baltic context (see Figure 12.15 and 12.16).

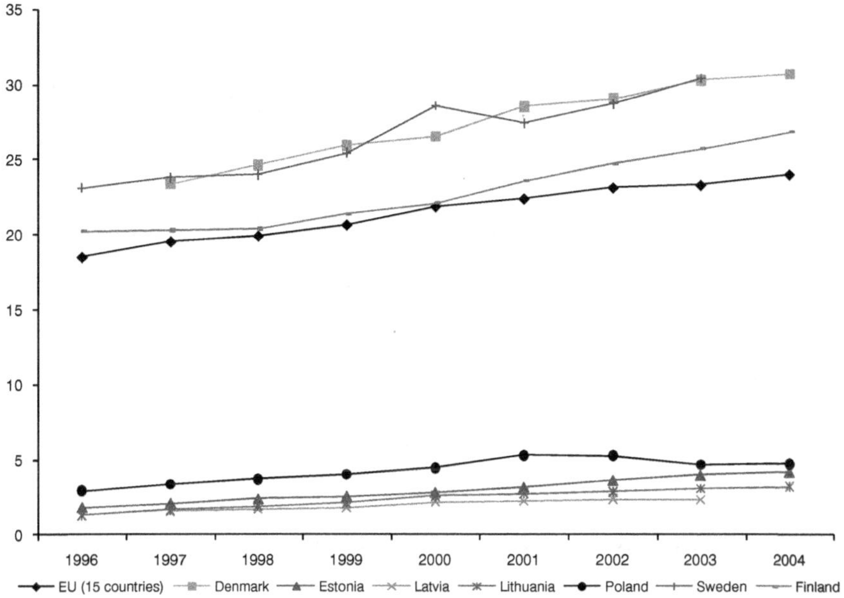

Figure 12.14 Average hourly labour costs, euro, 1996–2004 (source: EUROSTAT)

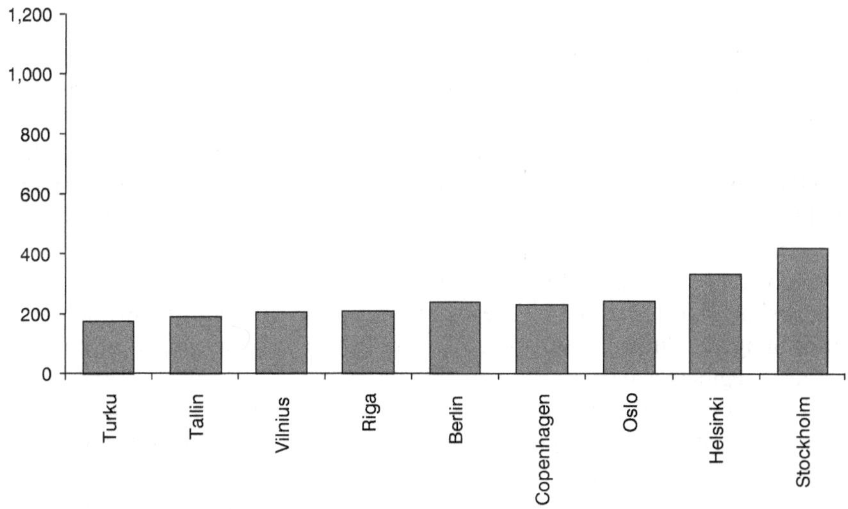

Figure 12.15 Prime office rents in northern Europe, 2006 (source: Turku Åbo, 2006)

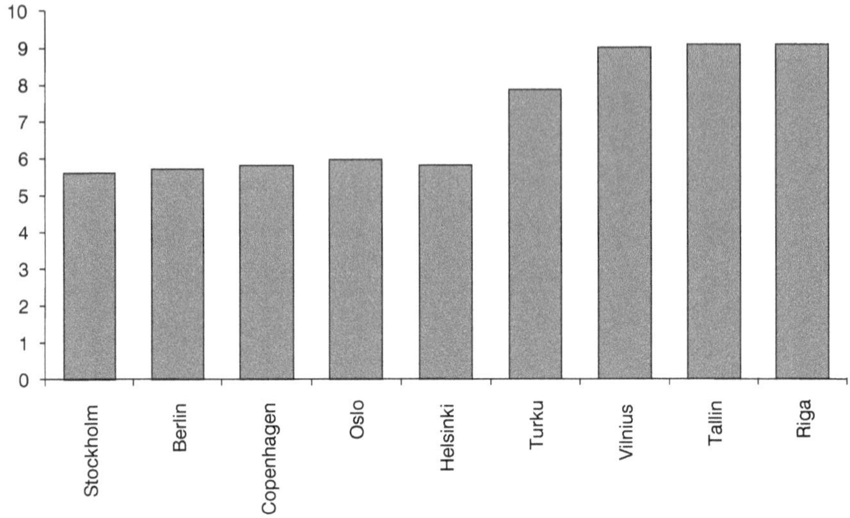

Figure 12.16 Prime office yields in northern Europe, 2006 (source: Turku Åbo, 2006)

Regarding soft factors, the adverse natural conditions propelled the Finnish towards innovation and collaborative behaviour. In the 1990s, Finland experienced an economic crisis, exacerbated by the fall of the Berlin wall and the collapse of the Russian market (which was the main export market for Finland). As a policy response, the country developed a national development strategy, with a strong focus on technology and ICT developments. Persistence, parsimony and honesty are usually highlighted as important characteristics of the Finns, and these traits are usually correlated with Finland's economic development level as a whole. The Finnish case is also shown as one of the prime examples of policy integration at different levels, especially in innovation policy. The Centres of Expertise programme (since 1994), has been shown to reinforce the national innovation system with strong regional linkages. Although led by the national government, it is designed and executed by means of strong regional partnerships between different agents (firms, universities, R&D institutes) with the support of politics and local/regional players.

In this context, Turku's social capital and institutional setting appears to be quite strong, as is the willingness of public and private institutions (at the local, regional and national levels) to cooperate. There are diverse examples of tacit/informal institutions, which tend to work for cooperative behaviours even without formalisation.

12.4 The shipbuilding cluster in Turku

The shipbuilding sector in Turku and its value chain are part of a complex cluster of interrelated firms, a large majority of which are located in the urban region,

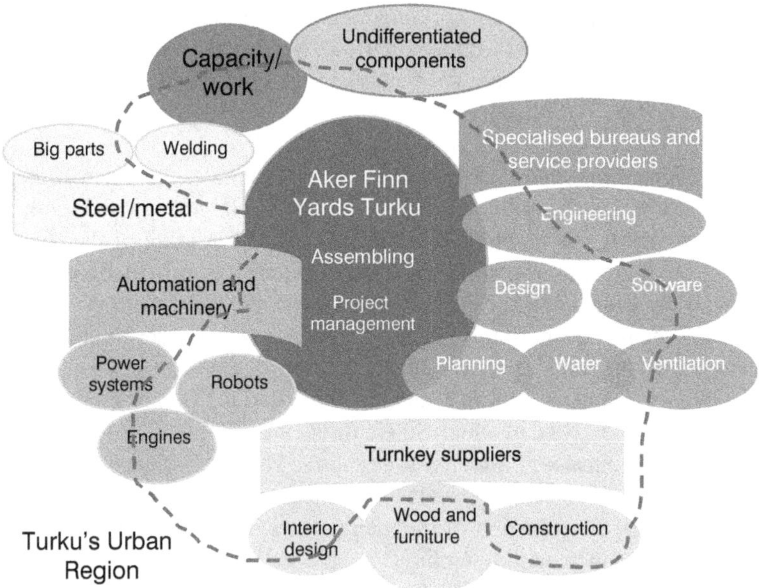

Figure 12.17 Shipbuilding macro cluster in Turku's urban region (source: own elaboration)

but have global networks. As depicted in Figure 12.17, the 'spider in the web', undoubtedly, is Aker Yards.

Following a global trend of mergers and acquisitions, Aker Yards merged with Kvaerner Masa Yards, today under the Aker Yards name. It has yards in several cities, but the yard in Turku is the largest. Aker used to have its headquarters in Helsinki. However, in early 2006, the headquarters moved to Turku, the company's largest and most important Finnish operations centre. Proximity to operations was a key argument for this move. This is curious in light of the international trend of spatial separation of management and production.

Turku's shipyard is one of the largest, most modern and innovative in Europe, producing breakthrough ships. Recent investments in the yard and development programmes have contributed to this. Today, the shipyard has a new singular building dock, equipped with a gantry crane capable of lifting 600 tonnes. It has increased the automation and indoor production facilities, including halls for steel production and block assembly. The yard has had periods of decline, the last one at the beginning of this century, when many jobs were cut. The present order book is impressive, however, with orders from Royal Caribbean International for 18 new ships (including the Genesis series). The total value of the order book of Aker Yards in Finland amounts to €4.3 billion. Recently, Turku's shipyard finished the first luxury cruise of the Freedom series (Freedom of the Seas), the largest and most innovative cruise ship ever built, with a capacity for 4,000 passengers and a crew of 1,500, an absolute ‘floating city’. Today, Aker directly employs roughly 3,700 persons in Finland (2,700 working in manufacturing, 460 in administration and 440 in management), half of them in Turku (interview source).

Aker’s core competences are concentrated and specialised. It focuses on the following activities:

- *Assembling, final outfit and testing*. The yard assembles the outsourced parts and ready-to-install/turnkey components, and the large parts and the hull of the ship, previously assembled, towards the final outfitting of the ship. The yard is an integrator of other organisations’ work (it is estimated that roughly 85% of the value added of the ship comes from the network of suppliers). In fact, a large part of the ship’s value added is increasingly developed by a vast network of suppliers and subcontracted firms, estimated to be roughly 500 companies (mainly small and medium firms that are highly innovative and knowledge intensive), of which 56 are turnkey suppliers.
- *Project management and coordination*. This task encompasses the coordination of the entire value chain. It involves the technical and organisational coordination of the many actors involved in the shipbuilding process. Aker increasingly employs people with specialised and high-level skills (planning, engineering, finance, logistics, ICT). It is evident that information management is an increasingly crucial ability for Aker.

Suppliers

The suppliers in the cluster can be classified in broad groups:

- *Automation and machinery suppliers*. These companies produce and supply customised parts for the ship, for instance engines and propulsion drives, power systems or other high-tech components, but also robots and production lines for the yard and for the shipbuilding process;
- *Specialised bureaus and service providers*. This group includes a vast range of specialised and ‘creative’ services, required for different stages of the value chain and ship’s production and assembling; it includes consultancy, planning, R&D, engineering and design bureaus, but also software design firms, as well as other providers, namely in terms of ventilation and water systems, crucial for the safety and quality of the ship and materials;
- *Turnkey suppliers*. A growing number of suppliers are ready-to-install suppliers, providing very specific and specialised parts of the ship, needed to satisfy the needs of increasing innovation, safety, comfort and leisure in the ship (e.g. cabins and bathroom units, furniture, restaurants, swimming pools, theatres, gambling rooms). Some of these firms are unique in the world, delivering finished components or finishing it in the yard;
- *Other components and work capacity*, not directly produced/employed by the shipyard, are needed to finish the ship. Besides diverse, undifferentiated components, the ships need a large amount of steel and metal work, namely

> related with the large parts of the ship (including the hull) and welders. In order to make capacity more flexible, some specialised welding and metal firms are also subcontracted, with qualified welders working directly in the yard. Moreover, capacity and undifferentiated work are also usually subcontracted, rendering the yard's structure more flexible.

Also other players, like knowledge institutes inside and outside Turku play an important role in the cluster's innovation dynamics.

The regional dimension

What is the role of proximity in this cluster, and how has it changed over time with globalisation? In this shipbuilding cluster, we found that proximity matters but to a varying degree, depending on the type of interaction.

As the scheme above suggests, a large majority of the firms and subcontractors that contribute to building a ship are actually found within the region. This does not mean that all suppliers and players (firms, universities) are found within Turku's boundaries: some of them are located in other cities nearby, like Tampere or Helsinki (e.g. design and engineering bureaus, logistics, R&D centres, Aker Yard's shared services), or even in other European countries (e.g. a few but important turnkey suppliers). Nevertheless, in this specific case, there is no doubt that proximity matters. The Turku 'shipbuilding network' is estimated to employ directly roughly 20,000 workers, 5,000 of them working in the yard (both blue and white collar).

Considering manufactured components, we found that the main suppliers and original equipment manufacturers of automation and machinery are located in the Turku region, as well as the main turnkey suppliers and specialised providers. This relation is also very strong for service providers and specialised bureaus, as well as for trade and logistics: the vast majority is found in Turku's service base. Two major reasons seem to underpin the 'proximity matters' reality for the value chain's services and manufactured components. First, the technological sophistication of the cluster product increases the need for frequent contact. Cultural proximity and face-to-face contacts facilitate this interaction. Second, proximity facilitates the interactions between the yard and its suppliers, contributing to an increased innovation pace and knowledge transfer. This is important to face the trend of reduction in the number of main suppliers but with stronger strategic relations. This holds particularly for suppliers of high-tech and more specific components. Representatives at Aker Yards speak of an 'extended enterprise' concept (yard/suppliers and supplier/supplier) in which Aker and its main suppliers form one 'virtual' enterprise.

It is important to note that the key suppliers are not only active in Turku's cluster: many of them have a strong international presence and are active in different value chains (e.g. the oil industry), and successfully export their expertise and products to other regions in the world (in particular in Asia's emerging markets). Several of the interviewed companies have opened sales offices in China to serve the local shipbuilding industry; others engaged in strategic partnerships. Some suppliers are subsidiaries of global companies with presence around the world. This gives them more resources, a larger corporate knowledge pool and gateways to new markets. Others have remained independent but have developed an international orientation, by exporting their products or by establishing partnerships with foreign partners. The trend of globalisation is thus evident not only in the yard segment but also in the supplier base.

When looking at R&D and innovation partners and at the delivery of specific skills and services, the functional region is larger than just Turku. There is substantial cooperation with universities and research institutes outside the region, mainly in Helsinki and Tampere. Companies consider those two cities as main 'knowledge bases' for their work, and to provide qualified staff and applied research. Aker Yards works closely with the Helsinki University of Technology, the only university in the country with a dedicated shipbuilding unit. Several firms indicated the importance of Tampere as a knowledge centre; the technical university is highly valued and several companies have strategic relations with it.

For more undifferentiated components and raw materials (like steel), as well as capacity supply (work), proximity becomes even less important; many materials and cheap labour increasingly come from lower wage countries in the Baltic region, like Poland or Latvia. Many workers from those countries stay in temporary housing in Turku. Several companies stressed that it is increasingly difficult to get skilled Finnish manual workers (especially welders) because the image of the sector is poor, and the national government encourages university education rather than mid-level technical education.

When looking at the clients of the cluster, of the shipyard and the other (specialised) firms and bureaus, we can also observe a strong degree of internationalisation. The yard's order book is logically international (the main client is Royal Caribbean International, USA), but other cluster firms are also well positioned in international markets.

12.5 Governance of the cluster

In this section, we will deal with the nature of interactions in the cluster. Also, we describe and discuss formal and informal platforms of knowledge exchange, as well as policy interventions by government agencies at different levels.

Coordination mechanisms in the cluster

One main feature of Turku is the strong tacit cooperation and institutionalised social capital within the shipbuilding players (e.g. yard, subcontractors). The yard and the main suppliers are active in competitive international markets, but at the same time they act in strong coordination at the local and regional levels by sharing resources and interacting towards a better and more competitive final product. Although the cooperative tradition in Finland as a whole explains part of the phenomenon, new market, network and firm dynamics play a crucial role.

The presence of a key player (Aker Yards) and its strategic outward-orientated policy certainly contributes to this tacit cooperation mechanism. Aker Yard's mission is 'to offer the right products at the right time – and at the right price. Based on our technological expertise and broad project experience … we have an excellent network of subcontractors and cooperation partners, and can take on demanding projects, regardless of size and complexity' (Aker Yards, 2005, p.3). In order to fulfil these aims, the preservation of a strong network is vital, and this actually tends to happen in the local and regional context. Aker Yards plays the 'spider in the web' role, organising the subcontractors and other institutions not only through contracts, but often tacitly. The presence of the strong network of firms organised around the yard helps to meet the market's demand for flexibility, efficiency and quality. In this respect, Turku markedly differs from other European yards with a dominant tradition of internalising different functions and parts of the value chain. The partnership endeavours and cooperation, where Turku's shipbuilding sector scores high, are ultimately led by market demand and by an understanding of the value added of cooperation. This reduces the need for formal or government-sponsored institutions to support cooperation and links (although they do exist).

We observed that relationships in the regional shipbuilding network are very stable; many firms have worked strategically together for many years. The presence of trust and stable relationships reduces uncertainly and facilitates knowledge transfers and joint innovation. Nevertheless, for the companies in the cluster, globalisation and rapid technological advancements increase the need to participate in different networks (national, international). Many players in the cluster are active in several networks, for instance in European research projects and in businesses around the world. The Baltic region is emerging as a relevant region (in between the 'global' and local' scales) for networking and cooperation, for skills, R&D and capacity. Some of our interviewees pointed out that the cluster should become even more globally networked (in terms of clients and knowledge base) and try to diversify into other markets (i.e. offshore and oil), in order to become less dependent on the shipyard's fortunes. However, with the current order book for the shipyard full, companies are tempted to dedicate all their resources to produce for the yard, and postpone further internationalisation and diversification efforts. In the long run, this may hurt the cluster as a whole.

In terms of ownership and decision-making of the industry, a large majority of the subcontractors are Finnish-owned independent firms; despite the yard restructuring, Aker Yards Turku maintains decision-making power, although the headquarters of the entire group are located in Oslo.

Coordinating actors

Notwithstanding the power of 'self-governance' between the main actors in the cluster, there are also several institutions and policies aimed to further improve the sectors' long-term competitiveness.

The importance of the shipbuilding sector can also be depicted by the large number of discussion platforms and forums that exist (shipbuilding, metal, marine industries) on different levels. At the regional level, the past cooperation and strategic relationships mean the actors know each other well, and can avoid the existence of highly formalised platforms. Sectoral platforms of cluster actors exist predominantly at the national level. The actors gather regularly for government advising and policy making. The Association of Finnish Maritime Industries (a branch of the Association of Technology Industries of Finland) is an important one. Representing shipyards, marine equipment manufacturers, turnkey suppliers and ship designers, it coordinates cooperation in industrial and economic policy, promotes sectoral networking and articulation for research and product development, deals with the application of the EU shipbuilding policy in Finland, and represents the entire marine technology sector in international associations. The yard and some of its suppliers are presently defining the Maritime Industries Strategic Agenda and preparing joint actions for 'creating better ships' at the national level. The presence of Aker Yards in Finland as a large player helps to structure the network in Finland towards internationalisation.

National policy actively promotes and supports innovation and technology sectors, and innovation funding is mainly national. Nevertheless, Finland's innovation policy has a strong regional dimension. It aims to support regional strengths through its Centres of Expertise Programme. This programme, more than a funding scheme, has been contributing towards the actors' organisation (firms, authorities, R&D and universities), with leverage effects in regional innovation systems (Lähteenmäki-Smith *et al.*, 2003). The Southwest of Finland Centre of Expertise targets the materials surface technology, among other technologies, that is used in metal and shipbuilding industries. An important development is the creation of the national Meritech centre in Turku, a research and training centre in maritime technology. The centre aims to combine the strengths, knowledge and resources of different actors (firms, knowledge institutions, local, regional and national government), develop intra- and inter-regional cooperation for the maritime sector, at different scales. There are plans to include a 'shipbuilding laboratory' to develop research for the sector and reinforce Turku as a national centre for shipbuilding excellence (in cooperation with other research facilities, like in the fields of ICT). The centre will support a Europe-wide network of experts and firms in the sector and showcase joint programme development and R&D, all towards keeping the European lead in shipbuilding design and technology. ICT Turku has strong competence of European R&D programmes and projects. The city of Turku can provide this experience as the basis for the further development of the extended enterprise for shipbuilding.

Another initiative is the Merike programme, supported by national policy (National Technology Agency, TEKES). It is devoted to developing maritime technology issues between different actors, with several projects coordinated from Turku. It contributes to knowledge networks of large companies and SMEs. The project, composed of several subprojects, has contributed towards the joint development of innovative and creative solutions, namely for large cruise lines (design, lifecycle support, environmental issues).

The Machine Technology Centre Turku provides up-to-date training for the metal industry (closely related to shipbuilding), including basic training and on-the-job training. It results from a joint initiative of the regional schools and firms – south-west Finland technology industries, Turku vocational adult education centre, Turku vocational institute and Turku Polytechnic – towards the practical training and continuing studies of the workers, as well as the support for business-specific necessities, including machinery outsourcing (City of Turku, 2006).

Local and regional level authorities (city of Turku and municipal institutes like the Turku Regional Development Centre) also consider the sector as a strategic one. Previously neglected in favour of other 'fancy' sectors (ICT and biotech), the local administration recognised the strong potential of metal and maritime industries in the region, not only for their contribution to employment and added value, but also to attract residents and jobs. Together with the city of Turku and other municipalities in the region, the Regional Development Centre internationally promotes the economy in general and specific sectors in particular (including the shipbuilding sector, together with ICT and biotech). One important sector is logistics. The aim is to develop Turku as an important hub in Northern Europe. Pilot Turku Ltd is a non-profit organisation that links different interested actors and gives assistance to firms to maximise its operations, in a context where timings and costs are essential. Pilot Turku Ltd, Regional Development Centre and Turku Science Park are all active in promoting the region through fairs and investment attraction, as well as business partnerships in the Baltic region. Moreover, they are developing specific endeavours in the shipbuilding sector, linking different actors (business, academia and public authorities), studying the sector's needs and developments and promoting better environment for business operations, including supporting entrepreneurship in connection with Turku's Science Park.

12.6 Conclusions and perspectives

Conclusions

Turku's shipbuilding cluster is a textbook example of a traditional manufacturing localized industry that has made the transition to a knowledge-intensive and globally competitive industry, based on high-level competences organised in a well-functioning network. In a heavily competitive environment, the cluster has managed to become a world leader in a niche market that is characterised by high product complexity. The skill of organising the network inside and outside the region is a key factor that explains the flexibility and innovation of the cluster.

The cluster actors have captured several opportunities offered by globalisation. First, the suppliers in the cluster are able to tap into the global knowledge base which makes them technologically 'state-of-the-art'. As shipbuilding is a multidisciplinary exercise, this is a key ability. Many suppliers have become part of multinational firms, and tap into the wide corporate knowledge base; others have developed particular competences to use new technologies and apply them to solve problems and provide practical solutions that make ships better. Second, the cluster benefits from emerging markets, mainly in Russia and Asia. It is increasingly able to 'sell' its competences, e.g. in the growing shipbuilding industry in China and the offshore oil industry in Russia. Third, globalisation enables the cluster actors to import raw materials and basic parts from emerging countries (mainly Eastern Europe) at very low prices. And finally, the cluster benefits from the internationalisation of labour flows: increasingly, the cluster depends on foreign (mainly manual) workers from the Baltic area. Thus, the Turku case study indicates that in the context of globalisation, networking, partnerships and markets are gaining prominence. Also lesser known regions are becoming increasingly important, both for skills and capacity sharing, in this case, the Baltic Sea area.

The case study confirms our hypothesis that the success of a manufacturing cluster depends on the quality of regional assets. The knowledge base and skills available in the region were considered one of the main sources of a firm's success. It is important to note that the 'relevant region' for knowledge and technology is not the Turku region but rather the city-network in the south of Finland (comprising Turku, Helsinki and Tampere).

Shipbuilding is a highly cyclical industry. However, the presence of an increasingly diverse and balanced economic base explains the region's resistance to ups and downs in the last ten years. The service base has been developing, but the manufacturing sector maintained its relevance for the economy, responsible for very large shares of the urban region's value added. Within this manufacturing base, sectors like ICT, biotech or high technology machinery gained prominence in the region and, together with the knowledge base, are sources of new competences in the region. Given the interdisciplinary nature of shipbuilding, the broad mix of industries and competences in the region are a core asset.

The fast growing ICT sector deserves special attention as contributor to the cluster, in two respects. First, there is the increased role of ICT in shipbuilding. Nowadays, ICTs are an integrated part of virtually every part of the ship, and the importance will grow in future. For building a ship, ICT expertise is thus indispensable. Another feature is the role of ICT in managing the production process. We observed an increasing interdependence in the cluster, reflected in the move towards an integrated 'virtual enterprise', in which companies will have access to each other's planning and inventory systems. This tendency will increase the need for sophisticated ICT solutions and expertise. The local ICT sector could bring an important contribution in this respect.

Finland generally performs less well when assessing issues like factor prices and accessibility. In fact, high taxation and labour costs hamper some FDI, as do the small internal markets and geographical distance to the centre of Europe. However, Turku still has some relative advantages in these factors, namely the lower office space and land prices, as well as good logistics conditions and on-going policies toward their improvement. Unlike in many other manufacturing sectors, for building cruise vessels proximity to consumer markets is not a factor at all; thus, for this activity the relatively peripheral position of Turku vis-à-vis large consumer markets does not pose any problems.

Moreover, the cultural and soft factors present in Turku have a strong influence in explaining the shipbuilding industry's permanence and success. The social capital, outward-looking attitude, networking and cooperative tradition between different actors (public-public, public-private, private-private) are in fact decisive in facing the present competitive challenges of the sector. Governance links in the sector (firms, academia, authorities) are to a large extent tacit and rooted, while also supported by platforms (not local) and different policy tiers. A main governance challenge for the future will be the continuity of the 'extended enterprise' model and the strategic links between the yard and the main suppliers, in order to meet capacity and develop new competitive edges. Another challenge is to cooperate more strategically with actors outside the Turku region, as this case study shows that 'knowledge links' in the shipbuilding cluster extend far beyond Turku.

The city and the Turku Regional Development Centre may have been late in recognising the strategic importance of the shipbuilding industry; they now take a positive role in improving the 'basics' of the local economy. Furthermore, these organisations, and also ICT Turku, play a positive role in supporting and fostering strategic networks, not only within the Turku region but also in the Baltic and even further abroad; one example is the organisation of business visits to firms or cities abroad in order to enable local entrepreneurs to develop business relations in this region.

Perspectives

The future looks bright for manufacturing in Turku, in particular for the shipbuilding industry and related branches: they have so far been able to respond effectively to the main challenges posed by globalisation. The shipbuilding cluster and its 'associated' firms are quite strong (locally and regionally), not only because they are constantly looking for new breakthroughs, through a strong network with the local yard, but also because they are increasingly international (in developed and emergent economies), and are gaining new skills.

However, the current full order book of the shipyard is putting strong pressure on the local and regional subcontractors and suppliers, and constrains them in their capacity to participate more actively in international markets (trade-off: supply locally or supply internationally) or diversify their client base. It should be taken into account that in the long run, the local and regional firms' internationalisation is an important asset for regional and cluster competitiveness; it may reduce the dependence of the region on the yard. This is important, as the cruise market is highly volatile and the heavy dependence on one company (Aker) is risky.

Another challenge is the lack of skilled manual workers. These people are crucial for the cluster. Increasingly, the vacancies are filled by people from Baltic states and Eastern Europe, but companies feel that this is not a long-term solution, with growing welfare and wage levels in these countries. On top of that, the image of the sector is rather poor: many Finnish youths are reluctant to work in a yard or a metal firm and prefer office jobs. The current policy of the Finnish government to increase the percentage of youths with an academic degree may further reduce the supply of skilled technicians, and may indirectly (and unwittingly) provide the signal that not having an academic degree is a major deficit.

What will be the impact of emerging economies on Turku's cluster? Emerging markets (China, Eastern Europe) are considered by the local firms much more as an opportunity to export know-how and competences. On the knowledge side, firms (albeit not all of them) have connected to global knowledge networks. On the supply side, the new markets enable cluster firms to import cheap semi-finished parts; this process reduces the need for manual work and improves efficiency of operations (it also reinforces the importance of efficient logistics). Thus, in general, globalisation requires new networking skills: the engagement and maintenance of networks with suppliers, knowledge providers and clients all over the world has become a crucial ability. Up to now many firms in the cluster have been shown to have these skills.

As long as the technical and organisational expertise in the cluster is superior and markets remain strong, globalisation is a blessing for the cluster. In the longer run, Asian players could be able to produce cruise vessels as well, at the expense of Turku's cluster. One reason for worry is the growing global overcapacity in the sector; this could be an incentive for some yards to explore new markets, including cruise ships, and produce them for below-margin prices. Most experts agree that the competitors, in the short run, cannot match the level of technical and organisational sophistication, but this may change over time. The pressure is high, and only a continuing process of innovation and cost reduction will secure the cluster's position.

Bibliography

Aker Yards (2005) *Preferred for innovation*, Corporate magazine, September.

Aker Yards (n.d.) Available online at: www.akeryards.com (accessed 26 June 2006).

Centres of Expertise Finland (n.d.) Available online at: www.oske.net/in_english (accessed 25 June 2006).

City of Turku (2006) *Creative Turku/SW Finland*, Annual report 2006.

Deltamarin (2006) *Corporate presentation,* March 2006.

Employment and Economic Development Centre of Southwest Finland (n.d.) *Statistics on employment, 1995–2005.*

European Comission (n.d.) Available online at: http://ec.europa.eu/ (accessed 26 June 2006).

Eurostat (n.d.) Available online at: http://ec.europe.eu/eurostat (accessed 26 June 2006).

Lähteenmäki-Smith, K. *et al.* (2003) *Regional competitiveness through expertise, mid term evaluation of the Finnish Centres of Expertise 1999–2002*, Helsinki: Ministry of the Interior.

Statistics Finland (n.d.) Available online at: www.stat.fi (accessed 17 May 2006).

Turku Åbo (2006) *Turku region real estate market review. real estate department*, March 2006.

Turku's Chamber of Commerce (2006) *Economic structures and outlook in southwest Finland, chamber of commerce Turku's economic review.* Available online at: www.turku.chamber.fi/english/ (accessed 23 May 2006).

Turku Science Park (n.d.) Available online at: www.turkusciencepark.com (accessed 26 April 2006).

Urban Audit (n.d.) Available online at: www.urbanaudit.org (accessed 26 June 2006).

13 Synthesis and conclusions

13.1 Introduction

This chapter contains a synthesis of the case studies described in this book. We draw general conclusions and lessons from the experiences in our case cities.

In section 13.2, we reflect on the development towards a network economy. It is meaningless to consider manufacturing as a 'stand-alone' activity: it has become part of networks and value chains that encompass urban regions, nations, continents or even the world. Based on a number of case studies, we show how companies engage in networks to develop new products (knowledge networks), and to produce and sell these products (production networks). Also, we look at interaction mechanisms: how do 'knowledge networks' relate to 'production networks'? How are these networks developed in space? We observe a move towards a 'project economy' in which companies organise themselves around industrial projects, in ever-changing network configurations, at varying geographical scales.

Next, in section 13.3, we argue that the industrial economy is moving towards a network economy in which projects play a key role. We analyse the role of proximity in this new network economy and we give insights how cities and regions can attract projects.

In section 13.4, we discuss the consequences of these trends for cities and regions. How are they affected by some fundamental changes in manufacturing industries? In this section, we make a distinction between five different city types: 1) 'traditional' industrial cities: Dortmund, Porto, Rotterdam, 2) advanced diversified cities: Munich, Paris, 3) post-communist industrial cities: Ostrava, 4) smaller industry towns: Eindhoven, Turku, and 5) emerging giants: Shanghai, São Paulo. Each city type has a different mix of 'foundations' (defined in this study as the economic base, knowledge base, quality of life, accessibility and factor conditions), and in each type of city, manufacturing plays a different role. A key question in this section is how can cities and regions prepare themselves to accommodate an increasingly knowledge-intensive manufacturing industry?

Finally, in section 13.5, we discuss a number of challenges and policy implications for cities. For each type of city, we address a number of critical issues: how important is manufacturing in the knowledge-based urban economy?

What challenges do city regions face? What are the policy implications on the European, national, regional and local level?

13.2 Manufacturing is part of the network economy

Our study confirms that manufacturing industries have undergone a process of fundamental change, with far reaching consequences for regions. The changes are so pronounced and have taken place in such a relatively short period of time, that it is justified to speak of a neo-industrial revolution. Internally, many industrial firms have redesigned themselves into connected sets of semi-independent business units. Externally, they engage in a variety of networks with different types of partners, in order to pool resources, to benefit from each other's capabilities, to share risks and to 'tap' from alternative knowledge sources. Knowledge networks and production networks are organised in new ways.

Manufacturing activities (plants, factories) have become embedded in knowledge networks and linked to service activities. Thus, we cannot understand the dynamics of manufacturing without studying the dynamics of the networks in which they are embedded.

In our study, we assume that urban economic development is strongly influenced by the dynamics of these networks. In this section, we present conclusions and findings about emerging industrial networks, and the development of these networks in space. To better understand these dynamics, in our cases we use the distinction between two types of networks: production networks (exploitation) and knowledge networks (exploration).

Pre-competitive networks

'Pre-competitive' networking is increasingly popular across industries

Our cases show that industrial companies increasingly engage in partnerships in the pre-competitive stages. Networking is an option to share risks and resources in a situation where outcomes are unsure. In our case studies, we identified several types of pre-competitive cooperation: cooperation projects with universities and academic institutes (often funded through European research framework programmes), business venturing and facility sharing.

Companies still tend to have their 'main academic partners' in their 'home' region

Pre-competitive partnerships often take the form of cooperation between corporate R&D departments and universities or other academic institutes. There is an evident mutual interest: the company can benefit from the research resources of the university, and the university gains additional financial resources to conduct research and produce publications.

Most of the companies in our study tend to have their 'main academic partners' in the original home region of the firms, despite their increasing global presence.

For instance, BMW mainly works with the Technical University of Munich; car producers SAIC and SMA (Shanghai) cooperate with two leading Shanghai universities Tongji and Jiaotong; in Paris, PSA and Renault have a number of partnerships with Paris-based public research institutes.

In several cases, knowledge-based partnerships stretch beyond the urban region. In the case of Turku, we found that for Aker Yards (the leading shipbuilding company), the universities of Tampere and Helsinki are more relevant than the local university in Turku, and the distance is not considered as a main barrier. Unilever (the food multinational that dominates Rotterdam's food industry) has its main research department in the Rotterdam region, but its academic partners are concentrated in Wageningen (a 100km drive), and some other Dutch university cities. There are hardly any links with academic institutions in the Rotterdam region. In Porto, the most dense and relevant networks for the metal-electro and machine building sectors stretch towards a large and diffuse urban-industrial region encompassing the cities of Guimarães, Braga and Aveiro.

In some cases, partner universities open 'branch plants' at the site of companies

The link between industrial companies and universities is becoming more intense, in all our case studies. In some cases, universities even open establishments near companies. One example is Audi, the German luxury carmaker. Its main academic partner, Munich Technical University, opened an establishment near the company's R&D department in Ingolstadt, some 80km from Munich. In Shanghai, Tongji university opened a facility at the main site of SMA (a carmaker), some 50km south of downtown Shanghai.

We may conclude that for multinational industrial companies, the quality of the regional and national public knowledge infrastructure is an increasingly valuable asset. For these types of partnerships, proximity matters, in the sense that frequent contacts can be maintained. An additional important explanatory factor for proximity is the personal contact between engineers in firms and university professors. Also, the national/regional bias in pre-competitive cooperation can be explained by the availability of large national and regional innovation and incentive programmes aimed to promote pre-competitive partnerships.

The Porto case shows that pre-competitive cooperation is relevant not only for multinationals or high-tech companies. During the 1990s, the regional footwear sector was hit by new competition from low-cost countries. The industry realised that it needed to shift from mass production towards the manufacturing of small series in specific niches, together with the development of own-brand products. However, the sector did not have the technology to shift to the new market needs. To address this problem, the Footwear Technological Centre, together with the Sector Association, set up a large project (FATEC – Footwear Active Technology) in cooperation with five R&D institutes, five machine builders and nine footwear firms as partners. The project delivered a diverse range of new high-tech equipments and machinery to manufacture small series of footwear: prototyping, laser cutting, logistics, automatic painting, robots, weaver technology, ICT solutions, etc.

Thanks to these outputs, several Portuguese footwear firms reinvented themselves and became competitive again, and the newly developed machinery has become an export product.

Networks for product development

Networks for product development and design are different from 'pre-competitive' networks in a number of respects. The commercial stakes for the cooperating partners are higher, and the outcome is more predictable, as the cooperation should lead to the creation of a new product with market potential. This asks for more detailed arrangements between the partners about the sharing of costs, future revenues, and property rights. Partners typically make arrangements about R&D and production as part of one deal.

Physical proximity is often not relevant for partner selection

Does distance matter in partnerships for product development? We found that in the selection of partners, physical proximity is not a dominant factor. Rather, 'institutional proximity' plays an important role: When a firm is part of an industrial alliance, there is often a preference for co-development with alliance partners. For instance, in developing car platforms and engines, Renault works almost exclusively with Nissan, its alliance partner. In the absence of alliances, partners are selected according to complementarities in competences and/or synergies in exploitation. PSA, for instance – not tied to any alliance – opts for a variety of partners, in various types of projects. For its new MPV platform, it works with Fiat; with BMW, it developed a new diesel engine to be used in small Peugeot cars and in BMWs Mini brand.

'Temporary proximity' or 'virtual proximity' are created to facilitate communication

The fact that companies often chose partners from far away does not mean, however, that proximity is irrelevant. On the contrary: joint product development is impossible without frequent interaction and tacit knowledge exchange. To achieve this, some companies create '*temporary physical proximity*', for instance by creating a joint team of engineers stationed at one location. We found many examples in our cases. In the BMW-PSA diesel engine project, one joint project team was created, based in BMW's technology centre in Munich. In São Paulo, a relevant part of the flex-fuel engine technology was developed through temporary visits of Brazilian engineers to other corporate R&D departments in Europe and the USA. Also DOC in Dortmund (a research centre owned by ThyssenKrupp and the Fraunhofer Institute) works in project teams with engineers from the Fraunhofer Institute, ThyssenKrupp and its customers (like BMW). In Turku, specialised Italian firms locate in temporary facilities in the shipyard in order to design and jointly assemble turnkey interior parts of cruise ships. In Porto, some machine building firms work with foreign laser technology firms, who test and adapt their solutions 'in situ'.

Alternatively, companies can create 'virtual proximity' to facilitate interaction over long distances. The Renault-Nissan alliance is a highly interesting case. Using highly sophisticated ICT systems, these firms manage to achieve co-development over a very large distance.

This indicates that the industrial knowledge economy is becoming a 'project' economy, in which multidisciplinary teams from different companies cooperate intensely towards new solutions, on a temporary basis.

Suppliers are increasingly important partners in product development

We found confirmation that 'main suppliers', or 'system suppliers' increasingly become an integrated part of the global knowledge networks of OEMs. Many of them have become multinational industrial conglomerates, sometimes specialised in one specific sector (Valeo, Knorr Bremse and Faurecia specialise in automotive systems), or with more broad competences (Bosch, Magneti Marelli, Siemens). They have a multitude of establishments (R&D and production locations) all over the world. Their knowledge intensity has become very high – witness their enormous R&D spending. In Paris, the share of suppliers in total automotive R&D spending increased from 1.8% in 1992 to 6.2% in 2000. Main suppliers create innovations (in-house or in-networks) and try to 'sell' them to OEMs, or alternatively, develop innovations in cooperation with OEMs in project teams. Due to their unique competences, some suppliers have gained power in the supply chain: rather than being 'exploited' or being in a subordinate position, they cooperate on an equal basis with OEMs.

In many of our cases, we observed that suppliers have opened R&D establishments near the R&D sites of their client, the OEM. In the innovation process, the cooperation between OEMs and suppliers is people-orientated, highly interactive and unpredictable; it is the 'logic of services'. Partners are selected on the basis of expertise, cost effectiveness and quality rather than spatial proximity. Here also, we see the project economy at work: OEMs and main suppliers create teams that work on a specific product; when it is ready, the team is dissolved. The physical location of the team depends on the situation.

Main suppliers, in turn, heavily rely on a network of lower tier suppliers. In these relationships, we see similar trends: a reduction of the number of sub-suppliers, an increasing involvement of suppliers in development, shorter lead times and a larger share of purchasing from low-cost countries. Valeo, a leading supplier is an example because it introduced the concept of 'VIP suppliers'. These best-in-class suppliers are integrated in projects from phase 0, with group-wide visibility. Valeo expects constant progress and innovation to ensure a high level of quality and competitiveness. VIP suppliers receive larger volume orders as Valeo continues to streamline its supplier base: it wants to reduce the number of suppliers by 25% per year.

Rising energy prices and climate change lead to new coalitions and networks

Industrial companies are deeply affected by two main trends: rising prices of energy and the global discussion about climate change. Governments set stricter rules (for instance on CO_2 emissions of cars), and are developing a wide range of incentives to save energy or promote alternatives for fossil fuels. Industrial firms have an incentive to invest in alternative energies, not only to save costs or comply to new rules but also to boost a 'green image' (which is increasingly important in today's consumer markets). These trends lead to new coalitions and product development networks. In particular in the car industry, we see new partnerships between car producers and firms or institutes specialised in energy solutions.

The most striking evidence comes from Brazil. In São Paulo, the new 'green arguments' associated with a long-standing tradition in the use and production of alternative bio-fuels – namely ethanol – has been leading to a large number of technological innovations in automotive engine systems and other complementary technologies. Nowadays, more than 75% of the cars in São Paulo run on a combination of gasoline and bio-fuels, under 'flex-fuel' systems. The development of these technologies involved OEMs and transnational tier-one suppliers, but increasingly their reach extends to local universities and other suppliers.

Another example is a recently established strategic cooperation in Paris between PSA and a knowledge institute specialised in nuclear energy. In the near future, manufacturing industries will increasingly rely on expertise in alternative energy sources and technologies (fuel cell, biomass, wind/sun/water), as well as technology to reduce emissions.

The dynamics of production networks

After a product has been developed, it is produced and sold. Here, cost efficiency becomes crucial, given pre-defined quality standards. In this 'exploitation stage', a multitude of networks and locational patterns can be observed. The location of production and the morphology of the production networks are far from uniform: determining factors are the volumes produced, the type of product, and firm-specific strategic considerations.

Assembly gradually shifts to low-cost countries, but there are notable exceptions

When deciding upon (re)location of assembly facilities, OEMs will consider unit labour costs, logistics, availability of skilled staff, proximity to markets, image, and sometimes also synergies with research and development.

Our cases confirm the overall trend of a relative reduction of assembly activities in developed countries, and a large growth in emerging economies. One reason is the large market potential in emerging economies, another one is factor costs (sometimes associated with the presence of good engineering skills – like in many Eastern European cities). French car manufacturers have recently opened new plants in Central and Eastern Europe (Renault in Romania, PSA in the Czech Republic) to benefit from lower production costs, to serve new growth markets as well as the saturated West European market. In the automotive industry, Central Europe has become a key location for manufacturing plants in the last decade. To name just a few recent plant openings in the region: Fiat in Bielsko Biala (Poland), Kia in Žilina (Slovakia), PSA-Toyota in Kolín (Czech Republic), Audi in Györ (Hungary), and Hyundai is building a large plant in the Ostrava region (Czech Republic). In their slipstream, suppliers open up establishments as well, and they often create more jobs than the OEMs. Similarly, São Paulo and its region concentrate production facilities of the major global OEMs. Some of them have already been there for almost a century (Volkswagen, GM and Ford), all targeting the fast growing Latin American market.

The gradual shift of production to low-cost countries does not hold in all cases. For car producers in the premium segment (BMW, Audi), the 'Made in Germany' label is one (but certainly not the only) argument to keep assembly in Germany. In the food industry, plants are often located near national and regional consumer markets. This sector is dominated by relatively small players that produce for national (or even local) markets. Also, there are more transport restrictions (especially for fresh products). Multinationals in the food industry (such as Unilever) however tend to rationalise their production activities on a European scale.

Another exception are family businesses with strong regional roots that are willing to keep production in the region. Often the managers of these firms have grown up in the region and are involved in many formal and informal networks. We have found various examples of this in the metal-electro sector in Dortmund and Porto. These family businesses can survive (price) competition from companies in low-cost countries by focusing on high-tech production and on small specialised niche markets. Although these family businesses are often SMEs, there are also larger companies with such strategies. An example is the Dutch VDL Group, which explicitly aims to keep production in the Eindhoven region.

In some industries, assembly is 'outsourced' to contract manufacturers, dedicated companies that specialise in assembly. Unilever, for instance, has outsourced 50% of its production to these types of companies, and also in the electronics industry, this is a common strategy. For example, TomTom (the developer of car navigation devices and software) focuses only on R&D and marketing, while it has sourced out the entire production function to contract manufacturers in low-cost countries. In other industries, such as shipbuilding, machine building and automotive, this strategy is less common. In the automotive industries, most OEMs still consider assembly as their core activity, given the complexity of the process, the functional link with product development and design, and the importance of quality. There are some exceptions, however; BMW has outsourced the production of its X3 model to Magna Steyr, a contract manufacturer in Austria. Volkswagen Buses in Brazil is another example and it is only responsible for brand and quality control. In some other sectors, like in machine building, development and assembly are very much intertwined, and also rely on the provision of after-care services as a core component of the product.

Suppliers are the main manufacturers

In many industries, suppliers have become the main manufacturers. In the car industry and the shipbuilding industry, they typically produce 80% of the value added of the final product. There is no single 'location behaviour' of suppliers. When deciding where to produce, a supplier takes several considerations into account: the location of its clients, the type of component involved, the type of contract it has with its clients (in contracts with suppliers, OEMs set a different time margin for the delivery of supplies to their assembly plants, depending on the type of component and specific logistics of the manufacturing process), the volumes to be delivered, the location of its sub-suppliers, logistics conditions of the site, costs and availability of skilled staff, etc. For some suppliers it makes sense to be located near the assembly plant of its client; others need to deliver, for instance, within 24 hours and have more freedom. 'Follow sourcing' is a common phenomenon: when an OEM opens a plant in a new market, its existing suppliers are often required to open a production facility in the same country or region, to ensure the delivery of high-quality components. For instance, Sungwoo High Tech, a Korean main supplier, opened a big plant in Ostrava to supply car parts to the factories of its clients Hyundai (in the Ostrava region) and KIA (Žilina, Slovakia).

Often, OEMs want to have 'local' suppliers in emerging markets

Many OEMs, when starting assembly operations in emerging markets, devote efforts to building up a local supplier base. This reduces their costs, as local suppliers are much cheaper. Also, they may use these suppliers to supply their factories in developed countries. Several OEMs in our study have set up programmes to raise the manufacturing abilities of firms in the host nation (Renault, PSA). Shanghai is a special case. There, the local government actively promoted the development of a strong local supplier base, based on the long-term vision that a strong supplier base is a necessary condition to build up a full-fledged automotive cluster. During the 1980s and early 1990s, Volkswagen was the only foreign car plant in the city, producing cars in a JV operation with state-owned company SAIC. The local government took a number of initiatives to ensure that local companies could learn basic manufacturing skills from Volkswagen and its German suppliers. The strategy has paid off: with the entrance of China to WTO and the influx of foreign investment in the automotive industry, Shanghai has emerged as a leading automotive cluster in China, and Shanghai-based Chinese suppliers are known for their skills, quality and reliability.

In some sectors, global sourcing 'hollows out' the regional manufacturing base

In the automotive clusters that we analysed, we gain the impression that spatial proximity of suppliers to assembly plants still makes sense but the relevance is decreasing. One of the reasons is the ongoing race for cost efficiency. Renault explicitly aims to have 70% of its French factories supplied from low-cost countries. System suppliers have similar ambitions: Valeo (a system supplier) aims at having 70% of its total purchases in low-cost countries by 2010. This 'hollows out' the local economy.

Many suppliers sell their products to far-away clients. Siemens, for instance, has a factory where it develops and produces different types of sensors for the automotive industry. The plant is located near Ostrava, Czech Republic, but its clients are car manufacturing plants spread all over Europe. Another example is Koninklijke Nedschroef Holding B.V. in the Eindhoven region. This company supplies fasteners for the automotive industry to various OEMs in Western Europe, which also export the fasteners further to overseas plants outside Europe. For light products, distance is not an important factor, but the quality requirements are increasingly tight.

However, in other cases, proximity to assembly is important. Some big car components, like fuel tanks and plastic systems need to be produced and assembled preferably in the same place – this is important because of transport costs, but also because it is a fragile product that might be broken on long journeys. Moreover, when just-in-time is essential, OEMs and key suppliers don't risk being distant. In a highly congested metropolis like São Paulo, many suppliers left the region (or redirected investments) to be closer to other growing OEM locations in Brazil to assure just-in-time deliveries.

In some cases, manufacturing cannot be separated from R&D

We have concluded that in many cases, 'fundamental' R&D is physically and functionally separated from production. There are notable exceptions, however. In Dortmund, for instance, we visited industrial SMEs that produce small batches of highly complex products. The managers explained that production and R&D are very much integrated. Furthermore, products are often tailor made. Also in Turku's shipbuilding cluster, the link between manufacturing and R&D is strong. Each ship is different, and built to order. Product innovations are incremental, and are typically developed in close cooperation with suppliers that are located on the yard's site. Process innovations are also significant for the sector to cut costs and improve quality, and in this respect, the yard works with machine builders from the region. Another example is truck assembler DAF, which mainly uses regional research partners because prototypes need to be tested and engineers need to walk around in the production plant to 'feel the production process'. Moreover, proximity of research partners speeds up the development of new models since there are no cultural or language barriers which might hinder the development of new trucks.

In a context of global competition, the manufacturing firms that we interviewed experience a strong pressure to cut costs and to achieve the most efficient size and operations configuration, catering for flexibility and financial discipline. Thus, many production facilities and functions have been moving to other locations. Aker Yards has yards in Romania and Brazil, the automotive OEMs have plants and suppliers in many different parts of the world, and also metal-electro firms, from large firms to SMEs, are actively developing facilities and sourcing components worldwide. Emerging economies are very attractive partners for production, not only for their still lower factor costs (labour, fiscal incentives, etc.), but also seen as opportunities to supply growing consumer and production markets.

Although outsourcing and off shoring trends are on the rise for many production functions, our research found that the proximity between physical production and other 'softer' functions and services remains of utmost relevance. In the automotive sector, new designs and innovations need to be tested 'in the plant', the creative staff and engineers need to see and touch production, and to interact with other departments for the integrated development of other solutions. This is the case, for example, in the automotive cluster of Munich, where both headquarters and production are located. The Finnish headquarters of Aker Yards was moved from the capital Helsinki to Turku, in order to locate closer to their main production facility. The presence of the yard is also relevant for other high-tech suppliers; they need to test their innovations in the assembling facilities. In the metal-electro sector in Porto, although many components are increasingly bought elsewhere (both undifferentiated and high-tech components), production is strongly linked to development, and the presence of the plant and production is essential for the development of new combinations and innovative machines.

In the automotive industry, R&D does not always follow manufacturing

In the automotive industry, OEMs produce large volumes of 'standardised' but technologically complex products to consumer markets worldwide. In the last decade, carmakers have opened plants in emerging or low-cost countries to serve local markets or benefit from cheaper labour costs or supplies. But often, the plants are just intended to produce existing models with technology developed elsewhere. An interesting example is the strategy of Hyundai, the Korean carmaker. This firm is building a large factory in the Ostrava region, but has its main European R&D and design centre in Germany. In Ostrava, the company does not plan to carry out any development activity: it just considers the region as an interesting location for production, given its central location and relatively cheap labour costs. Germany is considered a good location for R&D, given its large expertise in this field. Similarly, Mitsubishi uses the NedCar plant in the Eindhoven region as a 'regional' production hub for the Western European market, while it has its European R&D centre in the Frankfurt region where it benefits from the good intercontinental accessibility and the strong knowledge base.

But in other cases, we see that R&D follows manufacturing, albeit often in a gradual way. Often, around foreign plants, 'pockets' of R&D are emerging, starting from rather basic plant maintenance and engineering functions. But foreign bases can grow in terms of R&D significance, especially when OEMs invest in the local knowledge infrastructure and enter into partnerships with local knowledge institutes. Audi does this in Györ, Hungary, where it has a large engine plant; Renault invests in a technological centre in Pitesti, its main production location in Romania; GM has invested in a large technological centre in Shanghai, PATAC.

In the course of time, these centres may 'upgrade' and develop themselves as key knowledge centres for the automotive industry.

An important question for cities in Western Europe is whether R&D needs manufacturing nearby. Here, we found some differences between the Paris case, where the automotive industry is dominated by two 'volume car' producers PSA and Renault, and Munich, where two luxury car makers – BMW and Audi – play a key role. In the Munich case, both OEMs – Audi and BMW – stressed the importance of having higher level R&D functions, services and manufacturing on one site, and they have aimed not to reduce production capacity in Germany. They argue that co-location is required for communication between design teams, engineers and system suppliers. Product innovation also needs production process innovation, and therefore it makes sense to have the production plant at the same site. There are also other arguments to keep production in Germany. One is that 'made in Germany' is still a selling argument; also, some manufacturers have made deals with unions to guarantee production levels in exchange for longer working weeks.

In the Paris case, the link between R&D and manufacturing is becoming weaker. The strategy of Renault and PSA is to gradually reduce their production capacity in France, while opening new plants in emerging countries, mainly in Central and Eastern Europe. They increasingly 'tap' the growing technological competences there, and also invest in them (Renault Romania) in order to have high-quality but cost-effective local suppliers, and to use local engineering expertise to adapt existing models to local tastes. This does not imply that Paris will lose its R&D: it will specialise in higher segments, and Paris will develop as the 'brain' for the networks. The availability of skilled engineers is far more important than presence of factories. The Paris situation is in rather sharp contrast with firms in the Munich region (Audi and BMW): they have been more reluctant to open new factories abroad, far away from their domestic knowledge base. But this strategy can make a firm vulnerable. In 2007, for instance, BMW's profitability was hit by the low dollar. The USA is its main market, but it has just one factory there (with insufficient capacity to serve the US market), and that factory heavily depends on components from Germany.

In Brazil, the link between manufacturing and R&D is weak overall, but with remarkable exceptions. In the 1970s, Brazil designed (and produced) the first Volkswagen car outside Germany. However, during the 1990s, the link between manufacturing and R&D became rather weak. OEMs started to concentrate engineering and design functions in centralised R&D departments, just like transnational suppliers, who bought and merged with indigenous suppliers, hollowing out the developing R&D local departments. However, regarding specific bio-fuel technologies, Brazil – namely São Paulo – is a world R&D reference centre. To produce flex-fuel cars and systems, proximity between OEMs, suppliers and bio-fuel producers is key – they need to joint interact for new tests and incremental developments *in loco*.

13.3 Towards a project economy?

Synthesising our findings, we conclude that the 'industrial economy' is moving towards a network economy in which projects play an increasingly important role. The project is a temporary structure in which actors bring in complementary resources, and make arrangements to share costs and benefits.

Figure 13.1 shows that this 'project economy' can be observed in each of the three stages in the value chain (pre-competitive stage, the product development stage and the exploitation stage). In each stage, industrial companies increasingly operate in networks. These networks can be local, regional, national or international. Networks are rather loose, they can be passive, or just based on personal contacts or past cooperation experience. We can speak of a project when the network is being activated: this happens when a concrete goal is set. This goal can be anything, varying from the production of a new ship or car, a new joint research project, the construction of a new plant, the development of a new engine, etc. In a project, partners come together, often (but not always) selected and led by one 'lead partner'. When the goal of the project is achieved, the team typically dissolves. However, when cooperation is successful, and if there are repeated projects, it makes sense to create a new organisation in which the cooperation is structured and where rules and common objectives make collaboration smoother and more efficient. We call this a semi-permanent structure. Examples are joint research institutes between firms and universities, or strategic alliances among companies.

The car producers in our study see the development and production of a new model as a project. In the design stage, they organise competitions among internal and external design units, and select the best results. In the development and production stages, for each model, there is a new round of supplier selection, and those that best meet the demands (in terms of price and innovation) are selected to participate in the 'project'. Aker Shipyards works in a similar way. Evidently, some suppliers have a strong position thanks to unique capabilities. Also, reputation mechanisms play a role: when partners have worked together in projects in a satisfying way, they are more likely to team up again.

In inter-firm project teams, proximity matters to a varying degree

When a project is knowledge intensive and iterative (R&D projects are often like this), interaction and communication are crucial, and proximity helps. When the project leader has sufficient power in the network, the other team members will open (temporary) establishments close to the leader. This process can be observed in the automotive industry. Teams may also work together in cyberspace to some extent: Renault and Nissan increasingly do it. Alternatively, the partners can decide to locate the project team in a specific location. One example is the joint development of a new engine by PSA and BMW: the team was based in Munich, at BMW's technological centre. At the site of the Aker shipyard in Turku, we found a number of temporary establishments of supplying companies from a

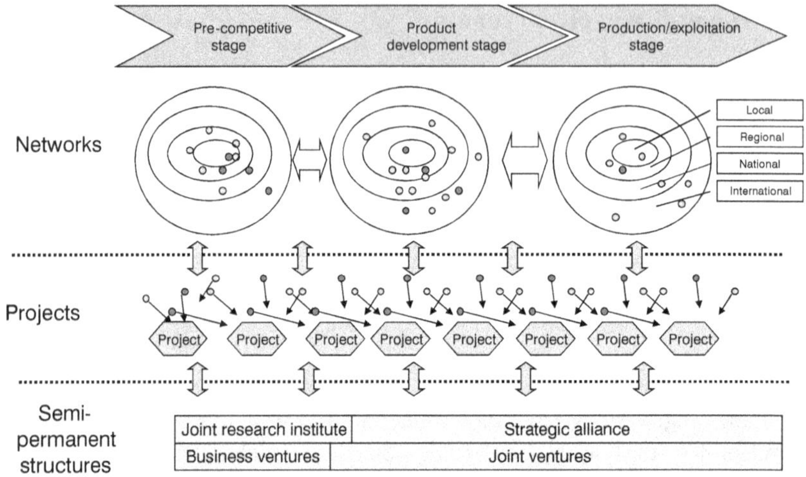

Figure 13.1 Networks, projects and semi-permanent structures (source: own elaboration)

variety of countries. The emergence of temporary structures has important human resources dimensions, as workers are increasingly asked to work abroad on a temporary basis.

How can regions and cities attract 'projects'?

We conclude that project teams, or other temporary structures, are becoming a more important aspect of the economy. For regions and cities, this raises the question of how to attract projects and other temporary forms.

In our case of São Paulo, the presence of a rather unique natural resource (sugar cane used to produce ethanol) and the accumulated know-how associated in related technologies set the conditions to attract related projects on a world scale. Generally, we conclude, tentatively, that cities with the best regional foundations (economic base, knowledge base, quality of life, accessibility and factor conditions) and with the presence of leading firms will attract the best projects. And these projects, in turn, make the foundations stronger because in some cases, temporary project structures are transformed into more permanent organisations. After some successful joint projects, Audi and the Technical University of Munich – helped by the local government – decided to create INI.TUM, an organisation based at Audi's research location in Ingolstadt. In the Netherlands, the Top Institute for Food and Nutrition (TIFN) was created in Wageningen, which is considered a world-class centre of excellence in food research. TIFN now coordinates pre-competitive research projects between industrial partners and universities.

Highly diversified and technologically advanced regions (such as Munich or Paris) are natural magnets for projects: they have many firms and competence centres that play a leading role in projects; they have a high quality of life, making it easier to attract temporary workers, and they have excellent international transport connections. The advantage of these 'diversified' cities is that they are a fertile ground to create 'new combinations' between sectors and activities that used to be separated. Less diversified cities should make sure that they become leaders in highly specialised niches, and create an 'image of excellence' in that specific area. This will help to make them a natural partner in international projects. Firms in emerging economies increasingly participate in international projects, but they are typically 'followers' in these projects.

Institutions and supportive organisations have a role to play, and may transform temporary networks in permanent institutions. In Turku, the development of the MERITECH centre is aimed to foster the already strong attractiveness of the region for shipbuilding and maritime technology projects. It aims to bring together different partners with research in the maritime industry, with Turku being the centre and many future projects co-funded by the European Union. In Porto, the presence and prestige of the research institutes of the University of Porto in different R&D European networks, on a temporary basis, contributed to attract to the university campus the first permanent structure of the Fraunhofer Institute outside Germany, in order to interact with the knowledge and industrial base of the region on a more regular basis.

13.4 Implications for urban regions

In the last section, we argued that firms increasingly operate in international networks. Manufacturing has become fully integrated in knowledge networks, and has undergone significant qualitative changes. Some networks are rather local or regional, but increasingly, networks are becoming international.

In this section, we turn our attention to urban regions. What is the impact of these changes on different types of urban regions? In our framework of analysis, we defined a number of 'regional assets' that are the foundations for regional competitiveness: the knowledge base, the economic base, accessibility, quality of life and factor conditions. Each region has a unique 'asset mix' that sets the margin for its development opportunities in the emerging network economy. For each region, a crucial question is: to what extent does the current asset mix meet the requirements of today's knowledge-based economy? In our case studies, we addressed this question. We asked companies, knowledge institutes, policymakers and intermediary institutes how they judged the local asset base, in relation to changes that occur in the industrial sector. In this section, we present the main findings of our analysis.

In our study, we covered ten city regions. Based on their different situations, we divide them in five groups:

1 'Traditional' industrial cities in transition (Rotterdam, Dortmund, Porto);
2 Post-communist industrial cities (Ostrava);
3 Advanced diversified cities (Munich, Paris);
4 Smaller industry towns (Eindhoven, Turku);
5 Emerging giants (Shanghai, São Paulo).

In Table 13.1, we show the characteristics of these city types. Note that the description is rather general, and major differences can exist between cities within one group. In the next sections, we elaborate issues in more detail.

De-industrialisation revisited

De-industrialisation continues, productivity continues to rise

In all our case studies manufacturing employment is decreasing as a share of total employment (see Figure 13.2). Ostrava is the only exception, where after a period of decline, the number of industrial jobs started to slightly increase again, namely as a result of important external industrial investments. The decline has been particularly fast in Dortmund, Porto and Shanghai. In Paris and Rotterdam, the share of industrial employment declined only slightly. In absolute terms, all our case studies have lost manufacturing jobs, except the emerging giants, Shanghai, and São Paulo (if we consider growing agglomerations outside the traditional metropolis). However, with fewer people, more value is created: in most cities, the share of manufacturing is relatively stable or has even increased (Shanghai). In all our case studies, the labour productivity has increased substantially in the last years. If in some cases, like Porto, that evolution was relatively modest, Shanghai almost doubled its industrial labour productivity (see Figure 13.3).

Manufacturing is a 'suburban' phenomenon'

In Europe, during the 1960s and 1970s, many plants left the city centres and moved towards peri-urban zones or newly created industrial estates and port areas far away from the city cores. In typical 'industrial' cities such as Porto, Dortmund and Ostrava, this fundamentally altered the spatial structure of the urban economy. Many 'traditional' manufacturing activities left the city in the 1960s and 1970s: some plants closed, others restructured, and many moved to industrial zones in the surrounding region because of space constraints and logistical reasons (especially in Porto). Not only did these cities lose many jobs, but also they had a legacy of worn-out and often polluted industrial estates in the core urban areas. Most Western European cities have cleaned up the old industrial sites and given them a new destination. In Central Europe, this process is still in full swing. In Ostrava, we found large, abandoned and polluted industrial sites right in the city centre. Foreign investors prefer to build new factories far outside the city centre, in greenfield sites. In São Paulo we could see a de-concentration of major industrial and automotive production from more traditional metropolitan locations towards other inner agglomerations and greenfields of the state, concentrating the largest share of the production. However, metropolitan São Paulo keeps an important share of automotive production of OEMs who locate there for decades. The inner city of São Paulo is mostly service based.

Table 13.1 Characteristics of the city types

	Role of manufacturing in local economy	*Knowledge base*	*Accessibility*	*Factor conditions*	*Quality of life*
'Traditional' industrial cities (Rotterdam, Dortmund, Porto)	Still significant, declining in terms of employment, moving towards high-tech	Low educational levels, labour market mismatches due to structural change, strong specialised universities that increasingly support industrial upgrading	Very good access for goods transport, but moderate international passenger transport connections	High prices of land and labour drive labour and space-intensive manufacturing out	Traditionally weaker on cultural and leisure amenities but improving, relatively poor housing stock
Post-communist industrial cities (Ostrava)	Dominant, engine of the regional economy, driver of services growth	Traditionally strong engineering knowledge base and technically skilled labour available	Moderate but improving freight connections, weak on international passenger transport	Low labour and land costs attract plants	Poorly developed service and leisure sectors, but rapid improvements; poor air quality and pollution
Advanced diversified cities (Munich, Paris)	Important though not dominant; wide range of industries, focus on 'upstream' functions	Strong and diverse knowledge infrastructure, large and high-educated workforce	Excellent (international) passenger connections, increasing problems for freight traffic due to congestion	Very expensive land and labour, very high cost of living making these cities unattractive for blue-collar workers	Excellent amenities
Smaller industry towns (Eindhoven, Turku)	Important but declining in terms of employment, strong upgrading processes	Specialised knowledge base (often matching the needs of local industrial specialisation), well educated labour force	Good freight connections, moderate on international passenger transport	High prices of land and labour drive labour and space-intensive manufacturing out	Good, but with some deficits in internationally appealing amenities
Emerging giants (Shanghai, São Paulo)	Very significant manufacturing sector, rapid technological upgrading, most growth comes from manufacturing related business services	Gateway for talent, top academic centres and concentration of specific know how.	Excellent in international freight and passenger conections, but growing metropolitan congestion	Rapidly increasing land and labour costs drive out space and labour-intensive plants	Excellent amenities, growing international appeal, but in growing jeopardy (air quality, safety)

Source: own elaboration

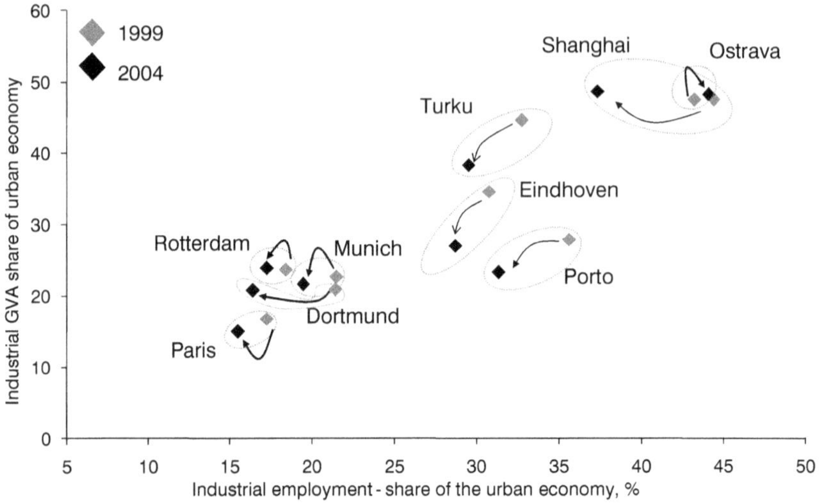

Figure 13.2 Industrial employment decline, 1999–2003 (source: own elaboration) (data from EUROSTAT – NUTS III level – and from Shanghai Statistical Yearbook)

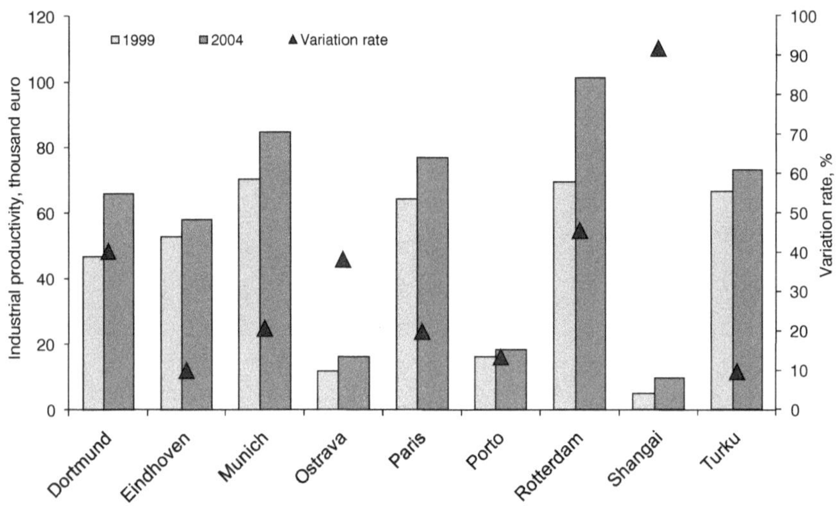

Figure 13.3 Industrial labour productivity, thousand euro and variation rate, 1999–2004 (source: own elaboration) (data from EUROSTAT – NUTS III level – and from Shanghai Statistical Yearbook). Note: São Paulo has been left out of the figure because of the lack of comparable data.

Our study confirms that manufacturing is still predominantly a peri-urban phenomenon in most cities. Figure 13.4 shows the significance of the manufacturing sector (value added) on three geographical levels: the core city, its region and the country. In most of the cases, the values for the core city are well below the regional levels (this is true for Munich, Dortmund, Paris and Porto), indicating that industrial activity largely takes place in the regions around the core city. Paris and Munich never had much heavy mass manufacturing within their borders.

But with the emerging knowledge-based economy, manufacturing becomes part of the urban knowledge economy

In our case studies, we found some signs of 're-urbanisation' of manufacturing, particularly small-scale, high-tech production. In Dortmund, many new small industrial firms have started up in the technology park near the university, and in Porto, we see similar trends. These modern 'factories' depend on highly skilled human resources and often have strategic links with knowledge institutes in the city. These types of manufacturing have become an integral part of the urban knowledge-based economy. Figure 13.5 illustrates the increasing knowledge intensity of production. It plots the development of the industrial value added created in each city (the vertical axis) and the R&D intensity of the urban economies (horizontal axis). It shows that the different types of industrial cities each have distinct characteristics. The 'traditional industrial' cities (Dortmund, Porto, Rotterdam) score relatively low on R&D, but they made significant

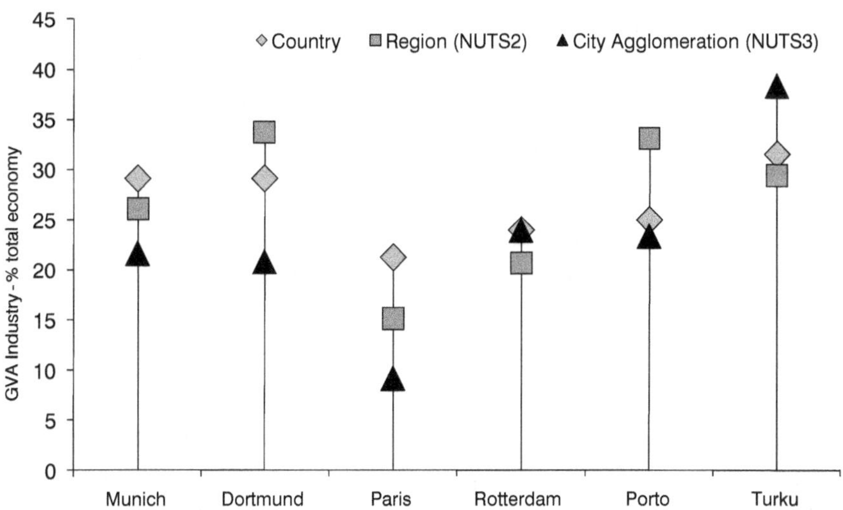

Figure 13.4 Industrial intensity in different spatial scales (2004, gross value added of industry, share in total economy) (source: own elaboration, based on EUROSTAT database). Note: this figure only includes the European cases. Comparable data for the other cases is lacking.

progress. On the other hand, the 'advanced diversified cities' (Munich and Paris) score very high on R&D intensity, but there is little progress (Munich) or even a slight decline (Paris). The figure also shows that the smaller industry towns (Eindhoven, Turku) still rely heavily on the industrial sector (they are high in the vertical axis) but this reliance has decreased strongly.

Manufacturing as a vital part of the regional economic base

As we have seen, in many urban regions, manufacturing is still a vital part of the regional economic base, but there are significant differences. In Ostrava, manufacturing is the driver of regional growth, with impacts on other sectors. Recently, the massive investments of foreign car makers, suppliers and other industrial firms give a boost to the regional economy, and good staff has become a scarce commodity. Also, the service sector is growing fast, thanks to these investments: there is a boom in the construction of new hotels and apartments, and many business service companies locate in Ostrava in the slipstream of industrial firms. This pattern is typical for post-communist countries. Under communism, they used to have an underdeveloped services sector. With the introduction of the market economy in the 1990s, the service sector began to expand rapidly, but mainly in the capital cities (like Prague, Budapest and Warsaw). Currently, secondary cities and industrial cities like Ostrava are catching up, thanks to their

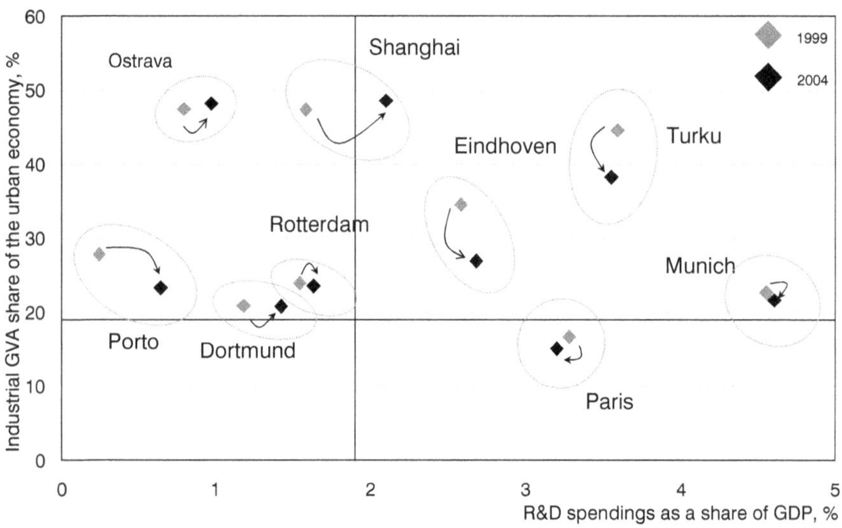

Figure 13.5 Evolution of the 'industrial intensity' (vertical axis, industrial GVA, share of the total urban economy) and 'technological intensity' (horizontal axis, R&D spending as share of GDP) (1999–2004) (source: own elaboration) (data from EUROSTAT – NUTSIII level for Industrial GVA; NUTSII level for R&D spending – and from Shanghai Statistical Yearbook). Notes: (1) Data on Ostrava corresponds to the period 2001–2003; (2) São Paulo has been left out of the figure because of the lack of comparable data.

attractiveness as manufacturing hubs, their improved accessibility and their relatively low factors costs.

In China, Shanghai's economy benefits from the development of China as 'workshop of the world'. The city itself has a big manufacturing sector, but first and foremost it has developed as one of the main service hubs for thousands of industrial companies that have opened plants in China. Many of its growing business services activities (banking, consultancy, legal services, headquarters functions) are linked to industrial companies, as is the case with metropolitan São Paulo.

In the traditional industrial cities in Western Europe (like Porto, Rotterdam, Dortmund, Turku), we see a different pattern. Here, manufacturing is not the driver of growth, but it is still an important economic sector, with significant spin-offs in other sectors. These cities try to develop new industries (services, ICT, biotech, logistics, creative industries) to compensate for the job losses in their manufacturing sectors, but it takes time before the local labour market is fully ready for the transition.

Finally, there are the highly diversified metropolises (Paris, Munich) in which manufacturing is still strong but not dominant. Industrial activity in these cities is typically high-tech, in the upstream parts of the value chain, and these cities are important services centres and 'brains' for internationally operating manufacturing firms.

Regional specialisation helps to develop particular niche industries

Many of our interviewees confirmed that industrial firms, even large ones, need to excel in a specific competence and serve well-defined niche markets in order to remain competitive. These competences are at least partly derived from the environment (the city or region) in which a firm is active. To some extent, co-location does still matter to realise cluster benefits. Specialised firms can function better when there is a specialised supplier base, a 'pool' of specialised labour and specialised knowledge institutes that conduct relevant research. Playing in such specialised sectors gives the region the advantage of specialisation and scale economies, since not many regions in the world can play in these niches. When specialised institutions interact (firms, suppliers, research facilities, organisations, etc.), regional competences become stronger. For example, many transnational suppliers in Brazil have been focusing their production and research capabilities in specific bio-fuel technologies, in which they are front runner worldwide.

However, regional specialisation implies a latent risk of lock-in, and may make the cluster more vulnerable to external shocks. The more specialised the economy, the less the possibility to develop 'new combinations' in case of fundamental changes. The small city of Turku is exposed to such risks, as it depends heavily on its dominant shipbuilding cluster. In 2002, the order book of the yard collapsed, due to the global economic downturn and the 9/11 aftermath, with large consequences for suppliers and the entire regional economy. By 2006, fortunes had reversed, and the order book was full. This posed another danger:

Box 13.1 Why are business services related to manufacturing growth?

Regions should be aware that fundamental changes, namely market, technology or legislation changes, may bring serious threats to specialised niche clusters. Despite the competitiveness and advantages of playing in market niches, it is important to support the diversification of the regional economy and facilitate the participation of local and regional players in other networks and value chains, outside the region. This contributes to reduce the risk of lock in and brings new knowledge to the region.

Several European Programmes support the international networking and cooperation of R&D institutes, firms and other associations, like the already four generations of the INTERREG initiative and the Framework Programme for Science and Technology. However, some regions are developing their own initiatives to support the internationalisation of their players, namely manufacturing entrepreneurs and R&D institutes. Some initiatives are relatively low cost and may reveal themselves as highly promising. The city of Turku and the Development Association support the networking between entrepreneurs in the Baltic region, willing to attract investment, but also to facilitate the entrance in new markets for the local firms. The north region of Portugal, where Porto belongs, developed a 'cooperation strategy', in order to support its own regional strategy (cluster development, manufacturing upgrade, insertion of its main cities is international thematic networks, etc). It aims namely to support firms and R&D institutes in developing pre-contacts in order to find the best partners for their businesses and research activities.

The yard pushes its suppliers so hard to deliver that they had neither time nor resources left to innovate and renew their product line-up, with dangerous longer term consequences.

Moreover, external changes in technology, markets or legislation may also yield risks for some clusters and regions. The successful luxury and high-segment automotive cluster in Munich is being challenged by new European legislation concerning CO_2 emissions. In the face of new legislation, more focus will be put on new technologies for filters, clean engines, etc., which is not the core competence of the Munich region, whose cluster relies on 'power' and designer cars.

Besides firms, research institutes also face a risk of lock-in. One example is the metal competences of some R&D institutes in the Porto region: when basic metal firms closed in the region, some institutes needed to find other applications for their competences, and diversified their activity and research towards new materials, energy and complex systems – here the quality and international contacts of the research interfaces was important to avoid lock-in in the region, and also to promote the development of new growth clusters in the region. They not only accessed new networks of research, but also inserted regional firms in those networks (like aerospace and aeronautics).

Some regional 'sector mixes' are highly productive

In some of our case studies, we observed a very fruitful interplay between different manufacturing sectors present in the same urban region. In Turku, the shipbuilding sector has been benefiting from the competences of the region in metal work – the shipyard interacts with different suppliers, who are involved in the increasingly demanding requests for state-of-the-art cruise ships. Producing new solutions and combinations, these suppliers are contributing to the strong innovative capacity of the yard. However, at the same time, suppliers gained new competences that they exploit elsewhere. In Porto, a similar situation was observed. Specialised machinery suppliers in the region developed different joint projects to steer the competitive upgrade of the footwear sector to allow it to produce smaller series. The outcome of the interaction was the development of specific machines and technologies to produce for footwear niches, and new technologies to be exploited in other sectors, like food or automotive.

In those cases, the 'technology suppliers' and the clients benefited from interaction towards the upgrade. The regional dimension and the diversity of the economic base were important for the achievement of new combinations, although all the sectors where 'traditional'. In both cases, trust, similar cultures and regular contact and interaction were important for the synergies and innovative processes towards upgrade.

For many of the urban regions in this study, the strength of their manufacturing sectors comes from the setting of interrelated activities. For example, the food cluster in Rotterdam is composed of very different components; not only production, but also the logistics and agriculture parts of the chain. In Turku, the strength of the shipbuilding cluster comes from the presence in the region of several specialised suppliers, namely for specific components and equipments. In São Paulo, the presence of OEMs, engines and bio-fuel producers led to new combinations and innovation. In Porto, the metal-electro sector is composed of very different subsectors that interact with other value chains in the region, namely the habitat or the footwear sectors. Although it lacks large OEMs for passenger cars, the Eindhoven region is a strong supplier of embedded systems to many OEMs due to the linkages the automotive industry has with the ICT, mechatronics and design industries. In Paris or Munich, the automotive sector is strongly linked to a network of specialised first-tier suppliers with powerful interactions with the OEMs. The varieties of competences available in these cities make them productive network environments.

It can be rewarding to link 'traditional' manufacturing to new growth clusters

In the last decade, many industrial cities have been trying to diversify their economy by fostering new, 'sexy' growth clusters. The most popular candidates are tourism, ICT, biotechnology, medical industries, nanotechnology and the creative industries. Too often, the policy focus on new industries comes at the expense of 'traditional' manufacturing industries that have been considered as outdated, old-fashioned and without growth potential.

Some cases suggest that it can be useful to link the development of promising growth sectors to 'traditional' regional manufacturing strengths. In the urban region of Porto, regional policy explicitly links the equipment and electrical machinery sector to new spearhead clusters like medical equipment, energy- or maritime-related activities, providing incentives for joint projects between firms and R&D institutes with competences in the field. Also, the city of Dortmund aims to link traditional industries with new key industries (ICT, logistics and materials science and technology). In Rotterdam, unexplored links were also found, namely between the food cluster and the creative industries, with the potential to interact in packaging and design issues. Local economic development agencies can help to explore these potentials and bring 'unusual' partners together.

Other manufacturing sectors, like automotive or shipbuilding, present good potential to foster links with leisure and tourism activities. Industrial tourism is a promising example. Car manufacturers were among the first to discover the potential. Audi and BMW have specific facilities (Audi Forum and BMW 'Welt') for visitors, such as museums and guided tours through the plants; Shanghai and Volkswagen in association with other partners develop different thematic tours and plant visits for visitors and the city has devoted specific policies for this aim.

For industrial clusters, the 'relevant region' is getting larger

In all of our case studies, industrial clusters stretched far beyond the borders of cities or city regions. On several occasions, interviewees indicated that they could very well maintain face-to-face contacts with relevant partners, as long as a return trip can be made in one day. In Rotterdam, the food cluster is strongly linked to the expertise of the Wageningen research centre (100km from Rotterdam) and many firms and producers locate in surrounding municipalities. The leading automotive companies in Paris have very frequent face-to-face interactions with main suppliers and facilities in Normandy and Bretagne. In Porto, the relevant region for some of the most interesting interactions is a larger area of 70km around Porto, which houses not only the most representative firms and SMEs but also research partners, like the University of Minho. In Turku, the shipbuilding cluster has strong connections within the urbanised part of south-west Finland (the triangle of Turku, Helsinki and Tampere), where some of the most important research centres are located. Both for Rotterdam and Turku, the knowledge infrastructure of the city is not the most adequate for the cluster, and the firms gather competences and skills in larger regions. In Dortmund, the industrial Ruhr area is the relevant region for many of the manufacturing relationships and research links. In Shanghai, the

Box 13.2 Options for promoting new combinations

Although the hard incentives for new combinations and large projects are usually found at the national level, regional and urban players are usually better able to identify the possibilities to promote new combinations between existent manufacturing activities and new growth clusters. Some options are opened to urban and regional managers in order to support the development of these combinations and promote the urban economy.

Some options are open in this respect: i) Jointly managed funds (city, lead firms, universities, etc.) can be created for the specific support of small but demonstrative projects, applying new competences and knowledge to manufacturing sectors (like, for example, industrial design); ii) launch of the most appropriate 'cluster or platform strategy', in order to link and foster the contacts between players with convergent resources for the new combinations pretended – afterwards, the network can be activated for large projects; iii) involve manufacturing players with the city development strategies – they may uncover other hidden potentials for economic urban development, like the development of integrated tourism products.

automotive cluster develops in a very wide area, along the Yangtze River Delta. Similarly, São Paulo's bio-fuel cluster stretches from metropolitan São Paulo to other new and non-contiguous urban agglomerations in the state, such as Campinas Sorocaba and São Jose dos Campos. Thus, in the 'project economy', larger urban regions are relevant spatial units, putting limits on city economic policy.

In some of our cases, the cross-border relation is relevant for the development of manufacturing clusters. Turku-based firms consider the Baltic Sea region as a resource of temporary labour, but also invest there in manufacturing plants and other facilities; Porto's equipment firms and R&D institutes, in order to achieve critical mass and access a wider economic base, develop several projects with Galician institutions (Spain).

The growing role of the knowledge base

The regional knowledge base is increasingly important for manufacturing industries

As a result of the upgrading process in all manufacturing industries and 'open innovation' strategies in all our case studies, the regional knowledge infrastructure has become a strong ally for modern manufacturing firms, and companies increasingly show an interest in investing in their 'knowledge environment'. The knowledge base provides skilled workers and qualifications, and many firms in our case regions keep the local universities and institutes as stable partners for applied research, the development of new products and processes, and as primary network nodes to engage in pre-competitive research.

There are persistent bottlenecks that hamper a more fruitful cooperation between firms and knowledge centres. Overall, however, evidence from our case studies suggests that many university research centres are increasingly willing to cooperate with private firms and manufacturing firms in R&D projects and education. One indication is that private funding sources are increasingly important for knowledge institutes, as a result of many R&D contracts and projects with manufacturing firms. Universities also increasingly recognise the intellectual value added of contacts with firms: it helps them to gather new knowledge from interactions with other sources and partners.

In Paris, Renault's Technocentre (12,000 employees) develops joint projects for basic and applied research with university teams of the University of Orsay and the Polytechnic school, but also with the INRETS (Research Institute for Transport). Moreover, its internal research facilities are actively cooperating with secondary education institutes to offer adapted and tailor-made courses. In the case of Rotterdam, diverse segments of food firms are very active in cooperation projects with the top research centre of Wageningen, both for basic and for more applied research, but also with TU Delft, namely for logistics and applied research. In Ingolstadt (near Munich), Audi has set up a joint research centre (INI. TUM) with the Technical University, supported by the city, in which they carry out research projects.

Joint PhD programmes are increasingly popular

The development of joint PhD programmes (in which firms and universities work together) is gaining ground. Typically, the student is involved in a project of interest for the firm. The supervision is done by a university professor and an expert in the company. We found this model becoming popular with OEMs in the automotive industries of Paris and Munich, but also in Porto, where the manufacturing fabric is composed mainly of SMEs. In the latter case, there is already evidence of researchers moving from the research institutes to firms during and after successful research projects, depicting the value that firms increasingly attribute to this type of qualification.

Productive cooperation can help to upgrade a local industry

In Porto, the interaction between firms and universities has been crucial for the upgrading of the metal-electro sector. A select number of technological institutes have gradually bridged the gap between firms and university. The Institute of Mechanical Engineering and Industrial Management is one of them. In a first stage (the late 1990s), the institute 'pushed' new technologies towards regional companies. Currently, firms actively and directly approach the institute for applied research projects. Several firms have even become shareholders of the institute and co-determine the institute's strategy.

Technological centres can help manufacturing firms in several ways, e.g. by providing certification of products and processes, quality assessments, product tests. In Porto, the metal-electro sector relies on an active technological centre for this purpose. In Turku, many SMEs in the metal sector (mainly shipbuilding suppliers), can use similar facilities. In both regions, these centres are usually linked with sector associations which promote several courses targeting the skills and professional qualifications of the sector for the lower and intermediate levels. These courses are relevant for SMEs who cannot afford their own training facilities, but also for large firms that complement their training department competences with external providers. In São Paulo, regarding 'automotive knowledge' cooperation between universities and firms, it is still early. However, some polytechnic institutes – like the Maua Institute of Technology – develop in close cooperation with the sector (namely for tests and prototypes), using that knowledge in their own post-graduate technical courses.

The knowledge base part two: human resources

All our cases reported shortages of skilled technical staff

Despite job losses in manufacturing, industrial companies across all our case studies complained about a lack of skilled staff, especially with intermediate technical qualifications. An important issue is the ageing of the workforce. On average, workers in manufacturing plants are older, and many are to retire in the coming years. At the same time, young people are reluctant to take their places and prefer jobs in other sectors. Young people saw their parents struggling in plants, and associate manufacturing with dirty and heavy work and long work days against low salaries. Although working conditions and salaries have improved strongly over the last decades, plants still have problems attracting workers.

The shortage of employees is also related to the higher qualification demands in the manufacturing sector. In the highly sophisticated manufacturing plants in Western Europe, non-qualified jobs have virtually disappeared, whereas the demand for highly qualified people has increased. Modern plants need technically skilled controllers, maintenance engineers, and experts in ICT, quality control, robotics and logistics. Manual work has been substituted by machines and robots. Furthermore, many labour-intensive tasks that used to be done by less educated people have been sourced out to low-cost countries in Central and Eastern Europe and Asia. Some industries still need substantial numbers of 'traditional' industrial workers, such as welders, but even here, we see a process of upgrading. The shipbuilding sector in Turku is a prominent example. Simple welding has been taken over by machines. Humans are to do the more complex parts.

For traditional 'industrial' cities (like Rotterdam or Dortmund), the upgrading of manufacturing work poses some challenges. The qualification level of the workforce in these cities is relatively low, and (youth) unemployment is high.

Box 13.3 Aligning the regional knowledge base with the needs of firms

The availability of regional knowledge infrastructure is crucial for SMEs, as a vehicle to gather knowledge competences and develop technologically, but also for large firms with own R&D facilities, who seek in the regional infrastructure a way to complement their competences.

Thus, having good universities, research institutions and other 'knowledge facilities' is a precious asset for any region and for their manufacturing base. Nevertheless, it may not be enough. Despite the increasing awareness of the potential gains from cooperation (and policy instruments to support it), there appears to be still room for better alignment between manufacturing and the regional knowledge infrastructure.

On the one hand, many cities and regions do not have within their boundaries the right partner for different types of projects. In Turku, firms active in shipbuilding gather competences in the south-west of Finland; in Rotterdam, the key knowledge player is in Wageningen, one hour's drive from Rotterdam. In this case, the instruments to foster interaction should be planned at the relevant scale – accessibility infrastructure increasingly allows longer distances, and local 'anchors' (a firm, a knowledge institution) can be valorised like this.

On the other hand, at the local level, there appears to be room for better interaction between firms and knowledge capabilities. Although Rotterdam knowledge infrastructure does not have the key competences for food innovation, its law school can provide key legal advice for the sector. In many other cities with design and art courses, the links with the (increasingly valued) industrial design is still largely unused.

Many lower-skilled people have become redundant. The manufacturing sector has ceased to be a refuge for people without formal education: it has become part of the urban 'knowledge-based economy'.

In Ostrava, the shortage of staff is also becoming a major issue. Foreign manufacturing firms open plants in the region (most recently Hyundai, and a number of suppliers) and lure away skilled workers from existing companies. As a result, wage levels are rapidly increasing. Furthermore, in the 1990s few people in the region believed that manufacturing would ever grow so fast, and many students opted for studies in management and economics instead of technology.

Several actions are being taken

Many of our case cities are taking action to tackle the shortage of educated workers in the manufacturing sector. In general, four types of solution are developed:

- *'Import' workers.* This happens in Turku, where the shortage of workers in the shipbuilding industry is addressed by attracting skilled manual workers from the Baltic and Central and Eastern Europe. Similarly, truck producer DAF, in the Eindhoven region, hires employees from Belgium, Germany and even the UK to solve the shortage of specialised workers.
- *Retraining workers from other industries.* In Dortmund, unemployed workers from other industries (mainly from the steel and mining industries) are retrained for jobs in the metal-electro sector.
- *Promotion programmes to improve the image of the manufacturing industry as a workplace.* The Chamber of Commerce in Dortmund (IHK zu Dortmund) organises encounters between youths and older workers, to help youths become enthusiastic about working in plants.
- *Vocational training programmes.* These programmes are organised by various actors. In many cities, business organisations offer apprenticeships and specific training courses, especially for youths. This is the case in Dortmund, Eindhoven and Porto. Through these courses, workers can obtain new professional skills or even an academic qualification.

National governments must not overstress the importance of academic qualifications

In several countries, national governments are determined to increase the share of the workforce with academic qualifications. Many of our interviewees, however, stressed that the government should recognise the importance of lower and medium technical qualifications as the foundation for manufacturing industries. Firms in Turku's shipbuilding cluster fear that the current national policy to have 50% of its young people achieve an academic degree will reduce the supply of skilled technicians. Also, this policy, unintentionally, provides the signal that not having an academic degree is a major deficit.

Manufacturing and regional image

Manufacturing industries affect the image of the region

As remarked in the previous section, 'manufacturing' still has a negative connotation with the general public, in particular among young people. People still associate manufacturing with black smoke, chimneys, pollution and dirty work, and images are hard to change. As a result, traditional manufacturing cities or regions suffer from a poor image. In several of our cases we found a substantial gap between the perception and reality. Dortmund and Rotterdam, for instance, are still widely regarded as 'industrial' cities, but in reality their economies have a lower share of manufacturing than Munich. Perceptions of places are deeply rooted in the collective memory and change only very slowly.

The 'visibility' of the industry plays a role in the perception. Rotterdam, with its large petrochemical complex, and Dortmund or Ostrava, surrounded by large industrial plants, are cities with very visible industries. In diversified cities such as Munich and Paris, the skyline is not dominated by chimneys or plants. They specialise in the upstream parts of the value chain (sales, R&D, marketing, headquarters, etc.), which contributes to an image of a city of knowledge rather than a city of manufacturing, despite the presence of large car plants in these regions.

But the image of the region also has an impact on manufacturing

In machine building and car manufacturing, 'made in Germany' still stands for quality and reliability. This is one of the reasons why the luxury car manufacturers in Bavaria (Audi and BMW) want to keep production in Germany. Some German cities or regions are associated with particular competence, for instance Solingen (knives), Jena (lenses), thanks to the brand names of world class companies from those regions. Also, the Czech Republic (and the Ostrava region in particular) has a long-standing reputation as a quality manufacturing hub, and has begun to regain its pre-communist eminence as one of Europe's manufacturing hubs. In food research and innovation, the Wageningen region is well known as a centre of world class competences. The positive image of the region helps companies located there: it gives them credibility and visibility, and they can associate themselves with the positive image of the region.

On the other hand, if a country or a region has a poor image, its companies suffer from it. Porto is a case in point. Firms in the region around Porto produce state-of-the-art cutting machines and other high-tech equipment for niche markets. However, in international markets, the country's image (Portugal has traditionally been an importer of machinery products, and has hardly been known to be a producer) is not an asset. To escape the image deficit, several Portuguese machine builders sell their products through subsidiaries in countries with a better manufacturing reputation.

Accessibility, quality of life and factor conditions

Goods transport is the lifeline of manufacturing industries but passenger traffic becomes more important

Transport connections are the lifeline for manufacturing plants. Many companies increasingly depend on inputs from far-away suppliers, and just-in-time production methods further increase the logistical complexities. As a general trend, deliveries become smaller, more frequent and take place over larger distances. These trends make heavily congested urban areas (like Paris, São Paulo or the Randstad area) less attractive for the types of manufacturing that require frequent and reliable deliveries.

Goods have to be transported, but so do people. For manufacturing industries, passenger connections become increasingly important. The trend of internationalisation (clearly visible in all the industries we studied) leads to more frequent passenger traffic. Company managers have frequent meetings with clients, suppliers, or 'sister' companies that belong to the same consortium. Also, the increasing knowledge intensity of production requires more face-to-face meetings between engineers and other experts. Many of the managers that we interviewed stressed the increasing importance of international air and train connections for their work. Diversified cities with traditionally very good international connections (like Munich and Paris) benefit from this trend, and it is no coincidence that they are very strong in the higher-level functions in manufacturing value chains, in which internationalisation has made the largest inroads. For traditional industrial cities, the increasing importance of passenger connections poses some challenges, as typically, they do not have large international airports. Thinking in terms of goods and volumes is almost in the genes of policymakers, but not in the best interest of the future of their industrial base.

As manufacturing becomes more knowledge intensive, quality of life becomes more important

Quality of life is a key asset to attracting or retaining the knowledge workers on which industrial cities increasingly depend. Diversified metropolitan areas (like Paris or Munich) score well in terms of amenities (leisure, culture, etc.), which helps them to attract and retain the best brains. Paris is home to 50% of all the engineers in France. Audi, which has its main location in the provincial town of Ingolstadt, located its design centre in Munich to benefit from the big city ambiance and the cultural 'vibe' there. On the other hand, these cities tend to lose manufacturing activity: skyrocketing land and real estate prices in this type of city makes them less suitable for space-intensive manufacturing plants.

'Traditional' industrial cities are less well endowed in terms of amenities. Many of them have grown fast during industrialisation, and developed as 'blue collar' working-class cities. They still struggle with a legacy of poor housing stock, fewer high-level amenities and a less developed 'intellectual climate', which makes it more difficult to attract the kind of workers needed for the knowledge economy.

Another aspect is the quality of the natural environment. Climate change offers big challenges for all cities to reduce pollution and CO_2 emissions. All should take action to make industries cleaner, but the challenge is greater for the industrial cities as they tend to pollute more. The challenges are particularly daunting for post-socialist industrial cities (Ostrava), and 'emerging giants' (such as Shanghai), where air quality is rather poor. These cities are in the middle of a transition towards a knowledge economy, but they will find it hard to attract the best brains if the air quality remains as it is now. Post-socialist cities like Ostrava face the additional challenge of addressing the problems of polluted industrial sites. Cleaning them up is costly and complex (the soil sometimes belongs to private owners, not the state), but necessary to safeguard the liveability of the urban region in the long run.

Metropolitan São Paulo faces an unstable attractiveness balance. On one hand, it concentrates top cultural and leisure amenities, and the massive adoption of flex-fuel car engines has been improving air quality, even in the context of high traffic congestion. On the other hand, high crime and resilient social disparities pose a severe threat to its long-term attractiveness. In this sense, metropolitan São Paulo faces sharp competition from other growing urban areas in the state, like Campinas – this urban agglomeration has top research centres and universities, a very good quality of life and is presently a growing R&D hub for bio-fuel-related technologies.

Manufacturing locations in cities

The changes in manufacturing industries are reflected in new spatial patterns in cities: new industrial landscapes are emerging. Manufacturing used to be a space-intensive and rather polluting activity, and this explains why many 'traditional' manufacturing locations are located far outside city cores. However, the emergence of the 'knowledge-based economy' has brought a much more differentiated landscape of manufacturing activity. Importantly, manufacturing industries have become cleaner, thanks to technological innovation and stricter regulations. Air quality and water quality has improved dramatically in most cities. Furthermore, overall, manufacturing has become less space intensive.

Still, there are large-scale plants that need a lot of land (as we observed in the automotive and shipbuilding industries), but increasingly, manufacturing plants have become smaller, cleaner and more knowledge intensive. Rather than large plots, they need the availability of skilled staff, good international passenger transport connections or to be in the vicinity of knowledge institutes. Another trend is the increasingly networked nature of manufacturing: industrial products are created by a large network of component suppliers, many of which are also involved in product development.

These developments led to a differentiation in the locational needs and requirements of manufacturing firms. This has important consequences for the location policy of local and regional authorities. In our study, we have encountered a surprising variety of industrial locations, with new forms constantly emerging. In Figure 13.6, we have depicted them along two lines: the degree of logistical integration (operational and logistical links between the companies at the park), and the degree of knowledge transfer among the partners at the park.

'Traditional' industrial zones still can be found in any type of city. Typically, they are created and run by local agencies or project developers, to accommodate industrial companies. They can be specialised and focusing on a specific industry: Shanghai, for instance, has a strict zoning policy based on industrial specialisations. But often these parks are generic and open to all kinds of firms. These parks are typically created for planning reasons, to regulate traffic pollution and congestion and several other nuisances caused by manufacturing space-intensive activities,

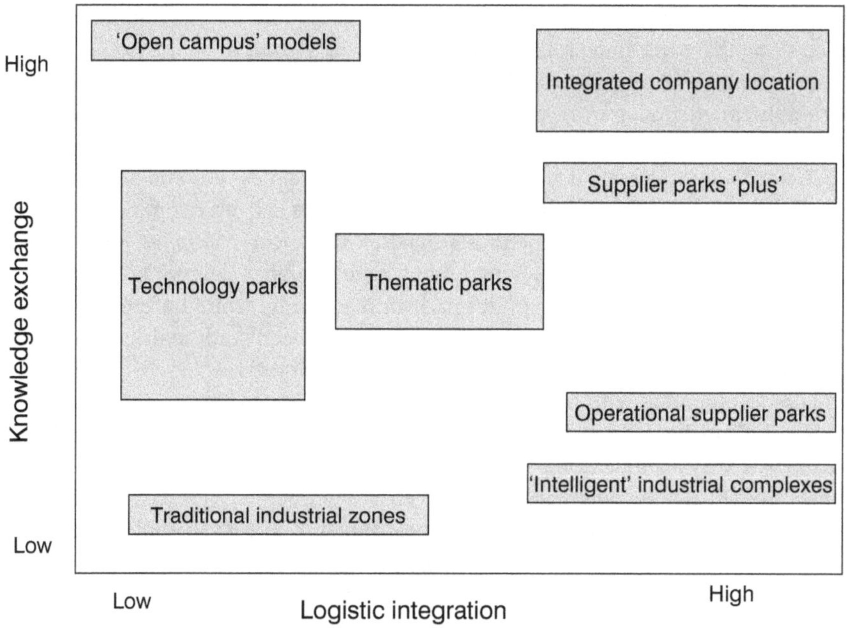

Figure 13.6 Industrial location types (source: own elaboration)

while facilitating specific infrastructure. All our case regions have this type of facility, although the very generic model is on the decline and replaced by more dedicated concepts.

Another type is the *'intelligent' industrial complex*. In these complexes, industries are functionally linked to each other in various ways. One example is the petrochemical complex in the Rotterdam port area, where petrochemical plants use common pipelines, and the by-products of firm A are the input for adjacent firm B. The waste heat from a power plant is used to heat a basin in which tropical shrimps are grown. In the light of environmental challenges, this type of complex has a bright future.

An increasingly popular type (since the 1990s) are *technology parks* or *university parks*. These complexes specialise in the accommodation of 'upstream' segments of the manufacturing value chain. Often, these parks are created to facilitate the interaction between firms and research institutions. Some offer shared facilities, such as lab space or test sites. For high-tech SMEs, these parks offer advantages, as they themselves have no resources for high-grade facilities. A good example is the Technology Park and Technology Centre in Dortmund that concentrate a high number of SMEs in electronics and nanotechnology, interacting with the knowledge institutes present in the park.

Thematic parks/areas are set up around a specific theme, an excellent example of which is Auto City in Shanghai. This is a large area, where all kinds of automotive-related activities are planned: plants, dealers, a race circuit, a second-hand market, many leisure facilities, and housing. The area is 'branded', i.e. everything there has some relationship with the car industry.

In *operational supplier parks*, suppliers are located near the site of the assembly plant and they are very common in the automotive industry. The rationale for these parks is that the co-location of suppliers and OEMs facilitates operational interaction: it reduces inventories, enables communication and increases the reliability of deliveries. Also, these parks can offer common services (e.g. workshops) and infrastructures (logistics facilities). *'Supplier parks plus'* are similar to 'normal' supplier parks but have a higher degree of knowledge interaction. The automotive supplier park in Ingolstadt, centred around Audi, is an example. On top of operational cooperation between Audi and its suppliers, this park also facilitates knowledge-intensive cooperation, such as co-engineering and co-development with suppliers, and training facilities.

Some firms have developed *integrated company locations*. Unlike the others, this is basically a single-company concept, in which several functions within one large firm (R&D, headquarters, production) are deliberately located on one site to facilitate interaction and integrated development. The best example is BMW's main Munich location; Aker Yards is another.

A relatively new approach is the '*open campus model*'. This concept is created to facilitate open innovation processes, in which several companies work together to develop a new product or process. An example is the high-tech campus of Philips, in Eindhoven. At the new campus, located near the highway and close to the city centre, Philips has concentrated several major research labs and installations. Other companies are welcome to locate there, as long as they cooperate strategically with Philips in (pre-competitive) innovation. The campus focuses on R&D and design, but it has prototyping and small-scale manufacturing facilities as well, in order to test products and production processes.

13.5 Challenges and policy implications

In this section, we present a number of key challenges for policy making. First we focus on challenges for different types of cities. What is the future of manufacturing, and what are the main policy challenges for urban policy? Second, we discuss implications for national policy. Finally, we argue the increasingly pro-active role of private firms in urban management.

The future of manufacturing: challenges for different types of cities

'Traditional' industrial cities

In the last decades, 'traditional' industrial cities (Rotterdam, Dortmund, Porto) have experienced heavy losses in manufacturing activity. This transition towards a service economy will continue but manufacturing is likely to remain resilient. Industrial activity in this type of city will evolve in two main segments. The first one is the 'knowledge-intensive' manufacturing of complex and specialised products, in small volumes, for global niche markets. This type of manufacturing relies on the local knowledge base, and behaves in the 'logics of services': it is people orientated, it has a strong service component and it is embedded in global knowledge networks. For this type, the physical 'production' cannot be separated from R&D and services. Local governments can promote this type of manufacturing by supporting cluster organisations, or by organising close partnerships between local businesses and local knowledge institutes to promote entrepreneurship and improve the match between the knowledge base and the industrial base.

'Traditional' industrial cities no longer have a monopoly on manufacturing industries. The knowledge-intensive niche markets increasingly face competition from other types of cities, such as the advanced diversified cities (e.g. Paris and Munich), and also from smaller industrial towns. This poses a number of challenges for this type of city. Most importantly, they need to catch up in terms of quality of life and accessibility (for people) to be able to support and develop their knowledge base.

Traditional industrial cities will also remain attractive for a second type of manufacturing: one in which proximity and access to consumer markets play a role. Examples are food production and distribution, but also value added manufacturing (i.e. re-working imported products for local markets). Cities with a favourable location (such as Rotterdam and Dortmund) have opportunities in this area. This type of manufacturing is less knowledge intensive (it uses more or less standardised technology), more space intensive and requires a less educated type of labour.

Post-communist industrial cities (Ostrava)

After a decade of decline, several post-communist industrial city regions in Central Europe are regaining their positions as manufacturing hubs: they attract foreign manufacturing plants on a large scale, and several indigenous industrial companies (former state companies as well as private start-ups) are growing. These regions are now fully integrated in the global production systems. It seems EU membership has greatly helped the industrial development of these regions, but some other points can be mentioned as well. The first is improved accessibility, thanks to large investments in roads, railways and airports. This enables the transport of goods and components to and from markets all over Europe. A second strong point is their still low labour costs, particularly compared to Western Europe. And third, these regions have an increasingly skilled labour force, which makes them attractive not only for simple assembly but also for complex operations and R&D activity. The role of manufacturing in these regions will probably evolve in a similar way to Western industrial cities. Productivity levels will increase and wages will rise, reducing the number of 'blue collar' workers, while increasing the knowledge component of manufacturing. Manufacturing will be a fully integrated part of the knowledge-based economy.

That also implies that in the longer run, quality of life becomes a key factor. And in this respect, these cities have a long way to go. Their challenge is not to attract as much FDI as possible, but rather to be more selective (focus on knowledge-intensive projects), and invest in quality of life. For sustainable development, these cities need attractive inner cities, cultural and leisure amenities, high-quality housing, a hospitality sector, and – last but not least – a clean and attractive natural environment. Another challenge for post-communist cities is to 'open up' more to the outside world: this implies improving passenger transport connections – increasingly important in the network economy – but also better foreign language skills, and welcoming more foreign students at local universities.

Advanced 'diversified' cities

As we have seen, manufacturing in 'advanced diversified cities' (Munich and Paris, in this study) is predominantly knowledge intensive, and this will remain the case. The main asset of these cities is their knowledge base: a highly skilled labour force and excellent knowledge institutes. They combine this with a diversified and modern economy, which can be conducive to the development of new types of networks and innovations. The manufacturing of physical goods plays a minor role in their knowledge-based economy because these cities specialise in R&D and services. They are the home of 'leaders' in the project economy, i.e. firms at the edge of the technological frontier that lead the development of innovations, and organise networks around them. Furthermore, they are 'decision centres' from which global production networks are controlled. Their excellent transport and communication connections reinforce this role. In these urban regions, mass production – if it still exists – will be further hollowed out, given high factor costs, and factories will be supplied from far-away regions. But this tendency will have no consequences for their knowledge base and their strengths in the upstream parts of the value chain. This does not mean that manufacturing will disappear altogether: their strong and diverse knowledge base makes them attractive environments for small-scale, sophisticated production that is linked to knowledge centres. In this respect, they will be new 'competitors' for industrial cities.

At the same time, these cities face new competition from emerging economies. Our study shows that urban regions in Central Europe and Asia are rapidly developing strategic R&D competences. Shanghai is emerging as a centre for automotive research, and in Ostrava, we found very advanced R&D activity in the development of sensors. For the advanced diversified cities, this may imply that they will have to specialise in a fewer number of core competences in which they are really world class.

Smaller industrial towns

The smaller industrial towns in our study (Turku and Eindhoven) have been very successful in constantly modernising their industry, and today they excel in specific industrial niche markets. Industrial companies in these towns have gone up in the value chain (now, they specialise in R&D, design, assembly and marketing), and have outsourced a large part of physical production. They benefit from growing global demand for sophisticated industrial products. These cities are strongly embedded in global production and knowledge networks. As a result, their economic fate depends on international economic developments. Traditionally, the economies of these cities were dominated by one large industrial company (Philips in Eindhoven, Aker Yards in Turku), but our research suggests that this dominance has decreased. One reason for this trend is the increased international orientation of suppliers: they no longer depend on assignments from the big local multinationals, but have successfully developed their own niche markets and exploit them internationally. For these cities, a key challenge is to keep talent in the region or attract it from outside. In this 'battle', they compete with all the other city types. A second issue is international accessibility: these cities score poorly compared to larger places, and this is increasingly seen as a disadvantage in the industrial network economy. A third, related, challenge for these cities is to effectively cooperate with cities nearby to overcome the disadvantage of a narrow and specialised knowledge base. Product innovation increasingly requires the combination of a variety of competences, and these cities don't have the scale to excel in everything. Strategic cooperation with nearby cities can help to overcome this deficit. To this end, Eindhoven has established networks with Leuven and Aachen.

Emerging giants

In our study, we analysed two emerging giants: Shanghai and São Paulo. Shanghai has experienced a remarkable development of its industrial base in the last decades, while at the same time it evolved as a major hub for services and R&D. In recent years, the city's manufacturing activities have upgraded substantially, and there is a tendency of migration of manufacturing to inland locations, where land is cheaper and cheap labour is more abundant. The significance of mass manufacturing is still high, but it will decrease further with rising productivity and pressure on wages. The industrial tradition of São Paulo is much older, but the recent trends are similar.

Thus, Shanghai and São Paulo are gradually becoming more like Munich and Paris and are beginning to specialise in the higher segments of the value chain. The cities' main attraction factor now is not cheap labour but rather their knowledge base and international orientation: they attract not only local talent, but talent from all over the world. R&D activity in Shanghai and São Paulo is rapidly becoming more sophisticated and world class (in some specific niches) thanks to enormous investments by different government tiers, as well as by private firms that consider them as good locations for R&D. These giants are becoming part of global innovation networks, although they do not play a truly leading role yet in cutting-edge projects (except in the case of bio-fuel-related research in São Paulo). To some extent, the cities have become new competitors for existing R&D centres such as Munich or Paris: they join the battle to attract the best brains and fundamental research activity.

On the other hand, the 'regional' economies (Asia and Latin America) are growing so fast that more R&D in emerging giants does not come at the expense of R&D in established centres, rather, it is becoming increasingly complementary. This implies that technological change will happen even faster because so many more people are active in it. It also reinforces the trend of specialisation, clearly visible in the case of São Paulo.

Environmental considerations (notably air pollution and climate change) will largely shape the future of Shanghai's manufacturing industries. The success of Shanghai's knowledge-based economy will depend on its quality of life because the city cannot afford to lose out in the battle for talent. Companies in the city will have to pay premium wages for skilled workers (if they can get them at all), especially if pollution is too bad, and if competing Asian cities take advantage. Thus, in principle, the city has a strong economic incentive to take environmental policy seriously. This problem cannot be solved by Shanghai alone as much of the air pollution is generated by industrialised cities and provinces in its hinterland (the Yangtze River Delta), and those cities have far fewer incentives to be strict on the environment.

Another challenge is to reduce pollution from car traffic in the city. In this respect, heavy government subsidies for the development and preferential treatment of 'clean cars' are an important step. Shanghai could become a leading city for clean technology given the size of its problems and the vast investments planned to develop clean technologies.

São Paulo has been significantly curbing its air pollution, namely through the use of cleaner technologies and fuels. However, congestion and safety are major threats for São Paulo manufacturing and research activities. For two decades now, OEMs and suppliers have been leaving the metropolis for greenfield locations elsewhere in the state, and even to other much less developed states. If the first rationale was mainly based on lower costs and less congestion, today many of these firms develop R&D outside São Paulo. Thus, metropolitan São Paulo is competing for private R&D firms that are choosing other manufacturing hubs, namely in the fields of bio-fuel research and production. A main challenge is to integrate these emerging R&D firms with the knowledge assets of the metropolis, namely its good (public) research centres and advanced service base.

Some implications for national policy

In the last section, we outline a number of challenges for different types of cities. What are the implications for national policies?

National governments need to adopt more integrated approaches towards the industry

National governments in Europe increasingly recognise that industrial development is linked to technological and scientific development. In several countries, different ministries are developing joint public agencies to integrate industrial innovation and technological development. In Portugal, the Ministry of Science and Technology and the Ministry of Economy have created ADI – Innovation Agency – a national public agency that manages a vast array of generic incentives with strong potential impacts on manufacturing development. Its action is not focused in any special sector, but provides support (technical and financial) to specific projects, and evaluates projects submitted by firms and knowledge institutions to ensure they meet certain strategic and quality standards. It not only supports R&D and technology transfer projects, but also the placing of qualified graduates in firms and joint firm–university PhD programmes. The state of São Paulo gives sound funding for R&D in specific fields (namely bio-fuels), but mainly for public partners and universities.

Another good option is to promote specific 'strong domains' in an integrated way, by linking the knowledge base to industrial strengths. For instance, the Dutch government has been developing industry-specific policies for food, mainly related to nutrition (Food and Nutrition Delta) at the national level. In China, the approach of the national government is different, and focuses on strategic sectors like automotive. The national government sets policy goals to strengthen the automotive industry (e.g. research areas, energy, organising models for technology transfer from outside OEMs present in China and JVs), guiding its future development. However, the implementation of the policy goals has a strong local and regional dimension. In Brazil, the federal government has been opting for a 'laissez-faire' strategy; it merely sets fiscal law, incentives and regulation, framing the private action. At the same time, there are no cluster or regionally based strategies for sectors like automotive. Opposite to the Chinese model, the main entry mechanism of transnational companies into Brazil was based on mergers and acquisition of indigenous firms, which became dependent on global headquarters elsewhere.

'Regionalised' industry policy may work, if designed with sufficient flexibility

Although the hard resources are still based at the national level, many national governments have recognised the need to give their policies a stronger regional dimension to better use their regions' assets. These policies tend to focus on the cluster – regionalised networks firms, R&D institutes, associations, etc. – rather than on individual firms or sectors. Our study suggests that the 'spatial reach' of networks is getting wider, with the implication that this type of approach should be sufficiently flexible. Policymakers struggle with the question of how functional relations in the cluster are related to the territorial dimension. Support structures should preferably cover large areas and have 'open membership'.

In Germany and Finland, the regional level is traditionally significant in economic and development policy. Recently, France has created a more bottom-up approach as well, through the initiative of 'competitiveness poles'. Here, the national government invites regions to submit cluster 'projects', in which firms, universities and public bodies would create collaborative networks, preferably based on existing strengths and competences (like in the automotive cluster in Paris). The projects are divided into three groups with a different status (the top being 'world class clusters') and the French government has a substantial budget to support initiatives within clusters, such as offering tax benefits, loans or direct subsidies. France has indeed been experiencing a change in the regional dimension of industrial and innovation policy – presently, regions are involved in funding university buildings and innovative SMEs.

Some other initiatives have a stronger top-down character, like the initiative ‘Peaks in the Delta’ from the Dutch government. In this initiative, the national government defines ‘top clusters’ that are eligible for support, appointing different regions as main targets of the initiative. The food sector is identified as an important sector in the east of the Netherlands and the region around Wageningen, but also for the southern part of the Randstad region. In Portugal, the initiative has been oscillating between a bottom-up and top-down approach, but the participation of regions has been different. The *Norte* region has been one of the most active in developing a strategy to identify traditional clusters to upgrade (for example, ‘fashion’ footwear and textiles) and for clusters to foster (ICT, maritime cluster, medical and pharmaceutical equipment). The region will benefit in the coming years from strong support from the EU regional policy. At the same time, the national government (also typically centralistic) is working on identifying ‘excellence poles’, but the articulation with the regional scale is not yet defined. Nevertheless, some ‘divisions of labour’ between the regional and national levels are already planned: the region will manage the funds for micro and small firms while the national government will manage the incentives for medium and large firms. Moreover, the central government’s direct incentives plan to give priority to projects inserted in other regional strategic clusters.

Companies increasingly recognise the importance of their ‘external environment’ for future competitiveness, and are prepared to invest in it

In our study, we discovered that companies increasingly recognise that their ‘external environment’ (in the broadest sense of the word) has a deep and fundamental impact on their competitiveness, and in many cases they are willing to invest in this environment.

Firms increasingly depend on their network partners, such as suppliers and knowledge institutes, and have an incentive to invest in these networks. We found various examples of companies that invest in these relations for their own benefit, e.g. Audi and INI.TUM (Munich), Unilever and WFI (Rotterdam). The growth of ‘open innovation’ approaches will make this trend more prevalent. In emerging countries, automotive firms ‘educate’ local suppliers to improve the quality.

Second, firms depend on the local/regional labour market. Especially when there are shortages of skilled staff, firms are willing to invest in regional education facilities, and the promotion of educational programmes. From the same perspective, companies want to improve the general image of a region, which may help to attract the much-needed human resources. For this reason, Philips (not one of our case studies) has become involved in inter-urban cooperation: the company was the main driver behind the intensified cooperation between the cities of Eindhoven, Leuven and Aachen, aimed at creating a ‘knowledge region’ of sufficient scale.

Third, with technological upgrading and internationalisation, the 'people-orientation' of manufacturing industries is increasing. Firms like to welcome their clients in an attractive environment that caters for face-to-face contact. This gives them an incentive to contribute to the 'quality of place' in the area where they are located.

'Corporate social responsibility' has become a major theme in the corporate world. Firms recognise that their image is absolutely crucial, not only to sell products but also to attract good workers. From this perspective, they are prepared to invest in social projects and urban revitalisation. Recently, climate change has emerged as a main issue, and several progressive industrial companies have committed themselves to reduce CO_2 emissions.

Index

For Product Safety Concerns and Information please contact our EU representative GPSR@taylorandfrancis.com
Taylor & Francis Verlag GmbH, Kaufingerstraße 24, 80331 München, Germany

www.ingramcontent.com/pod-product-compliance
Lightning Source LLC
LaVergne TN
LVHW011107300926
849160LV00008B/303

9780415710091